GLOSSARY OF GENETICS

GLOSSARIA
INTERPRETUM

Published by

AUSLANDS- UND DOLMETSCHERINSTITUT DER
JOHANNES GUTENBERG-UNIVERSITÄT
MAINZ IN GERMERSHEIM
DOLMETSCHER-INSTITUT DER UNIVERSITÄT HEIDELBERG
ÉCOLE D'INTERPRÈTES DE L'UNIVERSITÉ DE GENÈVE
ÉCOLE SUPÉRIEURE D'INTERPRÈTES ET DE
TRADUCTEURS DE L'UNIVERSITÉ DE PARIS
SCHOOL OF LANGUAGES AND LINGUISTICS
GEORGETOWN UNIVERSITY, WASHINGTON, D.C.
SCUOLA DI LINGUE MODERNE PER TRADUTTORI
ED INTERPRETI DI CONFERENZE, UNIVERSITÀ
DEGLI STUDI DI TRIESTE

under the General Editorship of
PROF. JEAN HERBERT
FORMER CHIEF INTERPRETER TO THE UNITED NATIONS

ELSEVIER PUBLISHING COMPANY
AMSTERDAM / LONDON / NEW YORK
1970

GLOSSARY OF GENETICS

IN

ENGLISH, FRENCH, SPANISH,
ITALIAN, GERMAN, RUSSIAN

Compiled and arranged by
FRANÇOISE BIASS-DUCROUX

in collaboration with
KLAUS NAPP-ZINN

Russian translation by
NIKOLAJ V. LUCHNIK

ELSEVIER PUBLISHING COMPANY
AMSTERDAM / LONDON / NEW YORK
1970

ELSEVIER PUBLISHING COMPANY
335 JAN VAN GALENSTRAAT, P.O. BOX 211, AMSTERDAM

AMERICAN ELSEVIER PUBLISHING COMPANY, INC.
52 VANDERBILT AVENUE, NEW YORK, N.Y. 10017

ELSEVIER PUBLISHING COMPANY LIMITED
RIPPLESIDE COMMERCIAL ESTATE
BARKING, ESSEX

Library of Congress Card Number: 68-54866

Standard Book Number: 444-40712-x

Printed in The Netherlands

FOREWORD

Now that conferences deal with such a multitude of questions and the attainments required of their interpreters have grown so diverse and specialized, we feel that there is room for a series of multilingual technical glossaries bearing on the principal subjects discussed at international conferences.

This undertaking is being supervised and co-ordinated by M. Jean Herbert, formerly Chief Interpreter to the United Nations. The purpose of its joint sponsorship by the Auslands- und Dolmetscherinstitut der Johannes Gutenberg-Universität Mainz in Germersheim, the Dolmetscherinstitut der Universität Heidelberg, the École d'Interprètes de l'Université de Genève, the École supérieure d'Interprètes et de Traducteurs de l'Université de Paris, the School of Languages and Linguistics, Georgetown University, Washington, D.C., and the Scuola di Lingue moderne per Traduttori ed Interpreti di Conferenze, Università degli Studi di Trieste, is to emphasize the international and scientific character of these glossaries. They are the work of language experts, including interpreters, some of whom are teachers or alumni of the sponsoring institutes.

The aim of these glossaries is to endow professional and apprentice interpreters with a terminological apparatus both scientifically planned and generally acknowledged in the various sectors of international activity. In broader terms, their intention is to enable experts to understand one another more readily, and to disseminate an accepted international terminology.

Each glossary will appear in several languages, to be chosen according to the frequency of their use at international conferences on that particular subject.

We have set out to make the glossaries as compact and efficient as possible. The usual order of terms is alphabetical, although a certain number are specially listed by their functions.

It is hoped that several glossaries will appear annually, dealing successively with subjects likely to arise at conferences where trained interpreters are employed.

DR. B. BEINERT, Direktor des Dolmetscherinstituts der Universität Heidelberg
PROF. DR. HANS JESCHKE, Direktor des Auslands- und Dolmetscherinstituts der Johannes Gutenberg-Universitat Mainz in Germersheim
PROF. CLAUDIO CALZOLARI, Preside della Facoltà di Economia e Commercio, Direttore della Scuola di Lingue moderne per Traduttori ed Interpreti di Conferenze dell'Università degli Studi di Trieste
MAURICE GRAVIER, Directeur de l'École supérieure d'Interprètes et de Traducteurs de l'Université de Paris
STEFAN F. HORN, Head, Division of Interpretation and Translation, School of Languages and Linguistics, Georgetown University, Washington, D.C.
NORBERT HUGEDÉ, Administrateur de l'École d'Interprètes de l'Université de Genève

TITLES PUBLISHED IN THE SERIES GLOSSARIA INTERPRETUM

PRÉFACE

La Génétique, ou science de l'hérédité, est au coeur de la biologie, puisqu'elle nous initie à la connaissance des mécanismes fondamentaux de la vie cellulaire, et même, depuis quelque temps, nous aide à connaître les propriétés des molécules qui sont à la base de ces mécanismes.

Il n'est pas un secteur des sciences de la vie qui ne puisse tirer parti des enseignements de la Génétique. Celle-ci, en effet, ne se borne pas à nous faire comprendre le mode de transmission des caractères organiques, à travers les générations. Elle éclaire les procédés essentiels de la différenciation sexuelle, et, collaborant avec l'embryologie, elle nous instruit sur le développement des êtres. Par les précisions qu'elle apporte sur la formation des races et des espèces, elle est un complément indispensable de la systématique, animale ou végétale. Par l'étude des phénomènes de mutation et par l'analyse des moyens dont dispose la sélection naturelle pour modifier, peu à peu, la composition des populations vivantes, elle contribue à nous fournir une représentation acceptable de l'évolution, et par là elle nous permet d'imaginer comment se sont réalisés, au cours des âges, les grands changements qui ont marqué l'histoire de la vie.

Enfin, elle constitue un chapitre important de la médecine dès lors que nous savons par elle qu'une foule de maladies et de tares sont dues à des modifications du patrimoine héréditaire, et plus précisément que certaines de ces maladies et de ces tares sont liées à des aberrations chromosomiques (altérations du nombre, cassures de chromosomes, etc.) qui se laissent apercevoir sous le microscope.

Un tel enrichissement du savoir ne saurait aller sans vastes conséquences dans le domaine pratique. Extrêmement fructueuses sont, dès maintenant, les applications de la Génétique en agriculture, en zootechnie, en pathologie humaine. Elles ne feront que s'amplifier sans cesse; déjà se laisse entrevoir le moment où il deviendrait possible d'agir directement sur les patrimoines héréditaires, et c'en est assez pour susciter des espoirs si hardis qu'ils ne peuvent pas ne pas côtoyer la crainte, tout au moins en ce qui concerne cette modification de l'homme par l'homme qui, en comblant nos rêves, nous affronterait aux plus ardus des problèmes.

Cette science de l'hérédité, comme toute science, est redevable de ses progrès à des chercheurs de tous pays, de toutes origines, de toutes langues, si bien qu'il n'est pas une seule question de Génétique dont l'étude ne doive s'appuyer sur une bibliographie internationale.

Il apparaissait donc comme nécessaire qu'on mît à la disposition des travailleurs, et plus généralement de tous ceux qui s'intéressent à la Génétique, un Glossaire qui leur offrît, pour chaque terme spécialisé, un ou plusieurs équivalents dans les principales langues d'Europe.

Dans l'ouvrage que j'ai la satisfaction de présenter en ces lignes, six langues — anglaise, italienne, française, espagnole, allemande, russe — ont été utilisées; et il est à peine besoin de faire ressortir les difficultés de toutes sortes (documentation, traduction, etc.) qu'ont dû surmonter Mme. Françoise Biass-Ducroux et ses collaborateurs pour mener à bien une telle entreprise, exigeant autant de soin, de patience, de conscience, de finesse linguistique, que de compétence scientifique.

Fruit de plusieurs années d'effort, ce Glossaire tient compte des dernières acquisitions de la Génétique: on y trouvera les termes correspondant aux notions les plus neuves de cette discipline.

Tel qu'il est, il constitue un précieux, un irremplaçable instrument de travail. Ceux qui l'ont composé ont bien servi la science, et aussi, en facilitant l'échange des informations et des pensées, ils ont heureusement concouru à établir cette liaison culturelle et spirituelle qui doit aider à la compréhension mutuelle des humains et à l'harmonieuse entente des peuples.

Jean Rostand
de l'Académie Française

INTRODUCTION

This glossary has been compiled to provide the user with a rapid and adequate translation of the terms in use today in the field of genetics. Those who have worked on it hope that it will fulfill its purpose. Much care has been taken to avoid a mere transcription of words from one language into the others. Each word has been given special attention to ensure provision of one equivalent or, where necessary, several equivalents corresponding to its specific meaning. Different spellings or synonyms have also been indicated; when neither an adequate equivalent nor a readily acceptable neologism could be found, the sense of the word has been conveyed in a short paraphrase. Much attention has at times been devoted to the translation of expressions or groups of words; in some cases it has been felt necessary to reproduce a literal transcription in the six languages in order to reassure the user struggling with an unfamiliar idiom. This has not always been possible, however, even in cases where a literal translation seemed to provide the obvious answer: in English, for instance, the term "lampbrush" chromosome does not convey the idea of "plumety" which is to be found in the Latin languages to describe this phenomenon. Within limits, then, literal translations or fairly succinct explanations have been given. But whenever a literal translation could be misleading or where no explanation could be given sufficiently briefly the corresponding space has been left blank. Unavoidable delays due to mail communications and technical difficulties are responsible for a certain lack of homogeneity in the choice of the terms to be found in this glossary; some of them are repeated, while others are missing. I am conscious of the inconvenience this may cause the reader and I offer my apologies in advance.

As the editor of the glossary, I wish in the first place to express my gratitude to the Springer-Verlag, publishers of the *Genetisches und Cytogenetisches Wörterbuch* by R. Rieger and A. Michaelis, who very kindly gave us permission to use this book as our basic work of reference.

Thanks are also due to Professor Alberto Pirovano of Milan and Rome for his untiring efforts to help us find the correct equivalents in Italian; to Dr. Manuel Alvarez-Luna of Mexico, whose close association with Professor Kihara of Japan has provided him with a vast store of up-to-date information on the most recent developments in genetics in the Far East, and who kindly revised the Spanish section; to Dr. J. Pourquié for revising the French section; and to Professor Heslot and Professor Valdeyron of the Institut Agronomique of Paris for their encouragement and help.

I also wish to acknowledge the kindness of Professor R. C. Clowes, formerly of the Microbiological Unit, Hammersmith Hospital, London, now at the Graduate Research Center of the Southwest, Dallas, in providing vital documents as well as a number of equivalents for very recent terms.

I take this opportunity to express my sincere gratitude to Mr. Jean Rostand, whose preface is the crowning reward of our work and zeal.

It would be impossible to attempt to mention here the names of all those whose advice and counsel helped me to carry out a task which proved far heavier than it originally seemed. However, I wish to emphasize how much I owe to Professor Klaus Napp-Zinn's moral support and active collaboration. Professor at

Cologne University, he most kindly assisted me in solving delicate problems, in spite of the fact that he had originally agreed to take part only as a translator for the German terms. I should like to thank also Professor Nikolaj V. Luchnik, Head of the Department of Biophysics, Institute of Medical Radiology of the Academy of Medical Sciences of the USSR at Obninsk, whose reliable translation into Russian will greatly contribute to the usefulness and the accuracy of this work. As translator into Russian of the Rieger-Michaelis Dictionary and as author of various other genetical books, he was able to perform this task most competently. Mme. Valentine Chassaignon de Cazes deserves my thanks for her careful transcription of the Russian manuscript. I also wish to convey my gratitude to Professor Vichyl and to Professor Vitale who so kindly agreed to read the proofs of the Russian and Italian texts.

Françoise Biass-Ducroux

VORBEMERKUNG

Als Bearbeiter für "Deutsch" halte ich es für zweckmässig, meine Arbeit kurz zu erläutern und einige Hinweise für die Benutzung dieses Wörterbuches bei Übersetzungen ins Deutsche zu geben. Was meine Arbeit betrifft, so hatte ich infolge unvorhergesehener Umstände, die ich nicht zu verantworten habe, keinen Einfluss auf die endgültige Auswahl der berücksichtigten Termini. Bei der Eindeutschung fremdsprachiger (vor allem englischer) Begriffe bemühte ich mich um möglichst weitgehende Berücksichtigung der beiden Forderungen, dass deutsche Arbeiten in möglichst gutem Deutsch abgefasst und dennoch fremdsprachigen Fachgenossen möglichst leicht verständlich sein sollten.

Die Kompromisslösungen sehen von Fall zu Fall verschieden aus. Manchmal wurde für Termini, die der Umgangssprache entstammen, das deutsche Gegenstück eingesetzt (z.B. laggard — Nachzügler). Bei anderen Gelegenheiten wurden die — meist griechischen oder lateinischen — Wortstämme fremdsprachiger Ausdrücke belassen und nur mit einer deutschen End-, eventuell auch Vorsilbe versehen. In diesen beiden Fällen sind vom Bearbeiter vorgeschlagene Übersetzungen, welche noch nicht allgemein gebräuchlich zu sein scheinen, durch das Zeichen # kenntlich gemacht. Bei der Beurteilung dessen, was bereits gebräuchlich ist, wurde in der Regel das *"Genetische und cytogenetische Wörterbuch"* von R. RIEGER und A. MICHAELIS (mit freundlicher Genehmigung der Verfasser und des Springer-Verlages) zu Rate gezogen.

In wieder anderen Fällen wurden fremdsprachige Termini, die schon mehr oder weniger dauerhaften Eingang in die deutsche Fachsprache gefunden haben, in ihrer fremdsprachigen Form belassen, wobei sie das grammatische Geschlecht, das sie in der Ursprungssprache haben, entweder beibehielten oder änderten. Letzteres gilt vor allem für englische Ausdrücke a) griechischer und lateinischer Herkunft, b) germanischer Herkunft, soweit es im Deutschen Wörter gleichen Stammes gibt (z.B. *der* "pool"). Verschiedentlich empfahl es sich auch, einen Terminus unübersetzt zu lassen, besonders wenn ein knapper fremdsprachiger Ausdruck im Deutschen nur durch eine umständliche Konstruktion hätte wiedergegeben werden können.

Bei Übersetzungen *ins* Deutsche sollte man etliche Ausdrücke vermeiden, nämlich einerseits antiquierte (⊙), andererseits sprachlich unbefriedigende (unkorrekte oder unsinnige) Wörter (!), z.B. Hybridwörter wie "monofaktoriell" oder verselbständigte Wortteile wie "Ploidie".

Die Schwierigkeiten einer befriedigenden Regelung für sie Schreibung mit C oder K bzw. Z sind bekannt. Zwar sind wir auch in dieser Hinsicht weitgehend dem Wörterbuch von RIEGER und MICHAELIS gefolgt, haben aber in stärkerem Masse der weit verbreiteten Tendenz Rechnung getragen, das C durch Z oder K zu ersetzen, und deshalb manchmal auch mehrere Schreibweisen angegeben. Im übrigen gilt der gewohnte Rat: Was man unter C, c vermisst, suche man unter K, k oder Z, z — und umgekehrt.

Ich bin mir bewusst, dass gewiss noch nicht immer die ideale Lösung für die vorstehend gekennzeichneten Probleme gefunden ist. Für den Fall, dass später eine Neuauflage dieses Wörterbuches erforderlich werden sollte, werde ich deshalb Verbesserungsvorschläge stets dankbar zur Kenntnis nehmen.

Klaus Napp-Zinn

LANGUAGE INDICATIONS AND

ABBREVIATIONS

f	French	*f*	feminine
e	Spanish	*m*	masculine
i	Italian	*n*	neuter
d	German	*pl*	plural
		adj.	adjective

SYMBOLS

⊙	!	#
not in common use	unadvisable	proposed translation
peu usité	à éviter	traduction proposée
poco frecuente	desaconsejable	término propuesto
poco usato	da evitare	termine proposto
wenig gebräuchlich	nicht empfehlenswert	vorgeschlagene Übersetzung
редко применяеться	не рекомендуется	предложенный перевод

CONTENTS

BASIC TABLE

A

1 ABERRATION
   f aberration *f*
   e aberración *f*
   i aberrazione *f*
   d Aberration *f*

   ABERRATION, POST SPLIT
   see 2271

2 ABERRATION RATE
   f taux *f* d'aberration
   e porcentaje *m* de
      aberración
   i percentuale *f* delle
      aberrazioni
   d Aberrationsrate *f*

3 ABIOGENESIS
   f abiogenèse *f;*
      génération spontanée *f*
   e abiogénesis *f*
   i abiogenesi *f*
   d Abiogenesis *f* (Θ);
      Abiogenese *f* (Θ);
      Urzeugung *f*

4 ABIOGENETIC
   f abiogénétique
   e abiogenético
   i abiogenetico
   d abiogenetisch

5 ABORTION
   f avortement *m*
   e aborto *m*
   i aborto *m*
   d Abort *m;*
      Ablast *m*

6 ABORTIVE
   f abortif
   e abortivo
   i abortivo
   d abortiv

7 ABORTIVE INFECTION
   f infection *f* abortive
   e infección *f* abortiva

   i infezione *f* abortiva
   d Abortiv-Infektion *f*

8 ABORTIVE MITOSIS
   f mitose *f* abortive
   e mitosis *f* abortiva
   i mitosi *f* abortiva
   d abortive Mitose *f*

9 ABORTIVE TRANSDUCTION
   f transduction *f* abortive
   e transducción *f* abortiva
   i trasduzione *f* abortiva
   d abortive Transduktion *f*

10 ACCELERATION
   f accélération *f*
   e aceleración *f*
   i accelerazione *f*
   d Akzeleration *f;*
      Acceleration *f*

11 ACCESSORY
   f accessoire
   e accesorio
   i accessorio
   d akzessorisch

12 ACCESSORY CHROMOSOME
   f chromosome *m* accessoire
   e cromosoma *m* accesorio
   i cromosoma *m* accessorio
   d akzessorisches
      Chromosom *n*

13 ACCESSORY PLATE
   f plaque *f* accessoire
   e placa *f* accesoria
   i placca *f* accessoria
   d akzessorische Metaphase-
      platte *f*

14 ACCLIMATATION;
   ACCLIMATIZATION
   f acclimatisation *f;*
      acclimatation *f*
   e aclimatación *f*

i   acclimatazione *f*
d   Akklimatisierung *f*

15   ACENTRIC
f   acentrique
e   acéntrico
i   acentrico
d   azentrisch

16   ACENTRIC-DICENTRIC
TRANSLOCATION
f   translocation *f*
acentrique-dicentrique
e   - -
i   traslocazione *f*
acentrica e dicentrica
d   asymmetrische Trans-
lokation *f*

17   ACENTRIC INVERSION
f   inversion *f* acentrique
e   - -
i   inversione *f* acentrica
d   azentrische Inversion *f*

18   ACHIASMATIC
f   achiasmatique
e   aquiasmático
i   achiasmatico
d   achiasmatisch

19   ACHROMASIE
f   achromasie *f*
e   acromasia *f*
i   acromasia *f*
d   Achromasie *f*

20   ACHROMATIC
f   achromatique
e   acromático
i   acromatico
d   achromatisch

21   ACHROMATIC FIGURE
f   fuseau *m* achromatique;
figure *f* achromatique
e   huso *m* acromático
i   fuso *m* acromatico
d   achromatische Figur *f*;
achromatische Spindel *f*

22   ACHROMATIN

f   achromatine *f*
e   acromatina *f*
i   acromatina *f*
d   Achromatin *n*

23   A-CHROMOSOME
f   chromosome *m* A
e   cromosoma *m* A
i   cromosoma *m* A
d   A-Chromosom *n*

24   ACQUIRED CHARACTER
f   caractère *m* acquis
e   carácter *m* adquirido
i   carattere *m* acquistato
d   erworbene Eigenschaft *f*

25   ACROCENTRIC
f   acrocentrique
e   acrocéntrico
i   acrocentrico
d   akrozentrisch

26   ACROSYNDESIS
f   acrosyndèse *f*
e   acrosíndesis *f*
i   acrosindesi *f*
d   Akrosyndese *f*

27   ACTIVATION
f   activation *f*
e   activación *f*
i   attivazione *f*
d   Aktivierung *f*

28   ACTIVATOR
f   activateur *m*
e   activador *m*
i   attivatore *m*
d   Aktivator *m*

29   ACTIVATOR-DISSOCIATION
SYSTEM
f   système *m* activateur-
dissociation
e   sistema *m* activador-
disociación (Θ)
i   sistema *m* attivatore-
dissociazione
d   Aktivator-Dissoziations-
System *n*

30 ACTIVITY SPECTRUM
  f spectre *m* d'activité
  e espectro *m* de actividad
  i spettro *m* d'attività
  d Aktivitätsspektrum *n*

31 ADAPTABILITY;
   ADAPTIVITY
  f adaptativité *f*
  e adaptabilidad *f;*
     adaptividad *f*
  i adattabilità *f*
  d Anpassungsfähigkeit *f;*
     Adaptibilität *f*

32 ADAPTATION
  f adaptation *f*
  e adaptación *f*
  i adattamento *m*
  d Adaptation *f;*
     Anpassung *f*

   ADAPTATION, ECOBIOTIC
   see 894

   ADAPTATION, ECOCLIMATIC
   see 895

33 ADAPTIVE MODIFICATION
  f modification *f* d'adaptation
  e modificación *f* adaptiva
  i modificazione *f* adattativa
  d adaptive Modifikation *f*

34 ADAPTIVE PEAK
  f pic *m* d'adaptation
  e cúspide *f* adaptiva
  i vertice *m* adattativo
  d Anpassungsgipfel *m*

35 ADAPTIVE RADIATION
  f radiation *f* adaptative
  e radiación *f* adaptiva
  i radiazione *f* d'adattamento
  d – –

36 ADAPTIVE TRAIT
  f caractère *m* adaptatif
  e carácter *m* adaptivo
  i carattere *m* adattativo

  d Anpassungsmerkmal *n*

37 ADAPTIVE VALUE
  f valeur d'adaptation
  e valor *m* de adaptación
  i valore *m* adattativo
  d Anpassungswert *m*

   ADAPTIVITY
   see 31

38 ADDITION
  f addition *f*
  e adición *f*
  i addizione *f*
  d Addition *f*

39 ADDITIVE EFFECT
  f effet *m* additif
  e efecto *m* aditivo
  i effetto *m* additivo
  d additiver Effekt *m;*
     additive Wirkung *f*

40 ADDITIVE FACTORS
  f facteurs *mpl* additifs
  e factores *mpl* aditivos
  i fattori *mpl* additivi
  d additive Faktoren *mpl*

41 ADDITIVE GENOTYPE VALUE
  f valeur *f* additive du
     génotype;
     valeur *f* héréditaire ;
     valeur *f* de reproduction
  e valor *m* aditivo del
     genotipo;
     valor *m* hereditario
  i valore *m* ereditario
     genotipico
  d – –

42 ADDITIVITY
  f additivité *f*
  e aditividad *f*
  i additività *f*
  d Additivität *f*

43 ADDOSPECIES
  f addoespèce *f*
  e adoespecie *f*

i   addospecie *f*
d   Addospecies *f*

44   ADHESION
f   adhésion *f*
e   adhesión *f*
i   adesione *f*
d   Adhäsion *f*

45   AFFINITY
f   affinité *f*
e   afinidad *f*
i   affinità *f*
d   Affinität *f*

46   AGAMETE
f   agamète *m*
e   agámeto *m*
i   agameto *m*
d   Agamet *m*

47   AGAMIC;
AGAMOUS
f   agame
e   ágamo
i   agamo
d   agam

48   AGAMOBIUM
f   agamobium *m*;
sporophyte *m*
e   agamobio *m*
i   agamobio *m*
d   Agamobium *n* (⊙);
Sporophyt *m*

49   AGAMOGENESIS
f   agamogénèse *f*
e   agamogénesis *f*
i   agamogenesi *f*
d   Agamogenese *f*

50   AGAMOGONY
f   agamogonie *f*
e   agamogonia *f*
i   agamogonia *f*
d   Agamogonie *f*

51   AGAMOSPECIES
f   espèce *f* agame
e   agamoespecie *f*

i   agamospecie *f*
d   Agamospecies *f*

52   AGAMOSPERMIC
f   agamospermique
e   agamospérmico
i   agamospermico
d   agamosperm

53   AGAMOSPERMY
f   agamospermie *f*
e   agamospermia *f*
i   agamospermia *f*
d   Agamospermie *f*

54   AGAMOUS;
AGAMIC
f   agame
e   agámico
i   agamo
d   agam

55   AG COMPLEX
f   complexe *m* AG
e   complejo *m* AG
i   complesso *m* AG
d   AG-Komplex *m*

56   "AGE AND AREA"THEORY
f   théorie *f* de l'âge et de
l'espace
e   teoría *f* de la edad y area
i   teoria *f* dell' età e dello
spazio
d   Alter-und-Areal-Theorie *f*

57   AGMATOPLOID (adj.)
AGMATOPLOID
f   agmatoploïde
agmatoploïde *m*
e   agmatoploide
agmatoploide *m*
i   agmatoploide
agmatoploide *m*
d   agmatoploid
Agmatoploide *f*

58   AGMATOPLOIDY
f   agmatoploïdie *f*
e   agmatoploidía *f*
i   agmatoploidismo *m*

d　Agmatoploidie *f*

59　AGMATO-PSEUDOPOLYPLOID
(adj.)
AGMATO-PSEUDOPOLYPLOID
f　agmatopseudoploïde
　　agmatopseudoploïde *m*
e　agmato-seudopoliploide
　　agmato-seudopoliploide *m*
i　agmatopseudopoliploide
　　agmatopseudopoliploide *m*
d　agmato-pseudopolyploid
　　Agmato-Pseudopolyploide *f*

60　AGMATO-PSEUDOPOLYPLOIDY
f　agmatopseudopolyploïdie *f*
e　agmato-seudopoliploidîa *f;*
　　agmato-pseudopoliploidîa *f*
i　agmato-pseudopoliploidismo *m*
d　Agmato-Pseudopolyploidie *f*

61　AGONISIS;
CERTATION
f　certation *f*
e　agonisis *f;*
　　certación *f*
i　certatio *f*
d　Zertation *f;*
　　Certation *f*

62　AKINETIC
f　acinétique
e　acinético
i　acinetico
d　akinetisch

63　ALLAESTHETIC
f　allaesthétique
e　alaestético
i　allaestetico
d　allaesthetisch

64　ALLAUTOGAMOUS
f　allautogame
e　alautógamo
i　allautogamo
d　allo-autogam

65　ALLAUTOGAMY
f　allautogamie *f*

e　allautogamia *f*
i　allautogamia *f*
d　Allo-autogamie *f*

66　ALLELE
f　allèle *m*
e　alelo *m*
i　allelo *m*
d　Allel *n*

ALLELE, VARIEGATION
see 2913

ALLELES, MULTIPLE
see 1876

ALLELES, POSITIONAL
see 2263

ALLELE SHIFT;
ALLELE TREND
f　modification *f* par la
　　présence d'un allèle
e　alelo-desvio *m;*
　　alelo-tendencia *f*
i　modificazione *f* dovuta alla
　　presenza d'un allelo
d　Änderung *f* der Allelen-
　　häufigkeit

67

68　ALLELIC
f　allélique
e　alélico
i　allelico
d　allel

69　ALLELISM
f　allélisme *m*
e　alelismo *m*
i　allelismo *m*
d　Allelie *f*

70　ALLELOBRACHIAL
f　allélobrachial
e　alelobraquial
i　allelobrachiale
d　allelobrachial

71　ALLELOMORPH
f　allélomorphe *m*
e　alelomorfo *m*

i   allelomorfo *m*
d   Allelomorph *n*

ALLELOMORPH, SPURIOUS
see 2655

72   ALLELOMORPHIC
f   allélomorphique
e   alelomórfico
i   allelomorfico
d   allelomorph

73   ALLELOMORPHIC SERIES
f   série *f* allélomorphique
e   serie *f* alelomórfica
i   serie *f* allelomorfica
d   allelomorphe Serie *f*

74   ALLELOMORPHISM
f   allélomorphisme *m*
e   alelomorfismo *m*
i   allelomorfismo *m*
d   Allelomorphie *f*

75   ALLELOTYPE
f   allélotype *m*
e   alelotipo *m*
i   allelotipo *m*
d   Allelotyp *m*

76   ALLIUM TEST
f   test *m* de l'ail
e   prueba *f* de alium
i   prova *f* dell' aglio
d   Allium-Test *m*

77   ALLOCARPY
f   allocarpie *f*
e   alocarpia *f*
i   allocarpia *f*
d   Allokarpie *f*;
    Allocarpie *f*

78   ALLOCHRONIC;
     ALLOCHRONOUS
f   allochrone
e   alocróno
i   allocronico
d   allochron

79   ALLOCYCLIC
f   allocyclique
e   alocíclico
i   allociclico
d   allozyklisch

80   ALLOCYCLY
f   allocyclie *f*
e   alociclia *f*
i   allociclia *f*
d   Allozyklie *f*

81   ALLODIPLOID (adj)
     ALLODIPLOID
f   allodiploïde
    allodiploïde *m*
e   alodiploide
    alodiploide *m*
i   allodiploide
    allodiploide *m*
d   allodiploid
    Allodiploide *f*

82   ALLODIPLOIDY
f   allodiploïdie *f*
e   alodiploidia *f*
i   allodiploidia *f*
d   Allodiploidie *f*

83   ALLODIPLOMONOSOME
f   allodiplomonosome *m*
e   alodiplomonosoma *m*
i   allodiplomonosoma *m*
d   Allodiplomonosom *n*

84   ALLOGAMOUS
f   allogame
e   alógamo
i   allogamo
d   allogam

85   ALLOGAMY
f   fécondation *f* croisée;
    allogamie *f*
e   alogamia *f*;
    fecundación *f* cruzada
i   allogamia *f*
d   Allogamie *f*;
    Fremdbestäubung *f*

86 ALLOGENE
    f allogène *m*
    e alógeno *m*
    i allogeno *m*
    d Allogen *n* (Θ);
       rezessives Allel n

87 ALLOGENETIC
    f allogénique
    e alogenético
    i allogenetico
    d allogenetisch

88 ALLOGENIC;
    ALLOGENOUS
    f allogène
    e alógeno
    i allogeno
    d allogenisch

89 ALLOHAPLOID (adj.)
    ALLOHAPLOID
    f allohaploïde
       allohaploïde *m*
    e alohaploide
       alohaploide *m*
    i allohaploide
       allohaploide *m*
    d allohaploid
       Allohaploide *f*

90 ALLOHAPLOIDY
    f allohaploïdie *f*
    e alohaploidía *f*
    i allohaploidismo *m*
    d Allohaploidie *f*

91 ALLOHETEROPLOID (adj.)
    ALLOHETEROPLOID
    f allohétéroploïde
       allohétéroploïde *m*
    e aloheteroploide
       aloheteroploide *m*
    i alloeteroploide
       alloeteroploide *m*
    d alloheteroploid
       Alloheteroploide *f*

92 ALLOHETEROPLOIDY
    f allohétéroploïdie *f*
    e aloheteroploidía *f*

i alloeteroploidismo *m*
d Alloheteroploidie *f*

93 ALLOIOBIOGENESIS;
    ALLOIOGENESIS
    f alloiogénèse *f*
    e aloiogénesis *f*
    i alloiogenesi *f*
    d Alloiogenesis *f*

94 ALLOLYSOGENIC
    f allolysogénique
    e alolisogénico
    i allolisogenico
    d allolysogen

95 ALLOMORPHOSIS
    f allomorphose *f*
    e alomórfosis *f*
    i allomorfosi *f*
    d Allomorphose *f* (#)

96 ALLOPATRIC
    f allopatrique
    e alopátrico
    i allopatrico
    d allopatrisch

97 ALLOPLASM
    f alloplasme *m*
    e aloplasma *m*
    i alloplasma *m*
    d Alloplasma *n*

98 ALLOPLOID (adj.)
    ALLOPLOID
    f alloploïde
       alloploïde *m*
    e aloploide
       aloploide *m*
    i alloploide
       alloploide *m*
    d alloploid
       Alloploide *f*

99 ALLOPLOIDY
    f alloploïdie *f*
    e aloploidía *f*
    i alloploidismo *m*
    d Alloploidie *f*

100 ALLOPOLYPLOID (adj)
ALLOPOLYPLOID
f allopolyploïde
allopolyploïde *m*
e alopoliploide
alopoliploide *m*
i allopoliploide
allopoliploide *m*
d allopolyploid
Allopolyploide *f*

ALLOPOLYPLOID, SEGMENTAL
see 2514

101 ALLOPOLYPLOIDY
f allopolyploïdie *f*
e alopoliploidía *f*
i allopoliploidismo *m*
d Allopolyploidie *f*

102 ALLOSOMAL
f allosomique
e alosómico
i allosomico
d allosomal

103 ALLOSOME
f allosome *m*
e alosoma *m*
i allosoma *m*
d Allosom *n*

104 ALLOSYNAPSIS;
ALLOSYNDESIS
f allosyndèse *f*
e alosíndesis *f*
i allosindesi *f*
d Allosynapsis *f;*
Allosyndese *f*

105 ALLOSYNDETIC
f allosyndétique
e alosindético
i allosindetico
d allosyndetisch

106 ALLOTETRAPLOID (adj)
ALLOTETRAPLOID
f allotétraploïde
allotétraploïde *m*
e alotetraploide
alotetraploide *m*
i allotetraploide

allotetraploide *m*
d allotetraploid
Allotetraploide *f*

107 ALLOTETRAPLOIDY
f allotétraploïdie *f*
e alotetraploidía *f*
i allotetraploidismo *m*
d Allotetraploidie *f*

108 ALLOTRIPLOID (adj)
ALLOTRIPLOID
f allotriploïde
allotriploïde *m*
e alotriploide
alotriploide *m*
i allotriploide
allotriploide *m*
d allotriploid
Allotriploide *f*

109 ALLOTRIPLOIDY
f allotriploïdie *f*
e alotriploidía *f*
i allotriploidismo *m*
d Allotriploidie *f*

110 ALLOTYPIC
f allotypique
e alotípico
i allotipico
d allotyp;
allotypisch

111 ALLOZYGOTE
f allozygote *m*
e alocigotó *m*
i allozigote *m*
d Allozygote *f*

112 ALTERNATE DOMINANCE;
ALTERNATIVE DOMINANCE
f dominance *f* alternée
e dominancia *f* alterna
i dominanza *f* alternata
d alternative Dominanz *f*

113 ALTERNATING DOMINANCE
f dominance *f* alternée
e dominancia *f* alternativa
i dominanza *f* alternativa
d wechselnde Dominanz *f (#)*

114 ALTERNATION OF
    GENERATIONS
    f alternance *f* de générations
    e alternación *f* de generaciones
    i alternazione *f* di generazioni
    d Generationswechsel *m*

115 ALTERNATIVE DISJUNCTION;
    DISJUNCTIONAL
    SEPARATION
    f disjonction *f* alternative
    e disyunción *f* alternativa
    i disgiunzione *f* alternativa
    d Alternativverteilung *f*

116 ALTERNATIVE INHERITANCE
    f hérédité *f* alternée
    e herencia *f* alternativa
    i eredità *f* alternativa
    d alternative Vererbung *f*

117 ALTRUISTIC
    f altruiste
    e altruisto
    i altruistico
    d fremddienlich

118 ALVEOLAR HYPOTHESIS
    f hypothèse *f* des alvéoles
    e hipótesis *f* de los alveolos
    i ipotesi *f* degli alveoli
    d Alveolar-Hypothese *f*

119 AMBISEXUAL
    f bissexuel;
      bissexué;
      ambo-sexuel
    e ambisexual
    i bisessuale
    d ambisexual (Ⓞ);
      zweigeschlechtig;
      monözisch

120 AMBIVALENT
    f ambivalent
    e ambivalente
    i ambivalente
    d ambivalent

121 AMEIOSIS
    f améiose *f*

    e ameiosis *f*
    i ameiosi *f*
    d Ameiose *f*

122 AMEIOTIC
    f améiotique
    e ameiótico
    i ameiotico
    d ameiotisch

123 AMICTIC
    f amictique
    e amíctico
    i amittico
    d amiktisch

124 A-MISDIVISION
    f A-misdivision *f*
    e división *f* errónea A
    i A-misdivisione *f*
    d A-Missteilung *f*

125 AMITOSIS
    f amitose *f*
    e amitosis *f*
    i amitosi *f*
    d Amitose *f*

126 AMITOTIC
    f amitotique
    e amitótico
    i amitotico
    d amitotisch

127 AMIXIA;
    AMIXIS
    f amixie *f*; amixis *f*
    e amixia *f*;
      amixis *f*
    i amissia *f*;
      amixi *f*
    d Amixie *f*; Amixis *f*

128 AMORPHIC
    AMORPHOUS
    f amorphe
    e amorfo
    i amorfo
    d amorph

129 AMPHEROTOKY

f   amphérotoquie *f*
e   anferotoquía *f*
i   amferotochia *f*;
    anferotochia *f*
d   Ampherotokie *f*

130   AMPHIASTER
f   amphiaster *m*
e   anfiaster *m*
i   amfiaster *m*
d   Amphiaster *n*

131   AMPHIASTRALMITOSIS
f   mitose *f* amphiastrale
e   mitosis *f* anfiastral
i   mitosi *f* anfiastrale
d   Amphiastralmitose *f*

132   AMPHIBIVALENT
f   amphibivalent *m*
e   anfibivalente *m*
i   amfibivalente *m*;
    anfibivalente *m*
d   Amphibivalent *n*

133   AMPHIDIPLOID (adj.)
    AMPHIDIPLOID
f   amphidiploïde
    amphidiploïde *m*
e   anfidiploide
    anfidiploide *m*
i   amfidiploide;
    anfidiploide
    amfidiploide *m*;
    anfidiploide *m*
d   amphidiploid
    Amphidiploide *f*

134   AMPHIDIPLOIDY
f   amphidiploïdie *f*
e   anfidiploidía *f*;
    doble diploidía *f*
i   amfidiploidismo *m*;
    anfidiploidismo *m*
d   Amphidiploidie *f*

135   AMPHIGAMY
f   amfigamie *f*
e   anfigamia *f*
i   amfigamia *f*;
    anfigamia *f*
d   Amphigamie *f*

136   AMPHIGENESIS
f   amphigénèse *f*
e   anfigénesis *f*
i   anfigenesi *f*; amfigenesi *f*
d   Amphigenese *f*

137   AMPHIGONY
f   amphigonie *f*
e   anfigonía *f*
i   anfigenia *f*;
    anfigonia *f*
d   Amphigonie *f*

138   AMPHIHAPLOID (adj.)
    AMPHIHAPLOID
f   amphihaploïde
    amphihaploïde *m*
e   anfihaploide
    anfihaploide *m*
i   amfiaploide;
    anfiaploide
    amfiaploide *m*;
    anfiaploide *m*
d   amphihaploid
    Amphihaploide *f*

139   AMPHIHAPLOIDY
f   amphihaploïdie *f*
e   anfihaploidía *f*
i   amfiaploidismo *m*;
    anfiaploidismo *m*
d   Amphihaploidie *f*

140   AMPHIKARYON
f   amphicarion *m*
e   anficarión *m*
i   - -
d   Amphikaryon *n*

141   AMPHILEPSY
f   amphilepsie *f*
e   anfilepsis *f*
i   amfilepsi *f*;
    anfilepsi *f*
d   Amphilepsis *f*

142   AMPHIMICTIC
f   amphimictique
e   anfimíctico
i   amfimittico;
    anfimittico
d   amphimiktisch

143 AMPHIMIXIA;
     AMPHIMIXIS
     f amphimixie *f*
     e anfimixis *f;*
        anfimixia *f*
     i amfimissi *f;*
        anfimixi *f;*
        amfimixi *f;*
        anfimissi *f*
     d Amphimixis *f;*
        Amphimixie *f*

144 AMPHIMUTATION
     f amphimutation *f;*
        mutation double *f*
     e anfimutación *f;*
        mutación *f* doble
     i amfimutazione *f;*
        anfimutazione *f*
     d Amphimutation *f*

145 AMPHIPLASTY
     f amphiplastie *f*
     e anfiplastia *f*
     i amfiplastia *f;*
        anfiplastia *f*
     d Amphiplastie *f*

146 AMPHIPLOID (adj.)
     AMPHIPLOID
     f amphiploïde
        amphiploïde *m*
     e anfiploide
        anfiploide *m*
     i amfiploide;
        anfiploide
        amfiploide *m;*
        anfiploide *m*
     d amphiploid
        Amphiploide *f*

147 AMPHIPLOIDY
     f amphiploïdie *f*
     e anfiploidía *f*
     i amfiploidismo *m;*
        anfiploidismo *m*
     d Amphiploidie *f*

148 AMPHITENE (adj.)
     AMPHITENE
     f amphitène
        amphitène *m*
     e anfiteno

anfiteno *m*
i amfitene;
   anfitene
   amfitene *m;*
   anfitene *m*
d Amphitän *n* (auch in
   Zusammensetzungen)

149 AMPHITHALLIC
     f amphithallique
     e anfitálico
     i amfitallico;
        anfitallico
     d amphithallisch

150 AMPHITOKY
     f amphitoquie *f*
     e anfitoquía *f*
     i amfitocchia *f;*
        anfitocchia *f*
     d Amphitokie *f*

151 AMPHOGENIC
     f amphogène
     e anfogénico
     i amfogenico;
        anfogenico
     d amphogen

152 AMPHOGENY
     f amphogénie *f*
     e anfogenía *f*
     i amfogenia *f;*
        anfogenia *f*
     d Amphogenie *f*

153 AMPHOHETEROGONY
     f amphohétérogonie *f*
     e anfoheterogonía *f*
     i amfoeterogonia *f;*
        anfoeterogonia *f*
     d Amphoheterogonie *f*

154 ANABOLY
     f anabolie *f*
     e anabolismo *m*
     i anabolismo *m*
     d Anabolie *f*

155 ANACHROMASIS
     f anachromasie *f*
     e anacromasia *f*
     i anacromasi *f*

d   Anachromasis *f*

156   ANAGENESIS
f   anagénèse *f*
e   anagénesis *f*
i   anagenesi *f*
d   Anagenese *f*

157   ANAGENETIC
f   anagénétique
e   anagenético
i   anagenetico
d   anagenetisch

158   ANALOGUE
f   analogue *m*
e   análogo *m*
i   analogo *m*
d   analog
      Analoge *n*

159   ANAMORPHIC
f   anamorphique
e   anamórfico
i   anamorfico
d   anamorph

160   ANAMORPHISM;
      ANAMORPHOSIS
f   anamorphisme *m*
e   anamorfosis *f*
i   anamorfosi *f*
d   Anamorphose *f*
      Entartung *f*

161   ANAPHASE
f   anaphase *f*
e   anafase *f*
i   anafase *f*
d   Anaphase *f*

162   ANAPHASE MOVEMENT
f   mouvement *m* anaphasique
e   movimiento *m* anafásico
i   movimento *m* anafasico
d   Anaphasebewegung *f*

163   ANAPHASIC
f   anaphasique
e   anafásico
i   anafasico

d   Anaphase-(in Zusammen-
      setzungen)

164   ANAPHRAGMIC
f   anaphragmique
e   anafrágmico
i   anafragmico
d   anaphragmisch (#)

165   ANAREDUPLICATION
f   anaréduplication *f*
e   anareduplicación *f*
i   anareduplicazione *f*
d   Anareduplikation *f*

166   ANASCHISTIC
f   anaschistique
e   anasquístico
i   anaschistico
d   anaschistisch (#)

167   ANASTOMOSE
f   anastomose *f*
e   anastomosa *f*
i   anastomosi *f*
d   Anastomose *f*

168   ANASTRALMITOSIS
f   anastralmitose *f*
e   anastralmitosis *f*
i   anastralmitosi *f*
d   Anastralmitose *f*

169   ANDRO-AUTOSOME
f   andro-autosome *m*
e   androautosoma *m*
i   andro-autosoma *m*
d   Andro-Autosom *n*

170   ANDRODIOECIOUS
f   androdioïque
e   androdioico
i   androdioico
d   androdioecisch;
      androdioezisch;
      androdiözisch

171   ANDRODIOECISM;
      ANDRODIOECY
f   androdioécie *f*
e   androdioecia *f*

i   androdioicismo *m*
d   Androdioecie *f;*
     Androdioezie *f;*
     Androdiözie *f*

172   ANDROECY
f   androécie *f*
e   androecia *f;*
     andrecia *f*
i   androicia *f*
d   Androecie *f;*
     Androezie *f;*
     Andrözie *f*

173   ANDROGAMY
f   androgamie *f*
e   androgamía *f*
i   androgamia *f*
d   Androgamie *f*

174   ANDROGENESIS
f   androgénèse *f*
e   androgénesis *f*
i   androgenesi *f*
d   Androgenese *f*

175   ANDROGENETIC
f   androgénétique
e   androgenético
i   androgenetico
d   androgenetisch

176   ANDROGENIC
f   androgène
e   androgénico
i   androgenico
d   androgen

177   ANDROGYNOUS
f   androgyne
e   androgino
i   androgino
d   androgyn

178   ANDROGYNY
f   androginie *f*
e   androginia *f*
i   androginia *m*
d   Androgynie *f*

179   ANDROHERMAPHRODITE (adj.)

ANDROHERMAPHRODITE
f   androhermaphrodite
     androhermaphrodite *n*
e   androhermafrodito
     androhermafrodito *m*
i   androermafrodito
     androermafrodito *m*
d   androhermaphroditisch
     Androhermaphrodit *m*

180   ANDROMEROGONY
f   andromérogonie *f*
e   andromerogonia *f*
i   andromerogonia *f*
d   Andromerogonie *f*

181   ANDROMONOECIOUS
f   andromonoïque
e   andromonoico
i   andromonoico
d   andromonoecisch;
     andromonoezisch;
     andromonözisch

182   ANDROMONOECY
f   andromonoécie *f*
e   andromonoecía *f*
i   andromonoicismo *m*
d   Andromonoecie *f;*
     Andromonoezie *f;*
     Andromonözie *f*

183   ANDROSOME
f   androsome *m*
e   androsoma *m*
i   androsoma *m*
d   Androsom *n*

184   ANDROSPOROGENESIS
f   androsporogénèse *f*
e   androsporogénesis *f*
i   androsporogenesi *f*
d   Androsporogenese *f*

185   ANDROSTERILE
f   androstérile
e   androestéril
i   androsterile
d   androsteril; besser:
     im männlichen Geschlecht
     steril (kurz:männlich-steril)

186 ANDROSTERILITY
  f androstérilité *f*
  e androesterilidad *f*
  i androsterilità *f*
  d Androsterilität *f*; besser:
     Sterilität im männlichen
     Geschlecht

187 ANEUPLOID (adj.)
     ANEUPLOID
  f aneuploïde
     aneuploïde *m*
  e aneuploide
     aneuploide *m*
  i aneuploide
     aneuploide *m*
  d aneuploid
     Aneuploide *f*

188 ANEUPLOIDY
  f aneuploïdie *f*
  e aneuploidía *f*
  i aneuploidismo *m*
  d Aneuploidie *f*

189 ANEUSOMATY
  f aneusomatie *f* (Θ)
  e aneusomatía *f*
  i aneusomazia *f*
  d Aneusomatie *f*

190 ANISOAUTOPLOID (adj.)
     ANISOAUTOPLOID
  f anisoautoploïde
     anisoautoploïde *m*
  e anisoautoploide
     anisoautoploide *m*
  i anisoautoploide
     anisoautoploide *m*
  d anisoautoploid
     Anisoautoploide *f*

191 ANISOAUTOPLOIDY
  f anisoautoploïdie *f*
  e anisoautoploidía *f*
  i anisoautoploidismo *m*
  d Anisoautoploidie *f*

192 ANISOCARYOSIS
  f caryose *f*
  e anisocariosis *f*

  i anisocariosi *f*
  d Anisokaryose *f*

193 ANISOCYTOSIS
  f anisocytose *f*
  e anisocitosis *f*
  i anisocitosi *f*
  d Anisozytose *f*

194 ANISOGAMETE
  f anisogamète *m*
  e anisogameto *m*
  i anisogameto *m*
  d Anisogamet *m*

195 ANISOGAMY
  f anisogamie *f*
  e anisogamia *f*
  i anisogamia *f*
  d Anisogamie *f*

196 ANISOGENOMATIC
  f anisogénomatique
  e anisogenomático
  i anisogenomatico
  d anisogenomatisch

197 ANISOGENOMIC
  f anisogénomique
  e anisogenómico
  i anisogenomico
  d anisogenomatisch

198 ANISOGENY
  f anisogénie *f*
  e anisogenia *f*
  i anisogenia *f*
  d Anisogenie *f*

199 ANISOPLOID (adj.)
     ANISOPLOID
  f anisoploïde
     anisoploïde *m*
  e anisoploide
     anisoploide *m*
  i anisoploide
     anisoploide *m*
  d anisoploid
     Anisoploide *f*

200 ANISOPLOIDY

f   anisoploïdie *f*
e   anisoploidía *f*
i   anisoploidismo *m*
d   Anisoploidie *f*

201   ANISOSYNDESIS
f   anisosyndèse *f*
e   anisosíndesis *f*
i   anisosindesi *f*
d   Anisosyndese *f*

202   ANISOSYNDETIC
f   anisosyndétique
e   anisosindético
i   anisosindetico
d   anisosyndetisch

203   ANISOTRISOMY
f   anisotrisomie *f*
e   anisotrisomía *f*
i   anisotrisomia *f*
d   Anisotrisomie *f*

204   ANNIDATION
f   annidation *f*
e   anidación *f*
i   annidazione *f*
d   Annidation *f*;
     Einnischung *f*

205   ANORMOGENESIS
f   anormogénèse *f*
e   anormogénesis *f*
i   anormogenesi *f*
d   Anormogenese *f*

206   ANORTHOGENESIS
f   anorthogénèse *f*
e   anortogénesis *f*
i   anortogenesi *f*
d   Anorthogenese *f*

207   ANORTHOPLOID (adj.)
      ANORTHOPLOID
f   anorthoploïde
     anorthoploïde *m*
e   anortoploide
     anortoploide *m*
i   anortoploide
     anortoploide *m*
d   anorthoploid

Anorthoploide *f*

208   ANORTHOPLOIDY
f   anorthoploïdie *f*
e   anortoploidía *f*
i   anortoploidismo *m*
d   Anorthoploidie *f*

209   ANORTHOSPIRAL
f   anorthospirale *f*
e   anortoespiral *f*
i   anortospirale *f*
d   Anorthospirale *f*

210   ANTEPHASE
f   antéphase *f*
e   antefase *f*
i   antefase *f*
d   Antephase *f*

211   ANTICIPATION
f   anticipation *f*
e   anticipación *f*
i   anticipazione *f*
d   Anticipation *f*

212   ANTIMORPH
f   antimorphe
e   antimorfo
i   antimorfo
d   antimorph

213   ANTIMUTAGENE
f   antimutagène *m*
e   antimutageno *m*
i   antimutageno *m*
d   Antimutagen *n*

214   ANTIMUTAGENIC
f   antimutagénique
e   antimutagénico
i   antimutagenico
d   antimutagen

215   ANTI-RECAPITULATION
f   anti-récapitulation *f* (#)
e   antirecapitulación *f* (#)
i   pre-conclusione *f*;
     antiricapitolazione *f* (#)
d   Antirekapitulation *f* (#)

216 ANTITHETICAL
   f antithétique
   e antitético
   i antitetico
   d antithetisch

217 APHASIC
   f aphasique
   e afásico
   i afasico
   d aphasisch

218 APOGAMETY
   f apogamétie f
   e apogametia f
   i apogametia f
   d Apogametie f

219 APOGAMIC;
    APOGAMOUS
   f apogame
   e apogámico
   i apogamico
   d apogam

220 APOGAMY
   f apogamie f
   e apogamia f
   i apogamia f
   d Apogamie f

221 APOHOMOTYPIC
   f apohomotypique
   e apohomotípico
   i apoomotipico
   d apohomotypisch

222 APOMEIOSIS
   f apoméiose f
   e apomeiosis f
   i apomeiosi f
   d Apomeiose f

223 APOMICTIC
   f apomictique
   e apomíctico
   i apomittico
   d apomiktisch

224 APOMICTOSIS

   f apomictose f
   e apomictosis f
   i apomitosi f
   d Apomiktosis f

225 APOMIXIA; APOMIXIS
   f apomixie f
   e apomixia f
   i apomissia f;
     apomixi f
   d Apomixis f

226 APOROGAMY
   f aporogamie f
   e aporogamia f
   i aporogamia f
   d Aporogamie f

227 APOSPORY
   f aposporie f
   e asporía f
   i asporia f
   d Aposporie f

228 ARCHALLAXIS
   f archallaxie f
   e arcalaxis f
   i arcallassi f
   d Archallaxis f

229 ARCHEBIOSIS
   f archébiose f
   e arquebiosis f
   i archebiosi f
   d Urzeugung f

230 ARCHETYPE
   f archétype m
   e arquetipo m
   i archetipo m
   d Urtyp m;
     Archetyp m

231 ARCHIPLASM; ARCHOPLASM
   f archiplasme m
   e arquiplasma m
   i archiplasma m
   d Archiplasma n;
     Archoplasma n

233 AROMORPHOSIS
f aromorphose *f*
e aromorfosis *f*
i aromorfosi *f*
d Aromorphose *f (#)*

234 ARRHENOGENIC
f arrhénogénique
e arrenogenético
i arrenogenetico
d arrhenogenetisch

235 ARRHENOGENY
f arrhénogénie *f*
e arrenogenía *f*
i arrenogenia *f*
d Arrhenogenie *f*

236 ARRHENOTOKY
f arrhénotoquie *f*
e arrenetoquía *f*
i arrenotochia *f*
d Arrhenotokie *f*

237 ARTIFACT
f artefact *m*
e artefacto *m*
i artefatto *m*
d Artefakt *n*

238 ARTIFICIAL
PARTHENOGENESIS
f parthénogénèse *f*
artificielle
e partenogénesis *f*
artificial
i partenogenesi *f*
artificiale
d künstliche
Parthenogenese *f*

239 ARTIOPLOID (adj.)
ARTIOPLOID
f artioploïde
artioploïde *m*
e artioploide
artioploide *m*
i artioploide
artioploide *m*
d artioploid

Artioploide *f*

240 ARTIOPLOIDY
f artioploïdie *f*
e artioploidía *f*
i artioploidismo *m*
d Artioploidie *f*

241 ASEXUAL
f asexué, asexuel
e asexual
i asessuale
d ungeschlechtlich *f;*
asexuell

242 ASEXUAL REPRODUCTION
f reproduction *f* asexuée
e reproducción *f* asexual
i riproduzione *f* asessuale
d ungeschlechtliche
Fortpflanzung *f*

243 ASSOCIATION
f association *f*
e asociación *f*
i associazione *f*
d Assoziation *f*

244 ASSORTATIVE MATING
ASSORTIVE MATING
f accouplement *m* par
affinité phénotypique (Θ);
homogamie *f*
e apareamiento *m* análogo
i omogamia *f*
d Paarung *f* ähnlicher
Individuen;
Homogamie *f*

245 ASSORTMENT
f assortiment *m*
e separación *f* (de genes o
cromosomas sexuales en
meiosis)
i separazione *f*
d Genverteilung *f*

246 ASTER
f aster *m*
e áster *m*

    i aster *m*
    d Aster *n*

247   ASTROCENTRE
    f astrocentre *m*
    e astrocentro *m*
    i astrocentro *m*
    d Astrosphäre *f*

248   ASTROSPHERE
    f astrosphère *f*
    e astroesfera *f*
    i astrosfera *f*
    d Astrosphäre *f*

249   ASYNAPSIS;ASYNDESIS
    f asynapsis *f*
    e asindesis *f*;
       asinapsis *f*
    i asinapsi *f*
    d Asynapsis *f*;
       Asyndese *f*

250   ASYNAPTIC
    f asynaptique
    e asináptico
    i asinaptico
    d asynaptisch

251   ASYNDESIS; ASYNAPSIS
    f asynapsis *f*
    e asinapsis *f*; asindesis *f*
    i asinapsi *f*
    d Asynapsis *f*; Asyndese *f*

252   ASYNGAMIC
    f asyngamique
    e asingámico
    i asingamico
    d asyngam

253   ASYNGAMY
    f asyngamie *f*
    e asingamia *f*
    i asingamia *f*
    d Asyngamie *f*

254   ATACTOGAMY
    f atactogamie *f*
    e atactogamia *f*
    i atactogamia *f*

    d Ataktogamie *f*

255   ATAVISM
    f atavisme *m*
    e atavismo *m*
    i atavismo *m*
    d Atavismus *m*

256   ATAVISTIC
    f atavique
    e atávico
    i atavico
    d atavistisch

257   ATELOMITIC
    f atélomitique
    e atelomítico
    i atelomíttico
    d atelomitisch

258   ATORSIONAL
    f paranémique
    e atorsional
    i atorsionale
    d untordiert (#)

259   ATRACTOPLASM
    f atractoplasme *m*
    e atractoplasma *m*
    i atrattoplasma *m*
    d Atraktoplasma *n*

260   ATRACTOSOME
    f atractosome *m*
    e atractosoma *m*
    i atrattosoma *m*
    d Atraktosom *n*

261   ATTACHED X-CHROMOSOME
    f chromosome X *m* attaché
    e cromosoma-X *m* agregado
    i cromosoma-X *m* attaccato
    d "attached" X-Chromosom *n*

262   ATTACHED X Y-CHROMOSOM
    f chromosome X Y *m* attaché
    e agregado X Y *m*
    i cromosomi-X e Y *mpl*
       attaccati
    d "attached" X Y-Chromosom

263 ATTACHMENT
   f fixation *f*
   e fijación *f*
   i fissazione *f*
   d a. Spindelfaseransatzstelle *f*
      b. Chromosomenfusion *f*

264 ATTENUATION
   f atténuation *f*
   e atenuación *f*
   i attenuazione *f*
   d Verdünnung *f*;
      Verschmälerung *f*

265 ATTRACTION
   f attraction *f*
   e atracción *f*
   i attrazione *f*
   d Attraktion *f*;
      Anziehung *f*

266 AUTOADAPTATION
   f autoadaptation *f*
   e autoadaptación *f*
   i autoadattamento *m*
   d Autoadaptation *f*

267 AUTO-ALLOPLOID (adj.)
   AUTO-ALLOPLOID
   f auto-alloploïde
      auto-alloploïde *m*
   e autoaloploide
      autoaloploide *m*
   i autoalloploide
      autoalloploide *m*
   d autoalloploid
      Autoalloploide *f*

268 AUTO-ALLOPLOIDY
   f auto-alloploïdie *f*
   e autoaloploidía *f*
   i autoalloploidismo *m*
   d Autoalloploidie *f*

269 AUTOBIVALENT
   f autobivalent *m*
   e autobivalente *m*
   i autobivalente *m*
   d Autobivalent *n*

270 AUTOGAMOUS
   f autogame
   e autógamo
   i autogamo
   d autogam

271 AUTOGAMY
   f autogamie *f*
   e autogamia *f*
   i autogamia *f*
   d Autogamie *f*
      Selbstbefruchtung *f*

272 AUTOGENESIS
   f autogénèse *f*
   e autogénesis *f*
   i autogenesi *f*
   d Autogenese *f*

273 AUTOGENETIC
   f autogénétique
   e autogenético
   i autogenetico
   d autogenetisch

274 AUTOGENIC
   f autogénique
   e autogénico
   i autogenico
   d autogenisch

275 AUTOGENOMATIC
   f autogénomique
   e autogenomático
   i autogenomatico
   d autogenomatisch

276 AUTOGENOUS
   f autogène
   e autógeno
   i autogeno
   d autogen

277 AUTOHETEROPLOID (adj.)
   AUTOHETEROPLOID
   f autohétéroploïde
      autohétéroploïde *m*
   e autoheteroploide
      autoheteroploide *m*
   i autoeteroploide
      autoeteroploide *m*
   d autoheteroploid
      Autoheteroploide *f*

278 AUTOHETEROPLOIDY

f   autohétéroploïdie *f*
e   autoheteroploidía *f*
i   autoeteroploidismo *m*
d   Autoheteroploidie *f*

279   AUTOMIXIS
f   automixie *f*
e   automixis *f*
i   automissi *f*
d   Automixis *f*

280   AUTOMUTAGENE
f   automutagène *m*
e   automutagen *m*
i   automutagene *m*
d   Automutagen *n*

281   AUTOMUTAGENIC
f   automutagénique
e   automutagénico
i   automutagenico
d   automutagen

282   AUTONOMOUS FACTOR
f   facteur *m* autonome
e   factor *m* autónomo
i   fattore *m* autonomo
d   autonomer Faktor *m*

283   AUTO-ORIENTATION
f   auto-orientation *f*
e   auto-orientación *f*
i   auto-orientazione *f*
d   Autoorientierung *f*

284   AUTOPARTHENOGENESIS
f   autoparthénogénèse *f*
e   autopartenogénesis *f*
i   autopartenogenesi *f*
d   Autoparthenogenese *f*

285   AUTOPLASTIC
f   autoplastique
e   autoplástico
i   autoplastico
d   autoplastisch

286   AUTOPLOID (adj.)
AUTOPLOID
f   autoploïde
autoploïde *m*

e   autoploide
autoploide *m*
i   autoploide
autoploide *m*
d   autoploid
Autoploide *f*

287   AUTOPLOIDY
f   autoploïdie *f*
e   autoploidía *f*
i   autoploidismo *m*
d   Autoploidie *f*

288   AUTOPOLYPLOID (adj.)
AUTOPOLYPLOID
f   autopolyploïde
autopolyploïde *m*
e   autopoliploide
autopoliploide *m*
i   autopoliploide
autopoliploide *m*
d   autopolyploid
Autopolyploide *f*

289   AUTOPOLYPLOIDY
f   autopolyploïdie *f*
e   autopoliploidía *f*
i   autopoliploidismo *m*
d   Autopolyploidie *f*

290   AUTOSEGREGATION
f   autoségrégation *f*
e   autosegregación *f*
i   autosegregazione *f*
d   Autosegregation *f*

291   AUTOSOMAL
f   aurosomal
e   autosomal
i   autosomale
d   autosomal

292   AUTOSOME
f   autosome *m*
e   autosoma *m*
i   autosoma *m*
d   Autosom *n*

293   AUTOSOMIC
f   autosomique
e   autosómico

i autosomico
d autosomisch

294 AUTOSUBSTITUTION
f autosubstitution *f*
e autosubstitución *f*
i autosostituzione *f*
d Autosubstitution *f*

295 AUTOSYNDESIS
f autosyndèse *f*
e autosíndesis *f*
i autosindesi *f*
d Autosyndese *f*

296 AUTOSYNDETIC
f autosyndétique
e autosindético
i autosindetico
d autosyndetisch

297 AUTOTETRAPLOID (adj.)
AUTOTETRAPLOID
f autotétraploïde
autotétraploïde *m*
e autotetraploide
autotetraploide *m*
i autotetraploide
autotetraploide *m*
d autotetraploid
Autotetraploide *f*

298 AUTOTETRAPLOIDY
f autotétraploïdie *f*
e autotetraploidía *f*
i autotetraploidismo *m*
d Autotetraploidie *f*

299 AUTOTRANSPLANTATION
f autotransplantation *f*
e autotransplantación *f*
i autotrasplantazione *f*
d Autotransplantation *f*

300 AUTOTRIPLOID (adj.)
AUTOTRIPLOID

f autotriploïde
autotriploïde *m*
e autotriploide
autotriploide *m*
i autotriploide
autotriploide *m*
d autotriploid
Autotriploide *f*

301 AUTOTRIPLOIDY
f autotriploïdie *f*
e autotriploidía *f*
i autotriploidismo *m*
d Autotriploidie *f*

302 AUXOCYTE
f auxocyte *m*
e auxocito *m*
i auxocito *m*
d Auxocyte *f*;
Auxozyte *f*

303 AUXOSPIREME
f auxospirème *f*
e auxoespirema *f*
i auxospirema *f*
d Auxospirem *n* (#)

304 AUXOTROPHIC
f auxotrophe
auxotrophique
e auxótrofo
i auxotrofico
d auxotroph

305 AZYGOGENIC
f azygogène
e acigogénico
i azigogenico
d azygogenetisch

306 AZYGOTE
f azygote *m*
e acigoto *m*
i azigote *m*
d Azygote *f*

B

307 BACK CROSS
f recroisement *m;*
rétrocroisement *m;*
croisement *m* en retour;
backcross *m*
e retrocruzamiento *m*
i rincrocio *m*
d Rückkreuzung *f*

308 BACK CROSS BREEDING
f reproduction *f* par
rétrocroisement
e retroreproducción *f*
i riproduzione *f* per
rincrocio
d Rückkreuzungszüchtung *f*

309 BACK CROSS PARENT
f parent *m* récurrent
parent *m* de backcross
e progenitor *m* recurrente
i genitore *m* ricorrente
d Rückkreuzungselter *m*

310 BACK CROSS RATIO
f proportion *f* de
rétrocroisements;
proportion *f* de backcross
e relación *f* de
retrocruzamiento
i proporzione *f* di rincroci
d Rückkreuzungsverhältnis *n*

311 BACKGROUND ADAPTATION
f adaptation *f* au milieu
e adaptación *f* al ambiente
i adattamento *m* all'ambiente
d Milieu-Anpassung *f*

312 BACK MUTATION
f mutation *f* reverse;
mutation *f* renversée;
mutation *f* de retour
e retromutación *f;*
mutación *f* invertida;

mutación *f* de retroceso
i mutazione *f* di ritorno
d Rückmutation *f*

313 BACK POLLINATING
f rétrocroisement *m;*
croisement *m* en retour;
backcross *m*
e retropolinización *f*
i rincrocio *m*
d Rückbestäubung *f*

314 BACTERIOCIN
f bactériocine *f*
e bacteriocina *f*
i batteriocina *f*
d Bacteriocin *n;*
Bakteriozin *n*

315 BACTERIOCINOGENESIS
f bactériocinogénèse *f*
e bacteriocinogénesis *f*
i bacteriocinogenesi *f*
d Bacteriocinbildung *f*

316 BACTERIOLYSIS
f bactériolyse *f*
e bacteriólisis *f*
i bacteriolisi *f*
d Bakteriolyse *f;*
Lysis *f*

317 BACTERIOLYTIC
f bactériolytique
e bacteriolítico
i batteriolitico
d bakteriolytisch;
lytisch

318 BACTERIOPHAGE
f bactériophage *m*
e bacteriófago *m*
i batteriofago *m*
d Bakteriophage *m;*
Phage *m*

319 BACTERIOPHAGE
INCOMPATIBILITY
f incompatibilité *f* du
bactériophage
e incompatibilidad *f* del
bacteriófago
i incompatibilità *f*
batteriofaga
d Bakteriophageninkompati-
bilität *f*

320 BACTERIOPHAGE
REDUCTION
f réduction *f* du
bactériophage
e reducción *f* del
bacteriófago
i riduzione *f* batteriofaga
d Bakteriophagenreduktion *f*

BACTERIUM, DEFECTIVE
LYSOGENIC see 715

321 BACULIFORM
f bacilliforme; en forme de
bâtonnet
e baculiforme
i baculiforme
d stabförmig;
stäbchenförmig

322 BALANCE
f équilibre *m*
e equilibrio *m*
i equilibrio *m*
d Gleichgewicht *n*;
Balance *f*

323 BALANCED
f équilibré
balancé
e equilibrado
i equilibrato
d balanciert; im Gleichgewicht befindlich

324 BALANCED GAMETES
f gamètes *mpl* équilibrés
e gametos *mpl* equilibrados
i gameti *mpl* equilibrati
d balancierte Gameten *mpl*

325 BALANCED LETHALS
f léthaux *mpl* balancés
e letales *mpl* equilibrados
i letali *mpl* equilibrati
d balancierte Letalfaktoren
mpl

326 BALANCED POLYMORPHISM
f polymorphisme *m* équilibré
e polimorfismo *m* equilibrado
i polimorfismo *m* equilibrato
d balancierter Polymorphismus *m*

327 BALANCED POLYPLOIDY
f polyploïdie *f* équilibrée
e poliploidía *f* equilibrada
i poliploidismo *m*
equilibrato
d balancierte Polyploidie *f*

328 BALANCE THEORY OF SEX
DETERMINATION
f théorie *f* de l'équilibre de
la détermination des sexes
e teoría *f* del equilibrio de
la determinación de los
sexos
i teoria *f* dell'equilibrio
della determinazione dei
sessi
d Balance-Theorie *f* der
Geschlechtsbestimmung

329 BALBIANI RING
f anneau *m* de Balbiani
e anillo *m* de Balbiani
i anello *m* di Balbiani
d Balbiani-Ring *m*

330 BALDWIN EFFECT
f effet *m* de Baldwin
e efecto *m* de Baldwin
i effetto *m* di Baldwin
d Baldwin-Effekt *m*

331 BALL METAPHASE
f métaphase *f* en forme de
boule (Θ)
e metafase *f* de bola
i metafase *f* a bolla

e Ballmetaphase *f*

332 BAND
  f bande *f*
  e banda *f*
  i banda *f*
  d Querscheibe *f*

333 BARRAGE
  f barrage *m*
  e barrera *f*
  i chiusura *f*
  d Barrage *f*

334 BARRELLIKE SPINDLE
  f fuseau *m* en forme de
    tonnelet
  e huso *m* en forma de barril
  i fuso *m* in forma di barile
  d Tonnenspindel *f*

335 BARRIER
  f isolation *f;*
    barrière *f*
  e isolación *f*
  i barriera *f;*
    impedimento *m*
  d Kreuzungsbarriere *f*

336 BASICHROMATIN
  f basichromatine *f*
  e basicromatina *f*
  i basicromatina *f*
  d Basichromatin *n*

337 BASIC .NUMBER
  f nombre *m* de base
  e número *m* básico
  i numero *m* basico
  d Basiszahl *f*

338 BASIDIUM (*pl*: ia)
  f baside *f*
  e basidio *m*
  i basidio *m*
  d Basidie *f*

339 BATHMIC FORCE
  f force *f* bathmique
  e energía *f* bátmica
  i forza *f* batmica

d bathmische Kraft *f* (#)

340 B-CHROMOSOME
  f chromosome B *m*
  e cromosoma B *m*
  i cromosoma B *m*
  d B-Chromosom *n*

341 BEHAVIOUR FLEXIBILITY
  f flexibilité *f* du comporte-
    ment
  e flexibilidad *f* de
    comportamiento
  i flessibilità *f* di
    comportamento
  d Verhaltensflexibilität *f*

342 BIGENERIC
  f bigénérique
  e bigenérico
  i bigenerico
  d bigenerisch (Θ);
    zwei Gattungen betreffend

343 BIGENERIC CROSS
  f croisement *m* bigénérique
  e cruza *f* intergenérica
  i incrocio *m* bigenerico
  d Gattungskreuzung *f*

344 BINUCLEATE
  f binucléé
  e binucleado
  i binucleato
  d binucleat;
    binukleat;
    zweikernig

345 BIOBLAST
  f bioblaste *m*
  e bioblasto *m*
  i bioblasto *m*
  d Bioblast *m*

346 BIOCHORE
  f biochore *m*
  e biocora *f*
  i biocora *f;*
  d Biochore *f;*
    Lebensraum *m*

347 BIOGENE *m*
   f biogène *m*
   e biógeno *m*
   i biogeno *m*
   d Biogen *n*

348 BIOGENESIS
   f biogénèse *f*
   e biogénesis *f*
   i biogenesi *f*
   d Biogenese *f*

349 BIOGENETIC
   f biogénétique
   e biogenético
   i biogenetico
   d biogenetisch

350 BIOGENIC ISOLATION
   f isolement *m* biogénique
   e aislamiento *m* biogénico
   i isolamento *m* biogenetico
   d biogenetische Isolation *f*

351 BIOGENY
   f biogénie *f*
   e biogenía *f*
   i biogenia *f*
   d Entstehung *f* des Lebens

352 BIOLOGICAL CONTROL
   f contrôle *m* biologique
   e control *m* biológico
   i controllo *m* biologico
   d biologische Bekämpfung *f*
      (von Schädlingen)

353 BIOLOGICAL ISOLATION
   f isolement *m* biologique
   e aislamiento *m* biológico
   i isolamento *m* biologico
   d biologische Isolierung *f*

354 BIOLOGICAL RACE
   f race *f* biologique
   e raza *f* biológica
   i razza *f* biologica
   d biologische Rasse *f*

355 BIOLOGICAL SPECTRUM
   (*pl* spectra)

   f spectre *m* biologique
   e espectro *m* biológico
   i spettro *m* biologico
   d biologisches Spektrum *n*
      (*pl*: Spektren)

356 BIOMASS
   f biomasse *f*
   e biomasa *f*
   i biomassa *f*
   d - -

357 BIOMECHANICS
   f biomécanique *f*
   e biomecánica *f*
   i biomeccanica *f*
   d Biomechanik *f*

358 BIOMETRICAL
   f biométrique
   e biométrico
   i biometrico
   d biometrisch

359 BIOMETRICS
   f biométrie *f*
   e biometría *f*
   i biometrica *f*
   d Biometrie *f*

360 BIOMETRY
   f biométrie *f*
   e biometría *f*
   i biometria *f*
   d Biometrie *f*

361 BIOMUTANT
   f biomutant *m*
   e biomutante *m*
   i biomutante *m*
   d Biomutante *f* (Θ)

362 BION
   f bion *m*
   e bion *m*
   i bion *m*
   d Bion *n*

363 BIOPHORE
   f biophore *m*

e bióforo *m*
i bioforo *m*
d Biophore *f*

364 BIOPLASM
f bioplasme *m*
e bioplasma *m*
i bioplasma *m*
d Bioplasma *n* (☉);
   Protoplasma *n*

365 BIOPOESIS
f biopoèse *f*
e biopoesis *f*
i biopoesi *f*
d Urzeugung *f*

366 BIOSOME
f biosome *m*
e biosoma *m*
i biosoma *m*
d Biosom *n*

367 BIOSYNTHESIS
f biosynthèse *f*
e biosíntesis *f*
i biosintesi *f*
d Biosynthese *f*

368 BIOSYSTEMATICS
f biosystématique *f*
e biosistemática *f*
i biosistematica *f*
d Systematik *f* der Lebewe-
   sen

369 BIOTIC
f biotique
e biótico
i biotico
d biotisch (#)

370 BIOTIC POTENTIAL
f potentiel *m* biotique
e potencial *m* biótico
i potenziale *m* biotico
d biotisches Potential *n*

371 BIOTIC RESISTANCE
f résistance *f* biotique
e resistencia *f* biótica
i resistenza *f* biotica

d biotische Resistenz *f*

372 BIOTOPE
f biotope *m*
e biotopo *m*
i biotopo *m*
d Biotop *m*

373 BIOTYPE
f biotype *m*
e biotipo *m*
i biotipo *m*
d Biotyp *m*

374 BIPARTITE
f biparti; bipartite
e bipartido
i bipartito
d a. Bipartit *n* (Chromosom)
   b. Halbseitenzwitter *m*
      (Gynandromorphe)

375 BIPARTITION
f bipartition *f*
e bipartición *f*
i bipartizione *f*
d Mosaikbildung *f*

376 BIREFRINGENT
f biréfringent
e birefringente
i birifrangente
d doppel(strahlen)brechend

377 BISEXUAL
f bissexué; bissexuel
e bisexual
i bisesso; bisessuale
d bisexuell;
   zweigeschlechtig

378 BITHALLIC
f bithallique
e bitálico
i bitallico
d bithallisch

379 BIVALENT
f bivalent *m*
e bivalente *m*
i bivalente *m*

d Bivalent *n*

BIVALENT, UNEQUAL
see 2890

380 BLASTOCYTE
f blastocyte *m*
e blastocito *m*
i blastocito *m*
d Blastocyte *f*; Blastozyte *f*

381 BLASTOGENESIS
f blastogénèse *f*
e blastogénesis *f*
i blastogenesi *f*
d Blastogenese *f*

382 BLASTOGENETIC
BLASTOGENIC
f blastogénique
e blastogenético
i blastogenetico
d blastogenetisch

383 BLASTOMERE
f blastomère *m*
e blastómero *m*
i blastomero *m*
d Blastomere *f*

384 BLASTOVARIATION
f blastovariation *f*
e blastovariación *f*
i blastovariazione *f*
d Blastovariation *f*

385 BLENDING CHARACTERS
f caractères *mpl* mêlés
e carácteres *mpl* mezclados
i caratteri *mpl* mescolati
d quantitative Merkmale *npl*

386 BLENDING INHERITANCE
f hérédité *f* à facteurs
multiples
e herencia *f* mezclada
i eredità *f* mista
d Vererbung *f* quantitativer
Merkmale

387 BLEPHAROPLAST

f blépharoplaste *m*
e blefaroplasto *m*
i blefaroplasto *m*
d Blepharoplast *m*

388 BLOCK
f bloc *m*
e bloque *m*
i blocco *m*
d Block *m*

389 BLOCK MUTATION
f mutation *f* chromosomique
e mutación *f* de bloques
i mutazione *f* di blocchi
d Blockmutation *f*

390 BLOCK OF GENES
f bloc *m* de gènes
e bloque *m* de genes
i blocco *m* di geni
d Genblock *m*

391 BLOCK PATTERN EFFECT
f effet *m* de combinaison en
bloc
e efecto *m* de combinación en
bloque
i effetto *m* di combinazione in
blocco
d Sperrmustereffekt *m*

392 BLOOD-LINE
f lignée *f*
e línea *f* sanguínea
i stirpe *f*
d Linie *f*

393 BLOOD RELATION
f rapport *m* consanguin
e relación *f* consanguínea
i relazione *f* consanguinea
d Blutsverwandtschaft *f*

394 BODY CELL
f cellule *f* somatique
e célula *f* somática
i cellula *f* somatica
d Körperzelle *f*

395 BOTTLE NECK PHENOMENON

f phénomène *m*
d'étranglement
e fenómeno *m* del cuello de
botella
i fenomeno *m* di
strozzamento
d "Bottleneck"-Phänomen *n*;
Flaschenhals-Phänomen *n* (#)

396 BOTTOM RECESSIVE
f récessif total *m*
e recesivo inferior *m*
i recessivo inferiore *m* (⊙)
d bottom recessive *n*

397 BOUQUET STAGE
f stade *m* du bouquet
e estado *m* de ramillete
i stadio *m* di fascio
d Bukettstadium *n*

398 BRACHYMEIOSIS
f brachyméiose *f*
e braquimeiosis *f*
i brachimeiosi *f*
d Brachymeiosis *f*

399 BRADYTELIC
f bradytélique
e braditélico
i braditelico
d bradytelisch

400 BRANCHED
f fourchu
e ramificado
i ramificato
d verzweigt

401 BRANCHED CHROMOSOME
f chromosome *m* fourchu
e cromosoma *m* ramificado
i cromosoma *m* ramificato
d verzweigtes Chromosom *n*

402 BRANCHING POINT
f point *m* de ramification
e punto *m* de ramificación
i punto *m* di ramificazione
d Verzweigungsstelle *f*

BREAK, HALF CHROMATID
see 1177

BREAK, ISOCHROMATID
see 1565

BREAKAGE, PARTIAL
see 2076

BREAKAGE, SECONDARY
see 716

403 "BREAKAGE FIRST"
HYPOTHESIS
f hypothèse *f* de la cassure
d'abord
e hipótesis *f* de "ruptura
inicial"
i ipotesi *f* della "prima
rottura"
d Bruch-Hypothese *f*

404 BREAKAGE-FUSION BRIDGE-
CYCLE
f cycle *m* "rupture-fusion-
pont" (⊙)
e ciclo *m* "ruptura-fusión-
puente" (⊙)
i ciclo *m* "rottura-fusione-
ponte" (⊙)
d Bruch-Fusionbrücken-
Zyklus *m*

405 BREAKAGE-REUNION
BIVALENT
f bivalent *m* "rupture-
réunion"
e bivalente *m* de ruptura y
reunión
i bivalente *m* di rottura e
riunione
d Bruch-Reunions-Bivalent *n*

406 BREAK CONSTRICTION
f constriction *f* de rupture
e constricción *f* de ruptura
i costrizione *f* di rottura
d Bruch-Einschnürung *f*

407 BREED
f race *f*

e   raza *f*
i   razza *f*
d   Brut *f;*
    Rasse *f;*
    Herkunft *f*

408   BREEDING
f   reproduction *f;* élevage *m*
e   reproducción *f;* cría *f*
i   riproduzione *f;*
    allevamento *m*
d   Zucht *f;* Züchtung *f*

409   BRIDGE-FRAGMENT
      CONFIGURATION
f   configuration *f* "fragment-
    pont" (Θ)
e   configuración *f* de
    fragmento-puente
i   configurazione *f* a ponte-
    frammento
d   Brücken-Fragment-
    Konfiguration *f*

410   BROCHONEMA
f   brochonème *f*
e   broconema *f*
i   broconema *f*
d   Brochonema *n*

411   BRUSQUE VARIATION

f   variation *f* brusque
e   variación *f* brusca
i   variazione *f* brusca
d   diskontinuierliche, alter-
    native, qualitative Varia-
    bilität *f*

412   BUD
f   bourgeon *m;* bouton *m*
e   yema *f*
i   germoglio *m;* gemma *f;*
    bottone *m*
d   Knospe *f*

413   BUD MUTATION
f   mutation *f* en bourgeon;
    mutation *f* bourgeonneuse
e   mutación *f* de yema;
    mutación *f* de sport
i   mutazione *f* gemmaria
d   Knospenmutation *f*

414   BUFFER GENE;
      BUFFERING GENE
f   gène *m* tampon
e   gen *m* amortiguador;
    gen *m* "Buffer"
i   gene *m* tampone;
    "buffer" *m*
d   Puffergen *n*

C

415 CACOGENESIS;
KACOGENESIS
f cacogénèse *f*
e cacogenesia *f*;
cacogénesis *f*
i cacogenesi *f*
d Kakogenese *f*

416 CACOGENIC; KAKOGENIC
f cacogénique
e cacogénico
i cacogenico
d kakogenisch

417 CACOGENICS
f cacogénie *f*
e cacogenia *f*
i cacogenia *f*
d Kakogenie *f*

418 CAENOGENESIS
f cénogénèse *f*;
caenogénèse *f*
e cenogénesis *f*
i cenogenia *f*
d Caenogenesis *f*

419 CAENOGENETIC
f cénogénétique;
caenogénétique
e cenogenético
i cenogenetico
d caenogenetisch

420 CANALIZATION
f canalisation *f*
e canalización *f*
i canalizzazione *f*
d Kanalisierung *f*

421 CANALIZED DEVELOPMENT
HYPOTHESIS
f hypothèse *f* du développement canalisé
e hipótesis *f* del desarrollo
canalizado

i ipotesi *f* dello sviluppo
canalizzato
d Hypothese *f* der kanali-
sierten Entwicklung

422 CANALIZING SELECTION
f sélection *f* canalisatrice
e selección *f* canalizante
i selezione *f* canalizzante
d kanalisierende Selektion *f*

423 CARPOXENIA
f carpoxénie *f*
e carpoxenia *f*
i carpoxenia *f*;
xenia *f* del frutto
d Carpoxenie *f*

424 CARYOGAMY
f caryogamie *f*
e cariogamia *f*
i cariogamia *f*
d Karyogamie *f*

425 CATAGENESIS
f catagénèse *f*
e catagénesis *f*
i catagenesi *f*;
attivazione *f* genetica
d Katagenese *f* (≠)

426 CATAGENETIC
f catagénétique
e catagenético
i catagenetico
d katagenetisch (≠)

427 CATAPHASE
f cataphase *f*
e catafase *f*
i catafase *f*
d Kataphase *f*

428 CATENATION
f caténation *f*
e catenación *f*

i   catenazione *f*
d   Kettenbildung *f*

429   CELL
   f   cellule *f*
   e   célula *f*
   i   cellula *f*
   d   Zelle *f*

430   CELL CONJUGATION
   f   conjugaison *f* cellulaire
   e   conjugación *f* celular
   i   coniugazione *f* cellulare
   d   Zellenkonjugation *f*

431   CELL DIVISION
   f   division *f* cellulaire
   e   división *f* celular
   i   divisione *f* cellulare
   d   Zellteilung *f*

432   CELL GRANULE
   f   ectosome *m*
   e   ectosoma *m*
   i   ectosoma *m*
   d   Ektosom *n*

433   CENOGENETIC;
      CAENOGENETIC
   f   cénogénétique;
      caenogénétique
   e   cenogenético
   i   cenogenetico
   d   caenogenetisch

434   CENTRAL BODY
   f   centrosome *m*;
      corpuscule central *m*
   e   centrosoma *m*;
      corpúsculo *m* central
   i   centrosoma *m*;
      corpuscolo *m* centrale
   d   Zentralkörper *m*;
      Centrosom *n*

435   CENTRALIZATION
   f   centralisation *f*
   e   centralización *f*
   i   centralizzazione *f*
   d   Zentralisation *f*

436   CENTRAL SPINDLE

   f   fuseau *m* central
   e   huso *m* central
   i   fuso *m* centrale
   d   Zentralspindel *f*

437   CENTRIC
   f   centrique
   e   céntrico
   i   centrico
   d   zentrisch

438   CENTRIC BREAKAGE
   f   rupture *f* centrique
   e   ruptura *f* céntrica
   i   rottura *f* centrica
   d   zentrischer Bruch *m*;
      Bruch *m* in Centromernähe

439   CENTRIC FUSION
   f   fusion *f* centrique
   e   fusión *f* céntrica
   i   fusione *f* centrica
   d   zentrische Fusion *f*

      CENTRIC REGION, SECONDARY
      see 2503

440   CENTRIOLE
   f   centriole *m*
   e   centríolo *m*
   i   centriolo *m*
   d   Centriol *n*

441   CENTROCHROMATIN
   f   centrochromatine *f*
   e   centrocromatina *f*
   i   centrocromatina *f*
   d   Centrochromatin *n*

442   CENTRODESM;
      CENTRODESMOSE;
      CENTRODESMUS
   f   centrodesmose *f*
   e   centrodesma *f*
   i   centrodesmosi *f*
   d   Centrodesmose *f*

443   CENTROGENE
   f   centrogène *m*
   e   centrogen *m*
   i   centrogeno *m*

d Centrogen *n*

444 CENTROMERE
f centromère *m*
e centrómero *m*
i centromero *m*
d Centromer *n*

CENTROMERE, MULTIPLE
see 1877

CENTROMERE, SEMILOCAL-
IZED see 2556

445 CENTROMERE DISTANCE
f distance *f* au centromère
e distancia *f* del centrómero
i distanza *f* centromerica
d Centromerabstand *m*

446 CENTROMERE
INTERFERENCE
f interférence *f* du
centromère
e interferencia *f* del
centrómero
i interferenza *f*
centromerica
d Centromer-Interferenz *f*

447 CENTROMERE
MISDIVISION
f fausse division *f* du
centromère
e misdivisión *f* del
centrómero
i falsa divisione *f* del
centromero
d Centromer-Missteilung *f*

448 CENTROMERE SHIFT
f déplacement *m* du
centromère
e desplazamiento *m* del
centrómero
i scivolamento *m* del
centromero
d Centromerdislokation *f*

449 CENTROMERIC
f centromérique

e centromérico
i centromerico
d centromerisch;
Centromer- (in Zusammen-
setzungen)

450 CENTROMERIC ATTRACTION
f attraction *f* des
centromères
e atracción *f* centromérica
i attrazione *f* centromerica
d Centromeranziehung *f*

451 CENTRONUCLEUS
f noyau *m* à chromocentre
e centronúcleo *m*
i centronucleo *m*
d Centronucleus *m*

452 CENTROPLASM
f centroplasme *m*
e centroplasma *m*
i centroplasma *m*
d Centroplasma *n*

453 CENTROPLAST
f centroplaste *m*
e centroplasto *m*
i centroplasto *m*
d Centroplast *m* (#)

454 CENTROSOME
f centrosome *m*
e centrosoma *m*
i centrosoma *m*
d Centrosom *n*

455 CENTROSPHERE
f centrosphère *f*
e centrosfera *f*
i centrosfera *f*
d Centrosphäre *f*

456 CENTROTYPE
f centrotype *m*
e centrotipo *m*
i centrotipo *m*
d Centrotyp *m* (#)

457 CEPHALOBRACHIAL
f céphalobrachial

e cefalobraquial
i cefalobrachiale
d cephalobrachial

458 CERTATION;
AGONISIS
f certation *f*;
agonisis *f*
e certation *f*;
agonisis *f*
i certatio *f*
d Certation *f*;
Zertation *f*

459 CHALAZA (*pl* CHALAZAS)
f chalaze *f*
e chalaza *f*
i calaza *f*
d Chalaza (*pl* Chalazae) *f*

460 CHARACTER
f caractère *m*
e carácter *m*
i carattere *m*
d Charakter *m*;
Merkmal *n*

461 CHECK-CROSS
f croisement *m* témoin;
check-cross
e cruza *f* de verificación
i incrocio *m* test
d Kontrollkreuzung *f*

462 CHEMOTYPE
f chémotype *m*
e quimiótipo *m*
i chemotipo *m*
d Chemotyp *m*

463 CHIASMA (*pl* CHIASMATA)
f chiasma *m*
(*pl* chiasmas ou chiasmata)
e quiasma *m*
i chiasma *m*
d Chiasma *n*
(*pl* Chiasmata)

CHIASMA, COMPENSATING
see 581

CHIASMA, COMPLEMENTARY
see 589

CHIASMA, DIAGONAL
see 750

CHIASMA, DISPARATE
see 844

CHIASMA, LATERAL
see 1662

CHIASMA, MULTIPLE
see 1878

CHIASMA, PARTIAL
see 2077

CHIASMA, TERMINAL
see 2786

464 CHIASMA FREQUENCY
f fréquence *f* des
chiasmata
e frecuencia *f* de los
quiasmas
i frequenza *f* dei chiasmi
d Chiasmafrequenz *f*

465 CHIASMA INTERFERENCE
f interférence *f* chiasmatique
e interferencia *f* quiasmática
i interferenza *f* chiasmatica
d Chiasmainterferenz *f*

CHIASMATA, COMPARATE
see 577

CHIASMATA, CROSSOVER
see 645

466 CHIASMA THEORY OF
PAIRING
f théorie *f* chiasmatique de
l'appariement
e teoría *f* quiasmática del
apareamiento
i teoría *f* chiasmatica
dell'appaiamento
d Chiasma-Theorie der
Paarung *f*

467 CHIASMA TYPE THEORY
f théorie f de la
chiasmatypie
e teoría f de los
quiasmátipos
i teoria f della chiasma-
tipia
d Chiasmatypie-Theorie f

468 CHIASMATYPY
f chiasmatypie f
e quiasmatipía f
i chiasmatipia f
d Chiasmatypie f

469 CHIMAERA
CHIMERA (pl CHIMERAS)
f chimère f
e quimera f
i chimera f
d Chimäre f

CHIMAERA, CHROMOSOMAL
see 518

CHIMAERA, HAPLOCHLAMY-
DEUS see 1189

CHIMAERA, MERICLINAL
see 1754

CHIMAERA, SECTORIAL
see 2511

470 CHLOROPLAST
f chloroplaste m
e cloroplastidio m;
cloroplasto m
i cloroplasto m
d Chloroplast m

471 CHONDRIOCONT;
CHONDRIOKONT
f chondrioconte m
e condrioconto m
i condrioconto m
d Chondriokont m

472 CHONDRIOGENE
f chondriogène m
e condriogén m

i condriogene m
d Chondriogen n

473 CHONDRIOGENIC
f chondriogénique
e condriogénico
i condriogenico
d chondriogenisch

474 CHONDRIOGENOTYPE
f chondriogénotype m
e condriogenotipo m
i condriogenotipo m
d Chondriogenotyp m

475 CHONDRIOID
f chondrioïde m
e condrioide m
i condrioide m
d Chondrioid n

476 CHONDRIOKINESIS
f chondriocinèse f
e condriocinesis f
i condriocinesi f
d Chondriokinese f

477 CHONDRIOKONT
CHONDRIOCONT
f chondrioconte m
e condrioconto m
i condrioconto m
d Chondriokont m

478 CHONDRIOLYSIS
f chondriolyse f
e condriolisis f
i condriolisi f
d Chondriolyse f

479 CHONDRIOMA
CHONDRIOME
f chondriome m
e condrioma m
i condrioma m
d Chondriom n

480 CHONDRIOMERE
f chondriomère m
e condriomero m
i condriomero m

d Chondriomere *f*

481 CHONDRIOMITE
f chondriomite *m*
e condriomito *m*
i condriomito *m*
d Chondriomit *m*

482 CHONDRIOSOMAL MANTLE
f manteau *m* chondriosomal
e manto *m* condriosomal
i mantello *m* condriosomico
d Chondriosomenmantel *m*

483 CHONDRIOSOME
f chondriosome *m*
e condriosoma *m*
i condriosomi *m.pl.*
d Chondriosom *n*

484 CHONDRIOSPHERE
f chondriosphère *f*
e condriosfera *f*
i condriosfera *f*
d Chondriosphäre *f*

485 CHOROGAMY
f chorogamie *f*
e corogamia *f*
i corogamia *f*
d Chorogamie *f*

486 CHROMASIE
f chromasie *f*
e cromasia *f*
i cromasia *f*
d Chromasie *f*

487 CHROMATIC
f chromatique
e cromático
i cromatico
d chromatisch

488 CHROMATIC INTERCHANGE
f échange *m* de chromatides
e intercambio *m* cromático
i scambio *m* di cromatidi
d Chromatidenaustausch *m*

489 CHROMATID

f chromatide *f*
e cromatidio *m*
i cromatidio *m*
d Chromatide *f*

CHROMATID, LAMPBRUSH
see 1658

CHROMATID, RING
see 2488

490 CHROMATID ABERRATION
f aberration *f* des
   chromatides;
   aberration *f* chromatidique
e aberración *f* cromatídica
i aberrazione *f* cromatidica
d Chromatidenaberration *f*

491 CHROMATID BREAK;
     CHROMATID BREAKAGE
f rupture *f* chromatidique;
   rupture *f* des
   chromatides
e ruptura *f* cromatídica
i rottura *f* cromatidica
d Chromatidenbruch *m*

492 CHROMATID BRIDGE
f pont *m* de chromatides;
   pont *m* chromatique
e puente *m* cromatídico
i ponte *m* di cromatidi
d Chromatidenbrücke *f*

493 CHROMATIDIC
f chromatidique
e cromatídico
i cromatidico
d chromatidisch;
   Chromatiden- (in Zusammensetzungen)

494 CHROMATID INTERFERENCE
f interférence chromatidique
e interferencia *f* cromatídica
i interferenza cromatidica *f*
d Chromatideninterferenz *f*

CHROMATIDS, NON-SISTER
see 1926

495 CHROMATID TIE
  f noeud *m* de chromatides
  e ligadura *f* cromatídica
  i nodo *m* cromatidico (Θ)
  d Chromatidenschlinge *f*

496 CHROMATIN
  f chromatine *f*
  e cromatina *f*
  i cromatina *f*
  d Chromatin *n*

  CHROMATIN, NUCLEOLAR
  ASSOCIATED
  see 1945

  CHROMATIN, SEX
  see 2573

497 CHROMATIN BRIDGE
  f pont *m* chromatique
  e puente *m* de cromatina
  i ponte *m* cromatidico
  d Chromatinbrücke *f*

498 CHROMATOID BODY
  f chromatoïde *m*
  e cromatoide *m*
  i cromatoide *m*
  d Chromatoidkörper *m* (#)

499 CHROMATOLYSIS
  f chromatolyse *f*
  e cromatolisis *f*
  i cromatolisi *f*
  d Chromatolyse *f*

500 CHROMIDIOSOME
  f chromidiosome *m*
  e cromidiosoma *m*
  i cromidiosoma *m*
  d Chromidiosom *n*

501 CHROMIDIUM
  f chromidie *f*
  e cromidio *m*
  i cromidio *m*
  d Chromidie *f*

502 CHROMIOLE
  f chromiole *f*

  e cromiolo *m*
  i cromiolo *m*
  d Chromiole *f* (#)

503 CHROMOCENTRE;
  CHROMOCENTER
  f chromocentre *m*;
     noeud *m* chromotique
  e cromocentro *m*
  i cromocentro *m*
  d Chromozentrum *n*

504 CHROMOCYTE
  f chromocyte *m*
  e cromocito *m*
  i cromocito *m*
  d Chromocyte *f*

505 CHROMOFIBRIL
  f chromofibrille *f*
  e cromofibrilla *f*
  i cromofibrilla *f*
  d Chromofibrille *f*

506 CHROMOFILAMENT
  f chromofilament *m*
  e cromofilamento *m*
  i cromofilamento *m*
  d Chromofilament *n*

507 CHROMOGENE
  f chromogène *m*
  e cromogén *m*
  i cromogene *m*
  d Chromogen *n*

508 CHROMOMERE
  f chromomère *m*
  e cromómero *m*
  i cromomero *m*
  d Chromomer *n*

509 CHROMOMERE SIZE
  GRADIENT
  f gradient de la taille du
     chromomère *m*
  e gradiente del tamaño del
     cromómero *m*
  i gradiente cromomero *m*
  d Chromomerengrössen-
     gradient *m*

510 CHROMOMERIC
     f chromomérique
     e cromomérico
     i cromomerico
     d chromomerisch;
        Chromomeren- (in Zu-
        sammensetzungen)

511 CHROMONEMA
     f chromonéma *m*
     e cromonema *m*
     i cromonema *m*;
        cromonemo *m*
     d Chromonema *n*

512 CHROMOPHOBE;
     CHROMOPHOBIC
     f chromophobe;
        chromophobique
     e cromofóbico
     i cromofobico
     d chromophob

513 CHROMOPLASMA
     f chromoplasme *m*
     e cromoplasma *m*
     i cromoplasma *m*
     d Chromoplasma *n*

514 CHROMOPLAST
     f chromoplaste *m*
     e cromoplasto *m*
     i cromoplasto *m*
     d Chromoplast *m*

515 CHROMOSIN
     f chromosine *f*
     e cromosina *f*
     i cromosina *f*
     d Chromosin *n*

516 CHROMOSOMAL
     f chromosomique;
        chromosomal
     e cromosómico
     i cromosomico
     d chromosomal;
        Chromosomen- (in Zu-
        sammensetzungen)

517 CHROMOSOMAL ABERRATION
     f aberration *f* chromosomique
     e aberración *f* cromosómica
     i aberrazione *f* cromosomica
     d Chromosomenaberration *f*

518 CHROMOSOMAL CHIMAERA;
     CHROMOSOMAL CHIMERA
     f chimère *f* chromosomique
     e quimera *f* cromosómica
     i chimera *f* cromosomica
     d Chromosomenchimäre *f*

519 CHROMOSOMAL FIBRES
     f fibres *f pl* chromosomiques
     e fibras *f pl* del huso
     i fibre *f pl* del fuso
     d Chromosomenspindelfasern
        f pl

520 CHROMOSOME
     f chromosome *m*
     e cromosoma *m*
     i cromosoma *m*
     d Chromosom *n*

     CHROMOSOME, A-
     see 23

     CHROMOSOME, ACCESSORY
     see 12

     CHROMOSOME, ATTACHED X-
     see 261

     CHROMOSOME, ATTACHED
     XY- see 262

     CHROMOSOME, B-
     see 340

     CHROMOSOME, BRANCHED
     see 401

     CHROMOSOME, GIANT
     see 1132

     CHROMOSOME, HETERO-
     TROPIC see 1314

CHROMOSOME, LAMPBRUSH
see 1659

CHROMOSOME, LINEAR
see 1672

CHROMOSOME, NLG
see 1917

CHROMOSOME, POLYMERIC
see 2231

CHROMOSOME, RESIDUAL
see 2459

CHROMOSOME, RING
see 2489

CHROMOSOME, SAT-
see 2496

CHROMOSOME, SEX
see 2574

CHROMOSOME, T-
see 2761

CHROMOSOME, W-
see 2928

CHROMOSOME, X-
see 2933

CHROMOSOME, Y-
see 2938

CHROMOSOME, Z-
see 2940

521 CHROMOSOME ARM
f   bras *m* chromosomique
e   brazo *m* cromosómico
i   braccio *m* cromosomico
d   Chromosomenarm *m*

522 CHROMOSOME BRIDGE
f   pont *m* chromosomique
e   puente *m* cromosómico
i   ponte *m* cromosomico
d   Chromosomenbrücke *f*

523 CHROMOSOME COMPLEX
f   complexe *m* chromosomique
e   complejo *m* cromosómico
i   complesso *m* cromosomico
d   Chromosomenkomplex *m*

524 CHROMOSOME CYCLE
f   cycle *m* chromosomique
e   ciclo *m* cromosómico
i   ciclo *m* cromosomico
d   Chromosomenzyklus *m*

525 CHROMOSOME MAP
f   carte *f* chromosomique
e   mapa *m* cromosómico
i   carta *f* cromosomica
d   Chromosomenkarte *f*

526 CHROMOSOME MOTTLING
f   coloration *f* segmentaire
    alternée (chromosome);
    mottling *m*
e   coloración *f* alternada de
    los segmentos cromosómic
    mottling *m*
i   colorazione *f* alternativa d
    segmenti cromosomici;
    mottling *m*
d   Chromosomenmottling *n*

527 CHROMOSOME NUMBER
f   nombre *m* chromosomique
e   número *m* cromosómico
i   numero *m* cromosomico
d   Chromosomenzahl *f*

CHROMOSOME NUMBER,
GAMETIC
see 1025

528 CHROMOSOME RING
f   chromosome *m* en anneau
e   cromosoma *m* anular
i   cromosoma *m* anulare;
    anello *m* cromosomico
d   Chromosomenring *m*

529 CHROMESOME SET
f   garniture *f* chromosomique
    génome *m* chromosomique
    lot *m* de chromosomes

e juego *m* de cromosomas
i corredo *m* cromosomico
d Chromosomensatz *m*

530 CHROMOSOME SHATTERING
f ruptures *f pl* chromosomiques multiples
e desmenuzamiento *m* cromosómico
i rottura *f* cromosomica
d multipler Chromosomen-
   bruch *m*

531 CHROMOSOME THREAD
f filament *m* chromosomique
e filamento *m* cromosómico
i filamento *m* cromosomico
d Chromosomenfaden *m* (!)

532 CHROMOSOMIC
f chromosomique
e cromosómico
i cromosomico
d chromosomal;
   Chromosomen- (in Zu-
   sammensetzungen)

533 CHROMOSOMIN
f chromosomine *f*
e cromosomina *f*
i cromosomina *f*
d Chromosomin *n*

534 CHROMOSOMOID
f chromosomoïde *m*
e cromosomoide *m*
i cromosomoide *m*
d Chromosomoid *n*

535 CHROMOTYPE
f chromotype *m*
e cromotipo *m*
i cromotipo *m*
d Chromotyp *m*

CIRCLE OF RACES
see 2403

536 CIS-ARRANGEMENT;
CIS-CONFIGURATION
f arrangement *m* en cis;

configuration *f* en cis
e cis-acomodo *m*;
   cis-configuración *f*
i configurazione-CIS *f* (Θ)
d Cis-Konfiguration *f*

537 CIS-TRANS TEST
f test *m* cis-trans
e prueba *m* cis-trans
i test *m* cis-trans
d Cis-Trans-Test *m*

538 CISTRON
f cistron *m*
e cistrón *m*
i cistron *m*
d Cistron *n*

539 CIS-VECTION EFFECT
f effet *m* de cis-vection
e efecto *m* CIS
i effetto *m* CIS
d Cis-Positionseffekt *m*

540 CLADE
f clade *f*
e - -
i - -
d Cladium *n* (≠)

541 CLADOGENESIS
f cladogénèse *f*
e cladogénesis *f*
i cladogenesi *f*
d Kladogenese *f*

542 CLEAVAGE
f clivage *m*
e escisión *f*
i scissione *f*
d Furchung *f*

CLEAVAGE, HOLOBLASTIC
see 1330

543 CLEAVAGE DELAY
f retard *m* de clivage
e retardo *m* de escisión
i ritardo *m* di scissione
d Furchungsverzögerung *f*

544   CLEAVAGE NUCLEUS
        f   noyau *m* de clivage
        e   núcleo *m* de escisión
        i   nucleo *m* di scissione
        d   Furchungskern *m*

545   CLEISTOGAMY
        f   cleistogamie *f*
        e   cleistogamia *f*
        i   cleistogamia *f*
        d   Kleistogamie *f*

546   CLINE
        f   cline *m*
        e   clina *f*
        i   clina *f*
        d   Kline *f*

547   CLON, CLONE, KLON
        f   clone *m*
        e   clon *m*
        i   clone *m*
        d   Klon *m*

548   C-MEIOSIS
        f   c-méiose *f*
        e   c-meiosis *f*
        i   c-meiosi *f*
        d   C-Meiose *f*

549   C-MITOSIS
        f   c-mitose *f*
        e   c-mitosis *f*
        i   c-mitosi *f*
        d   C-Mitose *f*

        C-MITOSIS, DISTRIBUTED
        see 853

        C-MITOSIS, EXPLODED
        see 972

        COADAPTATION, GENETIC
        see 1094

550   CODE READING
        DECODING
        f   décodage *m*
        e   lectura *f* del código
        i   lettura *f* del codice
        d   Entzifferung *f*
            (des genetischen Code)

551   CODOMINANCE
        f   codominance *f*
        e   codominancia *f*
        i   codominanza *f*
        d   Codominanz *f*

552   CODOMINANT
        f   codominant
        e   codominante
        i   codominante
        d   codominant

553   CODON
        f   codon *m*
        e   codon *m*
        i   codon *m*
        d   Codon *n*

554   COEFFICIENT OF
        INBREEDING
        f   coefficient *m* de
            consanguinité
        e   grado *m* de consanguinidad;
            grado *m* de inbreeding
        i   coefficiente *m* di
            consanguinità
        d   Inzuchtkoeffizient *m*

555   COEFFICIENT OF
        RELATIONSHIP
        f   coefficient *m* de parenté
        e   coeficiente *m* de parentesco
        i   coefficiente *m* di parentela
        d   Verwandtschaftskoeffizient *n*

556   COENOCYTE
        f   coenocyte *m*
        e   cenocito *m*
        i   cenocito *m*
        d   Coenocyte *f*

557   COENOGAMETE
        f   coenogamète *m*
        e   cenogameto *m*
        i   cenogamete *m*
        d   Coenogamet *m*

558   COENOGENESIS
        f   coenogénèse *m*
        e   cenogénesis *f*;
            cenogenia *f*

i   cenogenesi *f*
d   Coenogenese *f*

559   COENOSPECIES
f   coeno-espèce *f*
e   cenospecie *f*
i   cenospecie *f*
d   Coenospecies *f*

560   COENOZYGOTE
f   coenozygote *m*
e   cenocigoto *m*
i   cenozigoto *m*
d   Coenozygote *f*

COIL, PARANEMIC
see 2052

COIL, PLECTONEMIC
see 2178

561   COINCIDENCE
f   coïncidence *f*
e   coincidencia *f*
i   coincidenza *f*
d   Koinzidenz *f*

562   COLCHICINE
f   colchicine *f*
e   colchicina *f*
i   colchicina *f*
d   Colchicin *n*

563   COLCHIPLOID (adj.)
COLCHIPLOID
f   colchiploïde
colchiploïde *m*
e   colchiploide
colchiploide *m*
i   colchiploide
colchiploide *m*
d   colchiploid
Colchiploide *f*

564   COLCHIPLOIDY
f   colchiploïdie *f*
e   colchiploidia *f*
i   colchiploidismo *m*
d   Colchiploidie *f*

565   COLICIN
f   colicine *f*
e   colicina *f*
i   colicina *f*
d   Colicin *n*

566   COLICINOGENIC
f   colicinogène
e   colicinogénico
i   colicinogenico
d   colicinogen

567   COLICINOGENY
f   colicinogénie *f*
e   colicinogenia *f*
i   colicinogenia *f*
d   Colicinogenie *f*

568   COLLASOME
f   collasome *m*
e   colasoma *m*
i   collasoma *m*
d   Kollasom *n*

569   COLLATERAL INHERITANCE
f   hérédité *f* collatérale
e   herencia *f* colateral
i   eredità *f* collaterale
d   kollaterale Vererbung *f*

570   COLLOCHORE;
CONJUNCTIVE SEGMENT
f   segment *m* conjonctif;
collochore *m*
e   colocoro *m*;
segmento *m* conjuntivo
i   collocoro *m*;
segmento *m* congiuntivo
d   Konjunktionssegment *n* (#);
Collochore *f*

572   COMBINATION CHECKER
BOARD
f   échiquier *m* de croisement
e   tablero *m* de combinación
i   scacchiera *f* d'incrocio
d   Kombinationsquadrat *n*

573 COMBINING ABILITY
    f aptitude *f* à la combinaison
    e aptitud *f* por la combinación;
       aptitud *f* combinatoria
    i attitudine *f* combinatoria
    d Kombinationseignung *f*

574 COMMISCUUM
    f commiscuum *m*
    e comisco *m*
    i commiscuo *m*
    d Commiscuum *n*;
       Kommiskuum *n*

575 COMMUNAL CONNUBIUM
    f union *f* fortuite dans une
       communauté déterminée (Θ)
    e conubio *m* comunal
    i connubio *m* fortuito in una
       determinata comunità
    d - -

576 COMMUNITY
    f communauté *f*
    e comunidad *f*
    i comunità *f*
    d Gemeinschaft *f*

577 COMPARATE CHIASMATA *pl*
    f chiasmas *mpl* symétriques
    e quiasmas *mpl* acordes
    i chiasmi *mpl* compensati
    d einander kompensierende
       Chiasmata *mpl*

578 COMPARIUM
    f comparium *m*
    e comparium *m*
    i comparium *m*
    d Comparium *n*;
       Komparium *n*

579 COMPATIBILITY
    f compatibilité *f*
    e compatibilidad *f*
    i compatibilità *f*
    d Kompatibilität *f*;
       Verträglichkeit *f*

580 COMPATIBLE
    f compatible

    e compatible
    i compatibile
    d kompatibel;
       verträglich

581 COMPENSATING CHIASMA
    f chiasma *m* de compensation
    e quiasma *m* de compensación
    i chiasma *m* di compensazione
    d kompensierendes Chiasma *n*

582 COMPENSATION
    f compensation *f*
    e compensación *f*
    i compensazione *f*
    d Kompensation *f*

583 COMPENSATOR
    f compensateur *m*
    e compensador *m*
    i compensatore *m*
    d Kompensator *m*

584 COMPETENCE
    f compétence *f*
    e competencia *f*
    i competenza *f*
    d Kompetenz *f*

585 COMPETENT DONOR BAC-
    TERIUM CELL
    f bactérie *f* donatrice
       compétente
    e bacteria *f* donante
       competente
    i batteria *f* donatrice
       competente;
       cellula *f* donatrice
       competente
    d kompetentes Donor-Bakte-
       rium *n*;
       kompetente Donor-Zelle *f*

586 COMPETITIVE ABILITY
    f competitivité *f*
    e habilidad *f* competitiva
    i potere *m* competitivo;
       possibilità *f* competitiva
    d competitive ability; in
       einem anderen Sinne:
       Wettbewerbsfähigkeit *f*

587 COMPLEMENT
   f complément *m*
   e complemento *m*
   i complemento *m*
   d Komplement *n*

588 COMPLEMENTARY
   f complémentaire
   e complementario
   i complementare
   d komplementär

589 COMPLEMENTARY CHIASMA
   f chiasma *m* complémentaire
   e quiasma *m* complementario
   i chiasma *m* complementare
   d komplementäres Chiasma *n*

590 COMPLEMENTARY FACTORS
   f facteurs *mpl* complémen-
     taires
   e genes *mpl* complementarios;
     factores *mpl* complemen-
     tarios
   i geni *mpl* complementari;
     fattori *mpl* complementari
   d Komplementärgene *npl*

591 COMPLEMENTATION
   f complémentation *f*
   e complementación *f*
   i complementazione *f*
   d Komplementation *f*;
     Komplementierung *f*

592 COMPLETE
   f complet ; absolu
   e completo
   i completo
   d vollständig

593 COMPLETE LINKAGE
   f liaison *f* absolue;
     linkage *m* absolu
   e ligamiento *m* completo
   i linkage *m* completo;
     associazione *f* completa
   d vollständige Kopplung *f*

594 COMPLETE PENETRANCE
   f pénétrance *f* complète

   e penetración *f* completa
   i penetrazione *f* completa
   d vollständige Penetranz *f*

595 COMPLEX CHARACTER
   f caractère *m* complexe
   e carácter *m* complejo
   i carattere *m* complesso
   d komplexes Merkmal *n*

596 COMPOUND CROSSING-OVER
   f échanges *mpl* multiples;
     crossing-over *m* multiples
   e entrecruzamiento *m*
     compuesto
   i scambi *mpl* multipli
   d Mehrfachaustausch *m*

597 COMPOUND DETERMINERS
   f déterminants *mpl* composés
   e factores *mpl* determinantes
     compuestos
   i fattori *mpl* determinanti
     composti
   d compound determiners *mpl*

598 COMPOUND X-CHROMOSOMES
   f chromosomes X *mpl* composés
   e cromosomas-X *mpl* compuestos
   i cromosomi-X *mpl* composti
   d compound-X *n*

599 CONDENSATION;
   CONTRACTION
   f condensation *f*
   e condensación *f*
   i condensazione *f*
   d Kontraktion *f*;
     Kondensation *f*

600 CONDITIONAL DOMINANCE
   f dominance *f* conditionnelle
   e dominancia *f* condicional
   i dominanza *f* condizionale
   d konditionelle Dominanz *f* (#)

601 CONDITIONAL FACTORS *pl*
   f facteurs *mpl* conditionnels
   e factores *mpl* condicionales
   i fattori *mpl* condizionali
   d Konditionalfaktoren *mpl*

602 CONDITIONED DOMINANCE
  f dominance *f* conditionnée
  e dominancia *f* condicionada
  i dominanza *f* condizionata
  d bedingte Dominanz *f* (*#*)

603 CONDITIONED LETHAL
  f léthal *m* conditionné
  e letal *m* condicionado
  i letale *m* condizionato
  d bedingter Letalfaktor *m*

604 CONDITIONING
  f conditionnement *m*
  e condicionamiento *m*
  i condizionatura *m*
  d Induktion *f* von
    Dauermodifikationen

605 CONFIGURATION
  f configuration *f*
  e configuración *f*
  i configurazione *f*
  d Konfiguration *f*

606 CONGRESSION
  f congression *f*
  e congresión *f*
  i congressione *f*;
    adunanza *f*
  d Kongression *f*

607 CONGRUENT CROSSING;
    CONGRUENT HYBRIDISATION
  f croisement *m* congru
  e cruzamiento *m* congruente
  i incrocio *m* congruente
  d kongruente Kreuzung *f*

608 CONJUGANT
  f conjugant *m*
  e - -
  i coniugante *m*
  d Konjugant *m*

609 CONJUGATION
  f conjugaison *f*
  e conjugación *f*
  i coniugazione *f*
  d Konjugation *f*

610 CONJUNCTIVE SEGMENT;
    COLLOCHORE
  f segment *m* conjonctif;
    collochore *m*
  e colocoro *m*
  i collocoro *m*
  d Konjunktionssegment *n* (*#*);
    Collochore *f*

611 CONSTRICTION
  f constriction *f*;
    étranglement *m*
  e constricción *f*
  i costrizione *f*;
    coartazione *f*
  d Einschnürung *f*

CONSTRICTION, KINETIC
see 1638

CONSTRICTION, NUCLEOLAR
see 1946

612 CONSUBSPECIES
  f consub-espèce *f*
  e consubespecie *f*
  i consubspecie *f* (☉)
  d Consubspecies *f*

613 CONTACT HYPOTHESIS
  f hypothèse *f* du contact
  e hipótesis *f* del contacto
  i ipotesi *f* del contatto
  d Kontakthypothese *f*

614 CONTACT POINT
  f point *m* de contact
  e punto *m* de contacto
  i punto *m* di contatto
  d Kontaktpunkt *m*

CONTAMINATION, CRYPTIC
see 653

615 CONTINUOUS
  f continu
  e continuo
  i continuo
  d kontinuierlich

616   CONTRACTION;
       CONDENSATION
       f condensation *f;*
          contraction *f*
       e contracción *f*
       i contrazione *f*
       d Kontraktion *f;*
          Kondensation *f*

617   CONTROLLING ELEMENT
       f élément *m* de contrôle
       e elemento *m* de control
       i elemento *m* di controllo
       d Kontrollelement *n*

618   CO-NUCLEI;
       COMPLEMENTARY NUCLEI
       f noyaux *mpl* complémen-
          taires
       e núcleos *mpl* complemen-
          tarios
       i nuclei *mpl* complementari
       d Komplementärkerne *mpl*

619   CONVERGENCE
       f convergence *f*
       e convergencia *f*
       i convergenza *f*
       d Konvergenz *f*

620   CONVERGENT
       f convergent
       e convergente
       i convergente
       d konvergent

621   CONVERSION
       f conversion *f*
       e conversión *f*
       i conversione *f*
       d Konversion *f*

622   COOPERATION
       f coopération *f*
       e cooperación *f*
       i cooperazione *f*
       d Kooperation *f*

623   COORIENTATION
       f coorientation *f*
       e coorientación *f*

       i coorientazione *f*
       d Koorientierung *f*

624   CORE
       f noyau *m;* coeur *m*
       e núcleo *m* (Θ); centro *m* (Θ)
       i nucleo *m* (Θ); cuore *m* (Θ)
       d Kern *m*

625   CORRECTIVE MATING
       f croisement *m* correctif
       e apareamiento *m* correctivo
       i incrocio *m* correttivo (Θ)
       d korrektive Paarung *f* (≠)

626   CORRELATED PHENOTYPIC
       VARIATION
       f variation *f* phénotypique
          correspondante
       e variación *f* fenotípica
          correspondiente
       i variazione *f* fenotipica
          corrispondente
       d korrelierte Variation *f*
          phänotypischer Merkmale

       CORRELATION, INTRACLASS
       see 1543

       CORRELATION, MULTIPLE
       see 1879

       CORRELATION, PARTIAL
       see 2078

       CORRELATION, PHENOTYPIC
       see 2129

627   CORRELATION COEFFICIENT
       f coefficient *m* de corréla-
          tion
       e coeficiente *m* de correla-
          ción
       i coefficiente *m* di correla-
          zione
       d Korrelationskoeffizient *m*

628   COTYPE
       f cotype *m*
       e cótipo *m*
       i cotipo *m*

d Cotyp *m*

629 COUPLING
f attraction *f*
accouplement *m*
couplage *m*
e atracción *f*
acoplamiento *m*
i accoppiamento *m*
d Kopp(e)lung *f*
Paarung *f*

630 COUPLING PHASE
f phase *f* d'accouplement;
phase *f* d'attraction
e fase *f* de acoplamiento;
fase *f* de atracción
i fase *f* di accoppiamento
d Kopp(e)lungsphase *f*

631 CRISS-CROSS BREEDING
f croisement *m* alternatif;
criss-crossing *m*
e cruzamiento *m* alternativo;
criss-crossing *m*
i criss-crossing *m*;
incrocio *m* alternativo
d überkreuzte Brücke *f*

632 CRISS-CROSS BRIDGE
f pont *m* de croisement
e puente *m* de cruzamiento
i ponte *m* d'incrocio
d überkreutze Brücke *f*

633 CRISS-CROSSING
f croisement *m* alternatif
e cruzamiento *m* alternativo;
criss-crossing *m*
i criss-crossing *m*;
incrocio *m* alternativo
d Überkreuzzüchtung *f*

634 CRISS-CROSS INHERITANCE
f hérédité *f* alternative
e herencia *f* alternativa;
herencia *f* cruzada
i eredità *f* alternativa
d Überkreuzvererbung *f*

635 CRON
f cron *m* (⊖)
e cron *m* (⊖)
i cron *m* (⊖)
d Cron *n*

636 CROSS
f croisement *m*;
métissage *m*
e cruzamiento *m*;
mestizaje *m*
i incrocio *m*
d Kreuzung *f*

CROSS, FOUR-WAY
see 870

637 CROSS BACK, TO
f croiser en retour;
accoupler en retour
e retrocruzar
i rincrociare
d rückkreuzen

638 CROSS BREEDING
f croisement *m*
e cruzamiento *m*;
método *m* de crianza por
cruzamiento
i incrocio *m* (⊖)
d Kreuzungszüchtung *f*

CROSS BREEDING, ROTATION
see 2493

639 CROSS FERTILIZATION
f fertilisation *f* croisée;
fécondation *f* croisée
e fertilización *f* cruzada;
fecundación *f* cruzada
i fecondazione *f* incrociata;
fertilizzazione *f* incrociata
d Kreuzbefruchtung *f*

640 CROSS HOMOLOGY;
RESIDUAL HOMOLOGY
f homologie *f* résiduelle
e homología *f* residual
i omologia *f* residuale
d residuale Homologie *f*;
Resthomologie *f* (≠)

CROSSING, CONGRUENT
see 607

CROSSING, POLYALLELE
see 2205

641 CROSSING-OVER
f crossing-over *m;*
enjambement *m* (#)
e sobrecruzamiento *m;*
crossing-over *m*
i crossing-over *m;*
scambio *m*
d Crossing-over *n*

CROSSING-OVER, PARTIAL
see 2079

CROSSING-OVER, SISTER-
STRAND
see 2607

642 CROSSING-OVER MODIFIER
f modificateur *m* de
crossing-over
e modificador *m* de
sobrecruzamiento;
modificador de
crossing-over *m*
i modificatore *m* di scambio;
modificatore *m* di crossing-
over
d Crossing-over-Modifika-
tor *m*

643 CROSSING-OVER PERCEN-
TAGE
f pourcentage *m* de
crossing-over
e proporción *m* de
sobrecruzamiento;
percentaje *m* de
sobrecruzamiento
i proporzione *f* di crossing-
over;
percentuale *f* di scambio
d Crossing-over-
Prozentsatz *m*

644 CROSS LINKAGE
f cross-linkage *m*

e ligamiento *m* cruzado
i linkage *m* incrociato
d Cross-linkage *n;*
kreuzweise Verkettung *f*

645 CROSSOVER CHIASMATA
f chiasmas *mpl* de fusion
e quiasmas *mpl* de fusión
i chiasmi *mpl* di fusione
d Cross-over-Chiasmata *npl*

646 CROSSOVER HOMOZYGOTE
f homozygote *m* par crossing-
over
e homocigoto *m* por sobrecruzamiento
i omozigoto *m* per scambio
d Cross-over-Homozygote *f*

647 CROSSOVER INDUCER
f inducteur *m* de crossing-
over
e indúctor *m* de sobrecruzamiento
i induttore *m* di scambio
d Cross-over-Induktor *m*

648 CROSSOVER REDUCER
f réducteur *m* de crossing-
over
e reductor *m* de sobrecruzamiento
i riduttore *m* di scambio
d Cross-over-Reduktor *m*

649 CROSSOVER SUPPRESSOR
f suppresseur *m* de crossing-
over
e supresor *m* de
sobrecruzamiento
i sopressore *m* di scambio
d Cross-over-Suppressor *m*

650 CROSSOVER UNIT
f unité *f* de crossing-over
e unidad *f* de
sobrecruzamiento
i unità *f* di scambio
d Cross-over-Einheit *f*

651 CROSS-POLLINATION
 f pollinisation *f* croisée
 e polinización *f* cruzada
 i pollinazione *f* incrociata
 d Kreuzbestäubung *f*

652 CROSS REACTIVATION
 f réactivation *f* croisée
 e reactivación *f* cruzada
 i riattivazione *f* incrociata
 d Kreuzungsreaktivierung *f*

653 CRYPTIC CONTAMINATION
 f contamination *f* cryptique;
  contamination *f* latente
 e contaminación *f* críptica
 i contaminazione *f* criptica
 d latente Verunreinigung *f*

654 CRYPTIC GENETIC CHANGE
 f modification *f* génétique
  cryptique;
  modification *f* génétique
  latente
 e modificación *f* genética
  críptica
 i modificazione *f* genetica
  criptica
 d latente genotypische
  Veränderung *f*

655 CRYPTOCHIMERA
 f cryptochimère *f*
 e criptoquimera *f*
 i criptochimera *f*
 d Kryptochimäre *f*

656 CRYPTOENDOMITOSIS
 f crypto-endomitose *f*
 e criptoendomitosis
 i cripto-endomitosi *f*
 d Kryptoendomitose *f*

657 CRYPTOGONOMERY
 f cryptogonomérie *f*
 e criptogonomería *f*
 i criptogonomeria *f*
 d Kryptogonomerie *f*

658 CRYPTOHAPLOMITOSIS
 f crypto-haplomitose *f*

 e criptohaplomitosis *f*
 i cripto-aplomitosi *f*
 d Kryptohaplomitose *f*

659 CRYPTOMERE
 f cryptomère *m*
 e criptomero *m*
 i criptomero *m*
 d Kryptomere *f*

660 CRYPTOMERIC
 f cryptomérique
 e criptomérico
 i criptomerico
 d kryptomer

661 CRYPTOMERISM
 f cryptomérie *f*
 e criptomerismo *m*
 i criptomerismo *m*
 d Kryptomerie *f*

662 CRYPTOMITOSIS
 f cryptomitose *f*
 e criptomitosis *f*
 i criptomitosi *f*
 d Kryptomitose *f*

663 CRYPTOPOLYPLOID (adj.)
 CRYPTOPOLYPLOID
 f cryptopolyploïde
  cryptopolyploïde *m*
 e criptopoliploide
  criptopoliploide *m*
 i cripto-poliploide
  cripto-poliploide *m*
 d kryptopolyploid
  Kryptopolyploide *f*

664 CRYPTOPOLYPLOIDY
 f cryptopolyploïdie *f*
 e criptopoliploidía *f*
 i cripto-poliploidismo *m*
 d Kryptopolyploidie *f*

665 CTETOSOME
 f ctétosome *m*
 e ctetosoma *m*
 i ctetosoma *m*
 d Ktetosom *n*

666 CULL, TO
  f éliminer
  e eliminar
  i eliminare
  d eliminieren;
    ausmerzen

667 CUMULATIVE
  f cumulatif
  e acumulativo
  i cumulativo
  d kumulativ

668 CYANID EFFECT
  f effet *m* de cyanuration
  e efecto *m* de cianuración
  i effetto *m* di cianurazione
  d Cyanid-Effekt *m*

669 CYCLIC HYBRIDISATION
  f hybridation *f* cyclique
  e hibridación *f* cíclica
  i ibridazione *f* ciclica
  d zyklische Kreuzung *f*

670 CYCLOMORPHOSIS
  f cyclomorphose *f*
  e ciclomorfosis *f*
  i ciclomorfosi *f*
  d Cyclomorphose *f;*
    Zyklomorphose *f*

671 CYTASTER
  f cytaster *m*
  e citáster *m*
  i citaster
  d Cytaster *m*

672 CYTE
  f cyte *m*, cellule *f*
  e cito *m*
  i cito *m*
  d Zelle *f*

673 CYTOACTIVE
  f cytoactif
  e citoactivo
  i citoattivo
  d zellaktiv

674 CYTOBLAST
  f cytoblaste *m*

  e citoblasto *m*
  i citoblasto *m*
  d Cytoblast *m* (Θ);
    Zellkern *m*

675 CYTOCHIMAERA
  f cytochimère *f*
  e citoquimera *f*
  i citochimera *f*
  d Cytochimäre *f*

676 CYTODE
  f cytode *m*
  e citodo *m*
  i citode *m*
  d Cytode *f*

677 CYTODIERESIS (Θ);
    CYTODIAERESIS (Θ);
    MITOSE
  f cytodiérèse *f* (Θ);
    mitose *f*
  e citodiéresis *f* (Θ);
    mitosis *f*
  i citodieresi *f* (Θ);
    mitosi *f*
  d Cytodieresis *f* (Θ);
    Mitose *f*

678 CYTOGAMY
  f cytogamie *f*
  e citogamia *f*
  i citogamia *f*
  d Cytogamie *f*

679 CYTOGENE
  f cytogène *m*
  e citogén *m*
  i citogene *m*
  d Cytogen *n*

680 CYTOGENESIS
  f cytogénèse *f*
  e citogénesis *f*
  i citogenesi *f*
  d Cytogenese *f;*
    Zellteilung *f*

681 CYTOGENETIC
  f cytogénétique
  e citogenético

        i   citogenetico   i   citoma *m*
        d   cytogenetisch   d   Cytom *n*

682   CYTOGENETIC MAP   690   CYTOMERE
        f   carte *f* cytogénétique   f   cytomère *m*
        e   mapa *m* citogenético   e   citómero *m*
        i   carta *f* citogenetica   i   citomero *m*
        d   cytologische   d   Cytomere *f* (#)
            Chromosomenkarte *f*
                                          691   CYTOMICROSOME
683   CYTOGENETICS   f   cytomicrosome *m*
        f   cytogénétique *f*   e   citomicrosoma *m*
        e   citogenética *f*   i   citomicrosoma *m*
        i   citogenetica *f*   d   Mikrosom *n*
        d   Cytogenetik *f*
                                          692   CYTOMIXIS
684   CYTOGONY   f   cytomixie *f*
        f   cytogonie *f*   e   citomixis *f*
        e   citogonia *f*   i   citomixi *f*
        i   citogonia *f*   d   Cytomixis *f*
        d   Cytogonie *f*
                                          693   CYTOMORPHOSIS
685   CYTOKINESIS   f   cytomorphose *f*
        f   cytokinèse *f;*   e   citomórfosis *f*
            cytocinèse *f*   i   citomorfosi *f*
        e   citocinesis *f*   d   Cytomorphosis *f*
        i   citocinesi *f*
        d   Cytokinese *f;*   694   CYTOPLASM
            Zellteilung *f*   f   cytoplasme *m*
                                                 e   citoplasma *m*
686   CYTOLOGICAL   i   citoplasma *m*
        f   cytologique   d   Cytoplasma *n*
        e   citológico
        i   citologico   695   CYTOPLASMIC
        d   cytologisch   f   cytoplasmique
                                                 e   citoplásmico
687   CYTOLYSIN   i   cito-plasmico
        f   cytolysine *f*   d   cytoplasmatisch;
        e   citolisina *f*   Cytoplasma- (in Zusam-
        i   citolisina *f*   mensetzungen)
        d   Cytolysin *n*
                                          696   CYTOPLASMIC INHERITANCE
688   CYTOLYSIS   f   hérédité *f* extra-
        f   cytolyse *f*   chromosomique;
        e   citólisis *f*   hérédité *f* non-
        i   citolisi *f*   chromosomique;
        d   Cytolyse *f*   hérédité *f* cytoplasmique
                                                 e   herencia *f* citoplásmica
689   CYTOME   i   eredità *f* citoplasmica
        f   cytome *m*   d   (cyto)plasmatische Ver-
        e   citoma *m*   erbung *f;*
                                                     Plasmavererbung *f*

D

702 DALTONIAN
f daltonien
e daltoniano
i daltonico
d daltonisch; Dalton-
(in Zusammensetzungen)

703 DALTONISM
f daltonisme *m*
e daltonismo *m*
i daltonismo *m*
d Daltonismus *m*

704 DAM
f mère *f*
e madre *f*
i madre *f*
d Mutter *f* (von Säugetieren)

705 DAM-DAUGHTER
COMPARISON
f comparaison *f* mère-fille
e comparación *f* madres-
hijas
i comparazione *f* madre-
figlia
d Mutter-Tochter-
Vergleich *m*

706 DARWINISM
f darwinisme *m*
e darwinismo *m*
i darvinismo *m*
d Darwinismus *m*

707 DAUERMODIFICATION
f dauermodifikation *f*
e modificación *f* duradera;
modificación *f* persistente
i modificazione *f* durevole
d Dauermodifikation *f*

708 DAUGHTER CELL
f cellule fille *f*
e celula hija *f*
i cellula figlia *f*
d Tochterzelle *f*

709 DAUGHTER CHROMATID
f chromatide fille *f*
e cromatidio hijo *m*
i cromatidio figlio *m*
d Tochterchromatide *f*

710 DAUGHTER CHROMOSOME
f chromosome fils *m*
e cromosoma hijo *m*
i cromosoma figlio *m*
d Tochterchromosom *n*

711 DEADAPTATION
f désadaptation *f*
e desadaptación *f*
i disadattamento *m*
d Deadaptation *f;*
Anpassungsverlust *m*

712 DECODING
CODE READING
f décodage *m*
e lectura *f* del código
i lettura *f* del codice
d Entzifferung *f*
(des genetischen Code)

713 DECONJUGATION
f déliement *m* (Θ)
e deconjugación *f*
i slacciamento *m* (Θ)
d Dekonjugation *f*

714 DEFECTIVE
f défectif
e defectuoso
i difettivo
d defekt

715 DEFECTIVE LYSOGENIC
BACTERIUM
f bactérie *f* lysogène
défective
e bacteria *f* lisogénica
defectuosa
i bacterio *m* lisogeno difettivo
d - -

716 DEFERRED BREAKAGE;
 SECONDARY BREAKAGE
 f rupture *f* secondaire
 e ruptura *f* secundaria
 i rottura *f* secondaria
 d secondary breakage *n*;
 Sekundärbruch *m* (*#*)

717 DEFICIENCY
 f déficience *f*
 e deficiencia *f*
 i deficienza *f*
 d Defizienz *f*

718 DEFINITIVE NUCLEUS
 f noyau *m* définitif
 e núcleo *m* definitivo
 i nucleo *m* definitivo
 d definitiver Zellkern *m*

719 DEGENERATION
 f dégénération *f*;
  dégénérescence
 e degeneración *f*
 i degenerazione *f*
 d Degeneration *f*;
  Entartung *f*

720 DEGENERATION RESULTING
 FROM INBREEDING;
 INBREEDING DEPRESSION
 f dégénérescence *f* consanguine
 e degeneración *f* consanguínea
 i degenerazione *f*
  consanguinea
 d Inzuchtdegeneration *f*;
  Inzuchtdepression *f*

721 DEGRADATION
 f dégradation *f*
 e degradación *f*
 i degradazione *f*
 d Degradation *f*

722 DELAYED; LAGGING
 f retardé
 e retrasado
 i ritardato
 d verzögert; Nachzügler-
  (in Zusammensetzungen)

723 DELAYED DOMINANCE

 f dominance *f* retardée
 e dominancia *f* retrasada
 i dominanza *f* ritardata
 d verzögerte Manifestation *f*
  der Dominanz

724 DELAYED INHERITANCE;
 PREDETERMINATION
 f hérédité *f* retardée;
  prédétermination *f*
 e herencia *f* retrasada;
  predeterminación *f*
 i eredità *f* ritardata;
  predeterminazione *f*
 d verzögerte Vererbung *f*;
  Prädetermination *f*

725 DELETION
 f délétion *f*
 e deleción *f*
 i delezione *f*
 d Deletion *f*

726 DELETION HETEROZYGOTE
 f hétérozygote *m* pour une
  délétion
 e heterocigoto *m* por una
  deleción (Θ)
 i eterozigote *f* per una
  delezione (Θ)
 d Deletionsheterozygote *f*

727 DEME
 f dème *m*
 e demo *m*
 i demo *m*
 d -deme *m* (in Zusammensetzungen)

728 DENUCLEINATION
 f dénucléinisation *f*
 e desnucleinización *f*
 i denucleizzazione *f*
 d Denukleinisierung *f*

729 DEREPRESSION
 f dérépression *f*
 e derrepresión
 i - -
 d Derepression *f*

730 DERIVATIVE HYBRID
 f hybride *m* double

e híbrido *m* derivativo
i ibrido *m* derivativo
d Kreuzungsprodukt *n* zweier Bastarde

731 DERIVED
f dérivé
e derivado
i derivato
d abgeleitet

732 DESCENDANCE
f descendance *f*
e descendencia *f*
i discendenza *f*; progenie *f*
d Deszendenz *f*; Nachkommenschaft *f*

733 DESCENDANT
f descendant *m*; descendante *f*; rejeton *m*
e descendiente *m*
i discendente *m*
d Nachkomme *m*

734 DESMONE
f desmone *f*
e desmona *f*
i - -
d Desmon *n*

735 DESOXYRIBONUCLEIC ACID (DNA);
DEOXYRIBONUCLEIC ACID
f acide *m* désoxyribonucléique (ADN)
e ácido *m* desoxiribonucléico (ADN)
i acido *m* desossiribonucleico (ADN)
d Desoxyribonucleinsäure (DNS) *f*

736 DESYNAPSIS
f désynapsis *f*
e desinapsis *f*
i desinapsi *f*
d Desynapsis *f*

737 DETACHMENT
f détachement *m*

e separación *f*
i distacco *m*
d Detachment (engl.) *n*

738 DETERMINANT
f déterminant
e determinante
i determinante
d determinierend

739 DETERMINATE VARIATION
f variation *f* dans une direction déterminée; orthogénèse *f*
e variación *f* determinada
i variazione in una determinata direzione *f*; ortogenesi *f*
d gerichtete Variation *f*

740 DETRIMENTAL
f nuisible
e detrimental
i dannoso
d nachteilig

741 DEUTEROTOKY
f deutérotoquie *f*
e deuterotocia *f*
i deuterotochia *f*
d Deuterotokie *f*

742 DEUTHYALOSOME
f deuthyalosome *m*
e deutialosoma *m*
i deutialosoma *m*
d - -

743 DEVELOPMENTAL STABILITY; PHENOTYPIC STABILITY
f stabilité *f* de développement stabilité *f* phénotypique
e estabilidad *f* fenotípica
i stabilità *f* fenotipica
d phänotypische Stabilität *f*

744 DEVIATION
f déviation *f*
e desviación *f*
i deviazione *f*
d (mutationsbedingte Entwicklungs-)Abweichung *f*

745   DEVIATION DUE TO MODI-
      FYING GENES
      f   changement *m* influencé par
          des modificateurs
      e   desviación *f* debida a genes
          modificadores
      i   deviazione *f* dovuta a geni
          modificatori
      d   Abänderung *f* durch Modi-
          fikationsfaktoren

746   DEVOLUTION
      f   dévolution *f*
      e   devolución *f*
      i   devoluzione *f*
      d   Devolution *f (≠)*

747   DIAD; DYAD
      f   diade *f*
      e   diada *f*
      i   diade *f*
      d   Dyade *f*

748   DIADELPHOUS
      f   diadelphe
      e   diadelfo
      i   diadelfo
      d   diadelphisch *(≠)*

749   DIAGENIC; DIAGYNIC
      f   diagynique
      e   diagínico
      i   diaginico
      d   diagyn

750   DIAGONAL CHIASMA
      f   chiasma *m* en diagonale
      e   quiasma *m* diagonale
      i   chiasma *m* diagonale
      d   diagonales Chiasma *n*

751   DIAKINESIS
      f   diacinèse *f*
      e   diacinesis *f*
      i   diacinesi *f*
      d   Diakinese *f*

752   DIAKINETIC
      f   diacinétique
      e   diacinético
      i   diacinetico

      d   diakinetisch

753   DIALLEL
      f   diallèle
      e   dialelo
      i   diallelo
      d   diallel

754   DIANDRIC
      f   - -
      e   diándrico
      i   diandrico
      d   diandrisch

755   DIAPHOROMIXIS
      f   diaphoromixie *f*
      e   diaforomixis *f*
      i   diaforomixi *f*;
          diaforomissi *f*
      d   Diaphoromixis *f*

756   DIASCHISTIC
      f   diaschistique
      e   diasquístico
      i   diascittico
      d   diaschistisch *(≠)*

757   DIASTEM
      f   diastème *m*
      e   diastema *m*
      i   diastema *m*
      d   Diastema *n*

758   DIASTER
      f   diaster *m*
      e   diaster *m*
      i   diaster *m*
      d   Diaster *m*

759   DIBASIC
      f   dibasique
      e   dibásico
      i   dibasico
      d   dibasisch

760   DICARYON; DIKARYON
      f   dicaryon *m*
      e   dicarión *m*;
          dicarionte *m*
      i   dicarion *m*
      d   Dikaryon *n*

761 DICARYOTIC; DIKARYOTIC
f dicaryotique
e dicariótico
i dicariotico
d dikaryotisch

762 DICENTRIC
f dicentrique
e dicéntrico
i dicentrico
d dizentrisch

763 DICHLAMYDEOUS;
DICHLAMIDIUS
f dichlamyde
e diclamidio
i diclamidio
d dichlamyd;
dichlamydeisch

764 DICHOGAMIC;
DICHOGAMOUS
f dichogame
e dicógamo
i dicogamo
d dichogam

765 DICHOGAMY
f dichogamie f
e dicogamia f
i dicogamia f
d Dichogamie f

766 DICHONDRIC
f dichondrique
e dicóndrico
i dicondrico
d dichondrisch

767 DICLINIOUS
f dicline
e diclino
i diclino
d diklin

768 DICTYATE STAGE;
DICTYOTIC STAGE
f stade m dictyotique
e estado m dictiótico
i stadio m dictiottico
d retikuläres Stadium (der
Chromosomen) n

769 DICTYOKINESIS
f dictyocinèse f
e dictiocinesis f
i dictiocinesi f
d Diktyokinese f

770 DICTYOSOME
f dictyosome m
e dictiosoma m
i dictiosoma m
d Diktyosom n

771 DICTYOTIC STAGE;
DICTYATE STAGE
f stade m dictyotique
e estadio m dictiótico
i stadio m dictiottico
d retikuläres Stadium (der
Chromosomen) n

772 DIDIPLOID (adj.)
DIDIPLOID
f didiploïde
didiploïde m
e didiploide
didiploide m
i didiploide
didiploide m
d didiploid
Didiploide f

773 DIDIPLOIDY
f didiploïdie f
e didiploidía f
i didiploidismo m
d Didiploidie f

774 DIENTOMOPHILY
f dientomophylie f
e dientomofilia f
i dientomofilia f
d Dientomophilie f (#)

775 DIFFERENTIAL
f différentiel
e diferencial
i differenziale
d differentiell;
Differential- (in Zusam-
mensetzungen)

776   DIFFERENTIAL AND INTER-
      FERENCE DISTANCES
      f   distance *f* d'interférence
          et distance différentielle
      e   distancias *fpl* de inter-
          ferencia y diferenciales
      i   distanze *fpl* differenziali e
          d'interferenza
      d   Differential- und Interfe-
          renzabstand *m*

777   DIFFERENTIAL SEGMENT
      f   segment *m* différentiel
      e   segmento *m* diferencial
      i   segmento *m* differenziale
      d   Differentialsegment *n*;
          differentielles Segment *n*

778   DIFFERENTIAL SEX GENE
      f   gène *m* de différenciation
          sexuelle
      e   gen *m* sexual diferencial
      i   gene *m* di differenziazione
          sessuale;
          gene *m* sessuale
          differenziale
      d   Geschlechtsrealisator *m*

779   DIFFERENTIATION
      f   différenciation *f*
      e   diferenciación *f*
      i   differenziazione *f*
      d   Differenzierung *f*

780   DIGAMETIC
      f   digamétique;
          hétérogamétique
      e   digamético;
          heterogamético
      i   digametico;
          eterogametico
      d   digametisch;
          heterogametisch

781   DIGAMETY
      f   digamétie *f*
      e   digametia *f*
      i   digametia *f*
      d   Digametie *f*

782   DIGAMY S. (!)

      f   digamie *f* (!)
      e   digamia *f* (!)
      i   digamia *f* (!)
      d   Digamie *f* (!)

783   DIGENESIS
      f   digénèse *f* (Θ)
      e   digénesis *f* (Θ)
      i   digenesi *f* (Θ)
      d   Digenesis *f* (Θ);
          Generationswechsel *m*

784   DIGENIC
      f   digénique
      e   digénico
      i   digenico
      d   digen

785   DIGENOMIC
      f   digénomique
      e   digenómico
      i   digenomico
      d   digenomatisch

786   DIHAPLOID (adj.)
      DIHAPLOID
      f   dihaploïde
          dihaploïde *m*
      e   dihaploide
          dihaploide *m*
      i   diaploide
          diaploide *m*
      d   dihaploid
          Dihaploide *f*

787   DIHAPLOIDY
      f   dihaploïdie *f*
      e   dihaploidía
      i   diaploidismo
      d   Dihaploidie *f*

788   DIHETEROZYGOTE
      f   dihétérozygote *m*
      e   diheterozigoto *m*
      i   dieterocigoto *m*
      d   Diheterozygote *f*

789   DIHYBRID (adj.)
      DIHYBRID
      f   dihybride
          dihybride *m*

e dihíbrido
dihíbrido *m*
i diibrido
diibrido *m*
d dihybrid
Dihybride *f*

790 DIHYBRIDISM
f dihybridisme *m*
e dihibridismo *m*
i diibridismo *m*
d Dihybridie *f*

791 DIKARYON;
DICARION
f dicaryon *m*
e dicarión *m*;
dicarionte *m*
i dicarion *m*
d Dikaryon *n*

792 DIKARYOPHASE
f dicaryophase *f*
e dicariófase *f*
i dicariofase *f*
d Dikaryophase *f*;
Paarkernphase *f*

793 DIKARYOTIC;
DICARYOTIC
f dicaryotique
e dicariótico
i dicariotico
d dikaryotisch

794 DIMEGALY
f - -
e - -
i - -
d Dimegalie *f*

795 DIMERIC;
DIMERICAL
f dimérique
e dimérico
i dimerico
d dimer

796 DIMERY
f dimérie *f*
e dimería *f*

i dimeria *f*
d Dimerie *f*

797 DIMINUTION
f diminution *f*
e diminución *f*
i diminuzione *f*
d Verminderung *f*

798 DIMIXIS
f dimixie *f*
e dimixis *f*
i dimissi *f*
d Dimixis *f*

799 DIMONOECIOUS
f dimonoïque
e dimonoico
i dimonoico
d dimonoecisch;
dimonoezisch;
dimonözisch

800 DIMORPHIC
f dimorphe
e dimórfico
i dimorfico
d dimorph

801 DIMORPHISM
f dimorphisme *m*
e dimorfismo *m*
i dimorfismo *m*
d Dimorphismus *m*

DIMORPHISM, SEX
see 2578

802 DIOECIOUS
f dioïque
e dioico
i dioico
d dioecisch;
dioezisch;
diözisch

803 DIOECISM; DIOECY
f dioécie *f*
e dioecia *f*;
dioecismo *m*
i dioicismo *m*

d Dioecie *f*;
Dioezie *f*;
Diözie *f*

804 DI-OVAL TWINS
f jumeaux *mpl* bi-ovulaires
e gemelos *mpl* biovulares
i gemelli *mpl* bi-ovulari
d zweieiige Zwillinge *mpl*

805 DIPHYLETIC
f diphylétique
e difilético
i difiletico
d diphyletisch

806 DIPLOBIONT
f diplobionte *m*
e diplobionte *m*
i diplobionte *m*
d Diplobiont *m*

807 DIPLOBIONTIC
f diplobiontique
e diplobióntico
i diplobiontico
d diplobiontisch

808 DIPLOBIVALENT
f diplobivalent *m*
e diplobivalente *m*
i diplobivalente *m*
d Diplobivalent *n*

809 DIPLOCHLAMYDEOUS
f diplochlamyde
e diploclamidio
i diploclamidio
d dichlamyd;
dichlamydeisch

810 DIPLOCHROMOSOME
f diplochromosome *m*
e diplocromosoma *m*
i diplocromosoma *m*
d Diplochromosom *n*

811 DIPLOGENOTYPIC
f diplogénotypique
e diplogenotípico
i diplogenotipico
d diplogenotypisch

812 DIPLO-HAPLOID TWINNING
f formation *f* de jumeaux
diplo-haploïdes
e formación *f* de gemelos
diploides-haploides
i formazione *f* di gemelli
diplo-aploidi
d Entstehung *f* eines diploiden
und eines haploiden Zwillings

813 DIPLOHETEROECY
f diplohétéroecie *f*
e diploheteroecia *f*
i diploeteroicismo *m*
d Diploheteroecie *f*;
Diploheteroezie *f*;
Diploheterözie *f*

814 DIPLOID (adj.)
DIPLOID
f diploïde
diploïde *m*
e diploide
diploide *m*
i diploide
diploide *m*
d diploid
Diploide *f*

815 DIPLOIDISATION;
DIPLOIDIZATION
f diploïdisation *f*
e diploidización *f*
i diploidizzazione *f*
d Diploidisierung *f*

816 DIPLOID SET OF
CHROMOSOMES
f génome *m* diploïde
e genomio *m* diploide;
genoma *m* diploide
i genomio *n* diploide
d diploider Chromosomen-
satz *m*

817 DIPLOIDY
f diploïdie *f*
e diploidía *f*
i diploidismo *m*
d Diploidie *f*

818 DIPLOKARYOTIC (!);
TETRAPLOID
f diplocaryotique (!);
tétraploïde
e diplocariótico;
tetraploide
i diplocariotico;
tetraploide
d diplokaryotisch (!);
tetraploid

819 DIPLOMONOSOMIC
f diplomonosomique
e diplomonosómico
i diplomonosomico
d diplomonosom

820 DIPLONEMA
f diplonème *m*
e diplonema *m*
i diplonema *m*
d Diplonema *n* (#)

821 DIPLONT
f diplonte *m*
e diplonte *m*
i diplonte *m*
d Diplont *m*

822 DIPLONTIC
f diplontique
e diplóntico
i diplontico
d diplontisch

823 DIPLOPHASE
f diplophase *f*
e diplofase *f*
i diplofase *f*
d Diplophase *f*

824 DIPLOSIS
f diplosie *f*
e diplosis *f*
i diplosi *f*
d Diplosis *f*

825 DIPLOSOME
f diplosome *m*
e diplosoma *m*
i diplosoma *m*
d Diplosom *n*

826 DIPLOSPORY
f diplosporie *f*
e diplosporía *f*
i diplosporia *f*
d Diplosporie *f*

827 DIPLOTENE (adj.)
DIPLOTENE
f diplotène
diplotène *m*
e diplóteno
diploténico
i diplotene
diplotene *m*
d Diplotän *n*
(auch in Zusammenset-
zungen)

828 DIRECTIONAL SELECTION
f sélection *f* directionnelle
e selección *f* direccional
i selezione *f* direttoriale
d gerichtete Selektion *f*

829 DIRECT RELATIONSHIP
f parenté *f* directe
e parentesco *m* directo
consanguinidad *f*
i parentela *f* in linea retta
d Verwandtschaft *f* in gerade
Linie

830 DISASSORTATIVE MATING;
MATING UNLIKE TO UNLIKE
f accouplement *m* d'individus
dissemblables
e apareamiento *m* de oposició
i accoppiamento *m* fra
dissimili
d disassortative Paarung *f* (#)
Paarung *f* unähnlicher
Individuen

831 DISCONTINUOUS
f discontinu
e discontinuo
i discontinuo
d diskontinuierlich

832 DISCONTINUOUS VARIATION
DISCRETE VARIATION

f   variation *f* discontinue ;
   variation *f* discrète
e   variación *f* discontinua ;
   variación *f* discreta
i   variazione *f* discontinua;
   variazione *f* discreta
d   diskontinuierliche Variation *f;*
   alternative Variation *f*

833   DISDIERESIS
f   disdiérèse *f*
e   disdiéresis *f*
i   disdieresi *f*
d   Disdiärese *f*

834   DISGRESSIVE
f   complémentaire
e   disgregativo
i   disgressivo
d   komplementär

835   DISJUNCTION
f   disjonction *f*
e   disyunción *f*
i   disgiunzione *f*
d   Disjunktion *f*

DISJUNCTION, ALTERNATIVE
see 115

DISJUNCTION, HALF
see 1179

836   DISJUNCTIONAL SEPARA-
TION;
ALTERNATIVE DISJUNCTION
f   disjonction *f* alternative
e   disyunción *f* alternativa
i   disgiunzione *f* alternativa
d   Alternativverteilung *f*

837   DISLOCATED SEGMENTS
f   segments *mpl* disloqués
e   segmentos *mpl* dislocados
i   segmenti *mpl* dislocati
d   dislozierte Segmente *npl*

838   DISLOCATION
f   dislocation *f*
e   dislocación *f*
i   dislocazione *f*

d   Dislokation *f*

839   DISOMATY (Θ);
TETRAPLOIDY
f   disomatie *f* (Θ);
   tétraploïdie *f*
e   disomatía *f* (Θ);
   tetraploidia *f*
i   disomatia *f* (Θ);
   tetraploidismo *m*
d   Disomatie *f* (Θ);
   Tetraploidie *f*

840   DISOME
f   disome *m*
e   disoma *m*
i   disoma *m*
d   Disom *n*

841   DISOMIC
f   disomique
e   disómico
i   disomico
d   disom

842   DISOMY
f   disomie *f*
e   disomia *f*
i   disomia *f*
d   Disomie *f*

843   DISPARATE
f   disparate
e   dispar
i   disparato
d   ungleichartig

844   DISPARATE CHIASMA
f   chiasma *n* asymétrique
e   quiasma *m* dispar
i   chiasma *m* disparato
d   diagonales Chiasma *n;*
   nichtkompensierendes
   Chiasma *n*

845   DISPERMIC;
DISPERMOUS
f   dispermique
e   dispérmico
i   dispermico
d   disperm

846 DISPERMIC FERTILIZATION;
DISPERMY
f dispermie *f*;
fertilisation *f* dispermique
e dispermia *f*;
fertilización *f* dispérmica
i dispermia *f*;
fertilizzazione dispermica
d disperme Befruchtung *f*;
Dispermie *f*

847 DISPERSION STAGE
f stade *m* de dispersion
e estado *m* de dispersión
i stadio *f* di dispersione
d Zerstäubungsstadium *n*;
Dekondensationsstadium *n*

848 DISPIREME
f dispirème *m*
e dispirema *m*
i dispirema *m*
d Dispirem *n*

849 DISRUPTIVE SELECTION
f sélection *f* disruptive
e selección *f* disruptiva
i selezione *f* disruptiva
d disruptive Selektion *f*

850 DISSOCIATION
f dissociation *f*
e disociación *f*
i dissociazione *f*
d Dissoziation *f*

851 DISTAL
f distal
e distal
i distale
d distal

852 DISTANCE
f distance *f*
e distancia *f*
i distanza *f*
d Distanz *f*;
Abstand *m*

853 DISTRIBUTED C-MITOSIS
f c-mitose *f* distribuée

e c-mitosis *f* distribuída
i c-mitosi *f* distribuita
d - -

854 DITOKOUS
f - -
e - -
i - -
d ditok

855 DITYPE TETRADS
f tétrades *f pl* ditypes
e tetradas *f pl* ditipas
i tetradi *f pl* ditipiche
d ditype Tetraden *f pl*

856 D/I VALUES
f valeurs *m pl* D/I
e valores *m pl* D/I
i valori *m pl* D/I
d D/I-Werte *m pl*

DIVERGENCE,
ECOGEOGRAPHICAL
see 897

857 DIVISION
f division *f*
e división *f*
i divisione *f*
d (z.B. Zell-)Teilung *f*

DIVISION, HOMEOTYPIC
see 1348

DIVISION, HOMOTYPE
see 1402

DIVISION, NUCLEAR
see 1932

858 DIZYGOTIC
f dizygotique
e dicigótico
i dizigotico
d dizygotisch;
zweieiig

859 DIZYGOTIC TWINS
f jumeaux *m pl* dizygotiques
e gemelos *m pl* dicigóticos

i gemelli *mpl* dizigotici
d zweieiige Zwillinge *mpl*

860 D-MITOSIS
f d-mitose *f*
e d-mitosis *f*
i d-mitosi *f*
d D-Mitose *f*

861 DNA: DE(S)OXYRIBONUCLEIC ACID
f ADN: acide *m* désoxyribonucléique
e ADN: ácido *m* desoxiribonucléico
i ADN: acido *m* desossiribonucleico
d DNS: Desoxyribosenucleinsäure *f*

862 DOMINANCE
f dominance *f*
e dominancia *f*
i dominanza *f*
d Dominanz *f*

DOMINANCE, CONDITIONAL
see 600

DOMINANCE, DELAYED
see 723

DOMINANCE, INCOMPLETE
see 2080, 2084

863 DOMINANT
f dominant
e dominante
i dominante
d dominant

864 DOMINIGENE
f dominigène *m*
e dominigén *m*
i dominigene *m*
d Dominigen *n*

865 DONATOR
f donateur *m*
e donante *m*
i donatore *m*
d Donator *m*

866 DONOR (adj.)
DONOR
f donateur donneur *m*
e donante donor *m*
i donatore donatore *m*
d Donor *m*

867 DONOR CELL
f cellule *f* donatrice
e célula *f* donante
i cellula *f* donatrice
d Donor-Zelle *f*

868 DOSAGE COMPENSATION
f compensation *f* de dosage
e compensación *f* de dosis
i compensazione *f* di dosaggio
d Dosiskompensation *f*

869 DOSAGE INDIFFERENCE
f indifférence *f* au dosage
e indiferencia *f* de dosis
i indifferenza *f* di dosaggio
d Dosisindifferenz *f*

870 DOUBLE CROSS;
FOUR-WAY CROSS
f croisement double
e cruzamiento *m* doble cruzamiento *m* de cuatro líneas
i doppio incrocio *m*
d Doppelkreuzung *f*

871 DOUBLE CROSSING-OVER
f double crossing-over *m*
e doble sobrecruzamiento *m*
i doppio sovrincrocio *m*
d Doppel-Crossing-over *n*

872 DOUBLE DIPLOID (adj.)
DOUBLE DIPLOID
f amphidiploïde amphidiploïde *m*
e doble diploide; anfidiploide doble diploide *m*; anfidiploide *m*

i amfidiploide;
anfidiploide
amfidiploide *m*;
anfidiploide *m*
d amphidiploid
Amphihaploide *f*

873 DOUBLE DOMINANTS
f doubles dominants *mpl*
e doble dominantes *mpl*
i doppi dominanti *mpl*
d zwei dominante Komple-
mentärgene *npl*

874 DOUBLE F I
f double F I *f*
e doble F I *f*
i doppia F I *f*
d - -

875 DOUBLE HAPLOID (adj.)
DOUBLE HAPLOID
f amphihaploïde
amphihaploïde *m*
e doble haploide;
anfihaploide
doble haploide *m*;
anfihaploide *m*
i amfihaploide;
anfihaploide
amfihaploide *m*;
anfihaploide *m*
d doppelhaploid
Doppelhaploide *f*

876 DOUBLE REDUCTION
f double réduction *f*
e doble reducción *f*
i doppia riduzione *f*
d doppelte Reduktion *f*

877 DOUBLE SIZED RING
CHROMOSOME
f iso-anneau *m* double
e cromosoma *m* anular de
doble tamaño
i anello *m* cromosomico
raddoppiato
d Doppelring *m*

878 DOUBLING
f doublement *m*

e doblamiento *m*
i raddoppiamento *m*
d Verdoppelung *f*

879 DRIFT
f dérive *f*
e deriva *f*; drift *m*
i deriva *f*
d Drift *f*

DRIFT, RANDOM
see 2397

DRIFT, STEADY
see 2668

880 DRIVE
(MEIOTIC)
f mouvement *m* méiotique
e movimiento *m* meiótico
i impulso *m* meiotico
d meiotische Drift *f*

881 DUPLEX
f duplex
e duplexo
i duplex
d duplex

882 DUPLICATE
f doublé
e duplicado
i duplicato
d verdoppelt

883 DUPLICATE GENES
f gènes *mpl* doublés
e genes *mpl* duplicados
i geni *mpl* duplicati
d duplikate Gene *npl*

884 DUPLICATION
f duplication *f*
e duplicación *f*
i duplicazione *f*
d Duplikation *f*

885 DUPLICATIONAL POLYPLOII
f polyploïde *m* de duplication
e poliploide *m* duplicacional
i poliploide *m* da duplicazione
d Duplikationspolyploide *f*

886 DYAD, DIAD
   f   diada *f*; dyade *f*
   e   diada *f*
   i   diade *f*
   d   Dyade *f*

887 DYSGENESIS
   f   dysgénèse *f*
   e   disgenesia *f*;
       disgénesis *f*
   i   disgenesia *f*
   d   Dysgenese *f* (#)

888 DYSGENIC
   f   aneugénique;
       disgénésique
   e   disgénico
   i   desgenico
   d   dysgenisch

889 DYSPLOID (adj.)
    DYSPLOID

   f   dysploïde
       dysploïde *m*
   e   disploide
       disploide *m*
   i   disploide
       disploide *m*
   d   dysploid
       Dysploide *f*

890 DYSPLOIDION
   f   disploïdion *m*
   e   disploidión *m*
   i   disploidion *m* (Θ)
   d   Dysploidion *n* (Θ);
       dysploide Art *f*

891 DYSPLOIDY
   f   dysploïdie *f*
   e   disploidía *f*
   i   disploidismo *m*
   d   Dysploidie *f*

E

892 ECAD
f écade *f*
e écade *f*
i ecade *f*
d Ökade *f* (#)

893 ECESIS
f ecesis *f*
e ecesis *f*
i ecesi *f* (Ⓞ)
d - -

894 ECOBIOTIC ADAPTATION
f adaptation *f* écobiotique
e adaptación *f* ecobiótica
i adattamento *m* ecobiotico
d ökobiotische Anpassung *f*

895 ECOCLIMATIC ADAPTATION
f adaptation *f* écoclimatique
e adaptación *f* ecoclimática
i adattamento *m* ecoclimatico
d ökoklimatische Anpassung *f*

896 ECOCLINE
f écocline *f*
e - -
i - -
d Ökokline *f* (#)

897 ECOGEOGRAPHICAL DIVER-
GENCE
f divergence *f* écogéo-
graphique
e divergencia *f* ecogeo-
gráfica
i divergenza *f* ecogeo-
grafica
d ökogeographische Diver-
genz *f*

898 ECOPHENE
f écophène *m*
e ecofén *m*
i ecofene *m*
d Ökophän *n* (#)

899 ECOPHENOTYPE
f écophénotype *m*
e ecofenótipo *m*
i ecofenotipo *m*
d Ökophänotyp *m* (#)

900 ECOSPECIES
f écospecies *f*
éco-espèce *f*
e ecoespecie *f*
i ecospecie *f*
d Ökospecies *f*

901 ECOTYPE
f écotype *m*
e ecotipo *m*
i ecotipo *m*
d Ökotyp *m*

902 ECTOPIC PAIRING
f appariement *m* ectopique
e apareamiento *m* ectópico
i appaiamento *m* ectopico
d ektopische Paarung *f*

903 ECTOPLASM
f ectoplasme *m*
e ectoplasma *m*
i ectoplasma *n*
d Ektoplasma *n*

904 ECTOSOME
f ectosome *m*
e ectosoma *m*
i ectosoma *m*
d Ektosom *n*

905 ECTOSPHERE
f ectosphère *f*
e ectosfera *f*
i ectosfera *f*
d Ektosphäre *f*

906 ELECTOSOME
f electosome *m*
e electosoma *m*

i electosoma *m*
d Elektosom *n*

ELEMENT, CONTROLLING
see 617

907 ELIMINATION
f élimination *f*
e eliminación *f*
i eliminazione *f*
d Elimination *f*;
   Ausmerzung *f*

908 EMASCULATION
f émasculation *f*
e emasculación *f*
i emasculazione *f*
d Emaskulation *f*;
   Kastration *f*

909 EMBRYO-SAC
f sac *m* embryonnaire
e saco *m* embrional
i sacca *m* embrionale
d Embryosack *m*

910 ENDOGAMIC;
   ENDOGAMOUS
f endogamique
e endogámico
i endogamico
d endogam

911 ENDOGAMY
f endogamie *f*
e endogamia *f*
i endogamia *f*
d Endogamie *f*

912 ENDOGENOTE
f endogénote *m*
e endogenote *m*
i endogenote *m*
d Endogenote *f*

913 ENDOGENOUS
f endogène
e endógeno
i endogeno
d endogen

914 ENDOMITOSIS

f endomitose *f*
e endomitosis *f*
i endomitosi *f*
d Endomitose *f*

915 ENDOMITOTIC
f endomitotique
e endomitótico
i endomitotico
d endomitotisch

916 ENDOMIXIS
f endomixie *f*
e endomixis *f*
i endomissi *f*
d Endomixis *f*

917 ENDONUCLEAR
f endonucléaire
e endonuclear
i endonucleare
d endonukleär

918 ENDOPLASM
f endoplasme *m*
e endoplasma *m*
i endoplasma *m*
d Endoplasma *n*

919 ENDOPLASMIC
f endoplasmique
e endoplásmico
i endoplasmico
d endoplasmatisch

920 ENDOPOLYPLOID (adj.)
   ENDOPOLYPLOID
f endopolyploïde
   endopolyploïde *m*
e endopoliploide
   endopoliploide *m*
i endopoliploide
   endopoliploide *m*
d endopolyploid
   Endopolyploide *f*

921 ENDOPOLYPLOIDY
f endopolyploïdie *f*
e endopoliploidía *f*
i endopoliploidismo *m*
d Endopolyploidie *f*

922   ENDOSOME
      f   endosome *m*
      e   endosoma *m*
      i   endosoma *m*
      d   Endosom *n*

923   ENDOSPERM
      f   endosperme *m*
      e   endosperma *m;*
          endospermo *m*
      i   endosperma *m*
      d   Endosperm *n*

924   ENDOTAXONIC
      f   endotaxonique
      e   endotaxónico
      i   endotassonico
      d   endotaxonisch (#)

925   ENERGIC NUCLEUS
      f   noyau *m* quiescent
      e   núcleo *m* enérgico
      i   nucleo *m* energico
      d   Arbeitskern *m;*
          Ruhekern *m*

926   ENERGIC STAGE
      f   phase *f* métabolique
      e   estado *m* enérgico
      i   stadio *m* energico
      d   Ruhephase *f*

927   ENERGID
      f   énergide *f*
      e   enérgida *f*
      i   energide *m*
      d   Energide *f*

928   ENFORCED HETEROZYGOSIS;
      ENFORCED HETEROZYGOSITY
      f   hétérozygosis *f* forcée
      e   heterocigosis *f* permanen-
          te
      i   eterozigosi *f* permanente
      d   permanente Heterozygotie *f*

929   ENNEAPLOID (adj.)
      ENNEAPLOID
      f   ennéaploïde
          ennéaploïde *m*
      e   eneaploide

      eneaploide *m*
      i   enneaploide
          enneaploide *m*
      d   enneaploid
          Enneaploide *f*

930   ENNEAPLOIDY
      f   ennéaploïdie *f*
      e   eneaploidía *f*
      i   enneaploidismo *m*
      d   Enneaploidie *f*

931   ENVIRONMENT
      f   milieu *m;*
          environnement *m*
      e   medio *m* ambiente
      i   ambiente *m*
      d   Umwelt *f;*
          Umgebung *f;*
          Milieu *n*

932   ENVIRONMENTAL
      f   ambiant;
          influencé par le milieu
      e   ambiental;
          influenciado por el medio
      i   ambientale
      d   Umwelt-; Aussen- (in Zu-
          sammensetzungen)

933   ENVIRONMENTAL CONDI-
      TIONS
      f   conditions *fpl* du milieu
          extérieur
      e   condiciones *fpl* ambien-
          tales
      i   condizioni *fpl* dell'am-
          biente
      d   Umweltbedingungen *fpl*

934   ENVIRONMENTAL CORRE-
      LATION
      f   corrélation *f* due au milieu
      e   correlación *f* ambiental
      i   correlazione *f* dovuta
          all'ambiente
      d   umweltbedingte Korrelatio

935   ENVIRONMENTAL VARIATIC
      f   variation *f* influencée par
          le milieu;

71

variation *f* d'adaptation
e variación *f* ambiental
i variazione *f* dovuta
all'ambiente
d umweltbedingte Variation *f*

936 EPIGAMIC
f épigamique
e epigámico
i epigamico
d epigam (#)

937 EPIGENESIS
f épigénèse *f*
e epigénesis *f*
i epigenesi *f*
d Epigenese *f*

938 EPIGENETICS
f épigénétique *f*
e epigenética *f*
i epigenetica *f*
d Epigenetik *f* (Θ),
Entwicklungsmechanik *f*

939 EPIGENOTYPE
f épigénotype *m*
e epigenótipo *m*
i epigenotipo *m*
d Epigenotyp *m*

940 EPISOME
f épisome *m*
e episoma *m*
i episoma *m*
d Episom *n*

941 EPISOMIC TRANSISTORY
STADIUM
f état *m* épisomique tran-
sistoire
e estado *m* transistorio
episómico
i stadio *m* episomico
transistorio
d episomales Übergangs-
stadium *n* (#)

942 EPISTASIS;
EPISTASY
f épistasie *f*

e epistasis *f*
i epistasia *f*
d Epistasis *f*;
Epistasie *f*

943 EPISTATIC
f épistatique
e epistático
i epistatico
d epistatisch

944 EQUATIONAL
f équationnel
e ecuacional
i equazionale
d Äquations- (in Zusammensetzungen)

945 EQUATIONAL DIVISION;
EQUATION DIVISION;
HOMEOTYPIC DIVISION
f division *f* équationelle;
division *f* homéotypique
e división *f* celular somática;
división *f* homeotípica;
división *f* ecuacional
i divisione *f* equazionale;
divisione *f* omeotipica
d Äquationsteilung *f*;
homoiotypische Teilung *f* (Θ);
homöotypische Teilung *f* (Θ)

946 EQUATORIAL BODY
f cellule *f* équatoriale
e cuerpo *m* ecuatorial
i cellula *f* equatoriale
d Äquatorialkörper *m*

947 EQUATORIAL PLATE
f plaque *f* équatoriale
e placa *f* ecuatorial
i placca *f* equatoriale
d Äquatorialplatte *f*

EQUILIBRIUM, MUTATIONAL
see 1895

948 ERGATOMORPHIC MALE
f mâle *m* ergatomorphique
e macho *m* ergatomórfico
i maschio *m* ergatomorfico
d ergatomorphes Männchen *n*

949 ERYTHROPHILOUS
  f erythrophile
  e eritrófilo
  i eritrofilo
  d erythrophil

950 ETHEOGENESIS;
  ETHIOGENESIS
  f éthéogénèse *f*
  e eteogénesis *f*
  i - -
  d Etheogenesis *f*

951 EUCENTRIC
  f eucentrique
  e eucéntrico
  i eucentrico
  d euzentrisch

952 EUCHROMATIC
  f euchromatique
  e eucromático
  i eucromatico
  d euchromatisch

953 EUCHROMATIN
  f euchromatine *f*
  e eucromatina *f*
  i eucromatina *f*
  d Euchromatin *n*

954 EUCHROMOSOME
  f euchromosome *m*
  e eucromosoma *m*
  i eucromosoma *m*
  d Euchromosom *n*

955 EUGENIC
  f eugénique
  e eugenésico
  i eugenetico
  d eugenisch

956 EUGENICS
  f eugénique *f*
  e eugenesia *f*
  i eugenia *f*
  d Eugenik *f*

957 EUHETEROSIS
  f euhétérosis *f*

  e euheterosis *f*
  i eueterosi *f*
  d Euheterosis *f*

958 EUMEIOSIS
  f euméiose *f*
  e eumeiosis *f*
  i eumeiosi *f*
  d Eumeiose *f;*
    Eumeiosis *f*

959 EUMITOSIS
  f eumitose *f*
  e eumitosis *f*
  i eumitosi *f*
  d Eumitose *f;*
    Eumitosis *f*

960 EUPLOID (adj.)
  EUPLOID
  f euploïde
    euploïde *m*
  e euploide
    euploide *m*
  i euploide
    euploide *m*
  d euploid
    Euploide *f*

961 EUPLOIDY
  f euploïdie *f*
  e euploidía *f*
  i euploidismo *m*
  d Euploidie *f*

962 EVOLUTION
  f évolution *f*
  e evolución *f*
  i evoluzione *f*
  d Evolution *f*

  EVOLUTION, EXPLOSIVE
  see 973

963 EVOLUTION FACTORS
  f facteurs *mpl* d'évolution
  e factores *mpl* de
    evolución
  i fattori *mpl* di evoluzione
  d Evolutionsfaktoren *mpl*

964   EVOLUTIONARY SUCCESS
      f succès *m* évolutif
      e éxito *m* evolutivo
      i esito *m* evolutivo
      d Evolutionserfolg *m*

965   EVOLUTIVE
      f évolutif
      e evolutivo
      i evolutivo
      d evolutiv

966   EXAGGERATION
      f exagération *f*
      e exageración *f*
      i esagerazione *f*
      d Exaggeration *f*

967   EXHIBITION
      f exhibition *f*
      e exhibición *f*
      i esibizione *f*
      d Exhibition *f*

968   EX-MUTANT
      f ex-mutant *m*
      e ex-mutante *m*
      i ex mutante *m*
      d Exmutante *f*

969   EXOGAMY
      f exogamie *f*
      e exogamia *f*
      i exogamia *f*
      d Exogamie *f*

970   EXOGENOTE
      f exogénote *m*
      e exogenote *m*
      i esogenote *m*
      d Exogenote *f*

971   EXOGENOUS
      f exogène
      e exógeno
      i esogeno
      d exogen

972   EXPLODED C-MITOSIS
      f c-mitose *f* explosée
      e c-mitosis *f* explotada

      i c-mitosi *f* esplosa
      d explodierte C-Mitose *f*

973   EXPLOSIVE EVOLUTION;
      EXPLOSIVE SPECIATION
      f évolution *f* explosive
      e evolución *f* explosiva
      i evoluzione *f* esplosiva
      d explosive Evolution *f;*
         explosive Artbildung *f*

974   EXPONENTIAL DEATH
      PHASE
      f phase *f* de mort exponen-
         tielle
      e fase *f* de morte
         exponencial
      i fase *f* esponenziale di
         decesso
      d exponentielle
         Sterbephase *f*

975   EXPRESSIVITY
      f expressivité *f*
      e expresividad *f*
      i espressività *f*
      d Expressivität *f*

976   EXTENSION FACTOR
      f facteur *m* d'extension
      e factor *m* de extensión
      i fattore *m* d'estensione
      d Modifikationsgen *n*

977   EXTRACHROMOSOMIC
      f extrachromosomique
      e extracromosómico
      i extracromosomico
      d extrachromosomal

978   EXTRAMEDIAL HYBRID
      QUOTIENT
      f quotient *m* hybride extra-
         médial
      e cociente *m* híbrido
         extramedial
      i quoziente *m* ibrido extra-
         mediale (Θ)
      d - -

979   EXTRA-RADIAL
       f   extra-radial
       e   extra-radial
       i   extra-radiale
       d   extraradial

F

980 FACIES
f facies *m*
e facies *f*
i facies *f*;
aspetto *m*
d Facies *f*

981 FACTOR
f facteur *m*
e factor *m*
i fattore *m*
d Faktor *m*

982 FACTOR MAP
f carte factorielle *f*
e mapa *f* factorial
i carta *f* fattoriale
d Genkarte *f*

983 FACTOR-PAIR
f couple *m* de facteurs
héréditaires;
paire *f* de facteurs
héréditaires
e par *m* de factores
hereditarios;
pareja *f* de factores
hereditarios
i accoppiamento *m* fattoriale
d Genpaar *n*;
Allelenpaar *n*

984 FALSE LINKAGE
f fausse liaison *f*;
faux linkage *m*
e falso ligamiento *m*
i falso legame *m*
d falsche Kopp(e)lung *f*

985 FALSE PAIRING
f faux appariement *m*
e falso apareamiento *m*
i falso abbinamento *m*
d falsche Paarung *f*

986 FAMILIAL
f familial

e familiar
i familiare
d Familien- (in Zusammen-
setzungen)

987 FAMILY
f famille *f*
e familia *f*
i famiglia *f*
d Familie *f*

988 FAMILY SELECTION
f sélection *f* familiale;
sélection *f* interfamiliale
e selección *f* familiar
i selezione *f* familiare
d Familienauslese *f*

989 FECUNDATION;
FERTILIZATION
f fécondation *f*
e fecundación *f*
i fecondazione *f*;
fertilizzazione *f*
d Befruchtung *f*

990 FEMALE LINE OF BREEDING
f lignée *f* femelle;
lignée *f* maternelle;
filiation *f* maternelle
e línea *f* feminina;
línea *f* materna;
filiación *f* materna
i linea *f* materna;
linea *f* femminile
d weibliche Linie *f*;
mütterliche Linie *f*

991 FERTILE
f fertile
e fértil
i fertile
d fertil

992 FERTILISATION;
FECUNDATION

f   fertilisation *f;*
fécondation *f*
e   fertilización *f;*
fecundación *f*
i   fertilizzazione *f;*
fecondazione
d   Befruchtung *f*

FERTILISATION, DISPERMIC
see 846

FERTILISATION, PARTIAL
see 2081

993   FERTILIZING CAPACITY
f   pouvoir fécondant *m*
e   poder fecundante *m*
i   potere fertilizzante *m;*
potere fecondante *m*
d   Befruchtungsfähigkeit *f*

994   FESTOON ARRANGEMENT
f   disposition *f* en festons
e   arreglo *m* en festón
i   disposizione *f* a festoni
d   kettenförmige Chromoso-
menanordnung *f*

995   FIBRE
f   fibre *f*
e   fibra *f*
i   fibra *f*
d   Faser *f*

FIBRES, CHROMOSOMAL
see 519

996   FIBRIL
f   fibrille *f*
e   fibrilla *f*
i   fibrilla *f*
d   Fibrille *f*

997   FILAMENT
f   filament *m*
e   filamento *m*
i   filamento *m*
d   Filament *n*

998   FILIAL GENERATION
f   génération *f* fille ;

génération *f* des descen-
dants
e   generación *f* filial
i   generazione *f* filiale
d   Filialgeneration *f;*
Tochtergeneration *f*

999   FIRST DIVISION
f   première division *f*
e   primera división *f*
i   prima divisione *f*
d   erste Teilung *f*

1000   FISSION
f   scission *f;*
clivage *m*
e   fisión *f*
i   scissione *f*
d   Spaltung *f*

1001   FIXATION
f   fixation *f*
e   fijación *f*
i   fissazione *f*
d   Fixierung *f*

FIXATION, RANDOM
see 2398

1002   FIXATION RATE
f   taux *m* de fixation
e   proporción *f* de fijación
i   tasso *m* di fissazione
d   - -

1003   FIXED
f   fixé
e   estable
i   fissato
d   fixiert

1004   FLUCTUATION
f   fluctuation *f*
e   fluctuación *f*
i   fluttuazione *f*
d   Fluktuation *f*

1005   FORTIFICATION
f   renforcement *m*
e   fortificación *f*
i   fortificazione *f*
d   Verstärkung *f*

1006   FOUNDATION ANIMAL
        f   animal *m* fondateur ;
            animal *m* tête de ligne
        e   cabeza *f* de línea
        i   animale *m* prototipo
        d   Stammtier *n* ;
            Linienbegründer *m*

1007   FOUR-WAY CROSS
        f   croisement *m* double
        e   cruzamiento *m* doble
        i   doppio incrocio *m*
        d   doppelte Kreuzung *f*;
            Vierstrangaustausch *m*

1008   FRACTIONATION
        f   fractionnement *m*
        e   fraccionamiento *m*
        i   frazionamento *m*
        d   Fraktionierung *f*

1009   FRACTURE
        f   fracture *f*
        e   fractura *f*
        i   frattura *f*
        d   Bruch *m*

1010   FRAGMENT
        f   fragment *m*
        e   fragmento *m*
        i   frammento *m*
        d   Fragment *n*

1011   FRAGMENTATION
        f   fragmentation *f*
        e   fragmentación *f*
        i   frammentazione *f*
        d   Fragmentation *f*

        FRAGMENTATION,
        HALF CHROMATID
        see 1178

1012   FRATERNAL
        f   fraternel
        e   fraternal
        i   fraterno
        d   brüderlich

1013   FRATERNAL TWINS
        f   faux jumeaux *mpl*;
            jumeaux *mpl* bivitellins
        e   gemelos *mpl* isófanos

        i   gemelli *mpl* fraterni
        d   zweieiige Zwillinge *mpl*

1014   FREEMARTIN
        f   free-martin
        e   - -
        i   free martin *m*
        d   - -

1015   FREQUENCY
        f   fréquence *f*
        e   frecuencia *f*
        i   frequenza *f*
        d   Frequenz *f*;
            Häufigkeit *f*

1016   FRUCTIFICATION
        f   fructification *f*
        e   fructificación *f*
        i   fruttificazione *f*
        d   Fruktifikation *f*

1017   FULL-SIB: FULL-SISTER,
        FULL-BROTHER
        f   pleine soeur *f*;
            soeur *f* germaine ;
            plein frère *m* ;
            frère *m* germain
        e   hermana *f*;
            hermano *m*
        i   sorella *f*;
            fratello *m*
        d   Vollgeschwister:
            Vollschwester *f*,
            Vollbruder *m*

1018   FUNCTIONAL DIPLOID
        f   diploïde *m* fonctionnel
        e   diploide *m* funcional
        i   diploide *m* funzionale
        d   funktionelle Diploide *f*

1019   FUNDAMENTAL
        f   fondamental
        e   fundamental
        i   fondamentale
        d   fundamental

1020   FUNDAMENTAL NUMBER
        f   nombre *m* fondamental
        e   número *m* fundamental
        i   numero *m* fondamentale

d Grundzahl f;
Basiszahl f

FUSION, CENTRIC
see 439

1021   FUSION
f fusion f
e fusión f
i fusione f
d Fusion f

1022   FUSION NUCLEUS
f noyau m de fusion
e núcleo m de fusión
i nucleo m di fusione
d Fusionskern m

G

1023 **GAMETE**
f gamète *m*
e gameto *m*
i gameto *m*
d Gamet *m* ;
Geschlechtszelle *f*

GAMETES, BALANCED
see 324

1024 **GAMETIC**
f gamétique
e gamético
i gametico
d gametisch;
Gameten- (in Zusammen-
setzungen)

1025 **GAMETIC CHROMOSOME
NUMBER**
f nombre *m* de chromoso-
mes du gamète
e nombre *m* de cromosomas
gaméticos
i numero *m* cromosomico
del gameto
d gametische Chromoso-
menzahl *f*

1026 **GAMETIC INCOMPATIBILITY**
f incompatibilité *f* gamétique
e incompatibilidad *f*
gamética
i incompatibilità *f* gametica
d gametische (oder gametophytische) Inkompatibi-
lität *f*

1027 **GAMETIC LETHAL FACTOR**
f facteur *m* gamétique léthal
e factor *m* gamético letal
i fattore *m* gametico letale
d gametischer Letalfaktor *m*

1028 **GAMETIC MUTATION**
f mutation *f* gamétique
e mutación *f* gamética
i mutazione *f* gametica
d Gametenmutation *f*

1029 **GAMETIC NUMBER**
f nombre *m* gamétique
e número *m* gamético
i numero *m* gametico
d Gametenzahl *f*

1030 **GAMETIC STERILITY**
f stérilité *f* gamétique
e esterilidad *f* gamética
i sterilità *f* gametica
d gametische Sterilität *f*

1031 **GAMETOBLAST**
f gamétoblaste *m*
e gametoblasto *m*
i gametoblasto *m*
d Gametoblast *m*

1032 **GAMETOCYTE**
f gamétocyte *m*
e gametocito *m*
i gametocito *m*
d Gametocyte *f*

1033 **GAMETOGAMY**
f gamétogamie *f*
e gametogamia *f*
i gametogamia *f*
d Gametogamie *f*

1034 **GAMETOGENESIS**
f gamétogénèse *f*
e gametogénesis *f*
i gametogenesi *f*
d Gametogenese *f*

1035 **GAMETOGENIC;
GAMETOGENOUS**
f gamétogénique
e gametogénico
i gametogenico
d gametogen

1036 **GAMETOGENOUS;
GAMETOGENIC**
f gamétogénique
e gametogénico

i gametogenico
d gametogen

1037 GAMETOGONIUM
      (pl GAMETOGONIA)
      f gamétogonie *f*
      e gametogonia *f*
      i gametogonia *f*
      d Gametogonium *n*

1038 GAMETOID
      f gamétoïde *m*
      e gametoide *m*
      i gametoide *m*
      d Gametoid *n*

1039 GAMETOPHORE
      f gamétophore *m*
      e gametóforo *m*
      i gametoforo *m*
      d Gametophor *m*

1040 GAMETOPHYTE
      f gamétophyte *m*
      e gametofito *m*
      i gametofito *m*
      d Gametophyt *m*

1041 GAMETOPHYTIC
      f gamétophytique
      e gametofítico
      i gametofitico
      d gametophytisch

1042 GAMIC
      f gamique
      e gámico
      i gamico
      d - -

1043 GAMOBIUM
      (pl GAMOBIA)
      f gamobium *m*
      e gamobio *m*
      i gamobio *m*
      d Gamobium *n*

1044 GAMODEME
      f gamodème *m*
      e gamodemo *m*
      i gamodemo *m*
      d Gamodeme *m*

1045 GAMOGENESIS
      f gamogénèse *f*
      e gamogénesis *f*
      i gamogenesi *f*
      d Gamogenese *f*

1046 GAMOGENETIC;
      GAMOGENIC
      f gamogénétique
      e gamogenético
      i gamogenetico
      d gamogenetisch

1047 GAMOGENIC;
      GAMOGENETIC
      f gamogénétique
      e gamogenético
      i gamogenetico
      d gamogenetisch

1048 GAMOGONY
      f gamogonie *f*
      e gamogonia *f*
      i gamogonia *f*
      d Gamogonie *f*

1049 GAMOLYSIS
      f gamolyse *f*
      e gamólisis *f*
      i gamolisi *f*
      d Gamolyse *f*

1050 GAMONE
      f gamone *f*
      e gamón *m*
      i gamono *m*
      d Gamon *n*

1051 GAMONT
      f gamonte *m*
      e gamonte *m*
      i gamonto *m*
      d Gamont *m*

1052 GAMOPHASE
      f gamophase *f*
      e gamófase *f*
      i gamofase *f*
      d Gamophase *f*

1053 GAMOTROPISM
      f gamotropisme *m*
      e gamotropismo *m*
      i gamotropismo *m*
      d Gamotropismus *m*

1054 GEITONOCARPY
      f géitonocarpie *f*
      e geitonocarpía *f*
      i geitonocarpia *f*
      d Geitonocarpie *f*

1055 GEITONOGAMY
      f géitonogamie *f*
      e geitonogamia *f*
      i geitonogamia *f*
      d Geitonogamie *f*

1056 GEITONOGENESIS
      f géitonogénèse *f*
      e geitonogénesis *f*
      i geitonogenesi *f*
      d Geitonogenese *f*

1057 GEMINUS
      (pl GEMINI)
      f geminus (pl gemini) *m* (#);
         couple *m* bivalent
      e gémino (pl gémini) *m* (#)
      i gemino *m* (#)
      d Geminus *m*

1058 GEMMATION
      f gemmation *f*
      e gemación *f*
      i gemmazione *f*
      d Gemmatio *f*;
         Knospung *f*

1059 GEMMIPAROUS
      f gemmipare
      e gemíparo
      i gemmiparo
      d gemmipar

1060 GEMMULE
      f gemmule *f*
      e gemula *f*
      i gemmula *f*
      d Gemmula *f*;
         Brutknospe *f*

1061 GENE
      f gène *m*
      e gen *m* ;
         gene *m*
      i gene *m*
      d Gen *n*

      GENE, DIFFERENTIAL SEX
      see 778

      GENE, INHIBITING
      see 1492

      GENE, INVISIBLE
      see 1556

      GENE, KEY
      see 1633

      GENE, MAJOR
      see 1702

      GENE, MINOR
      see 1804

      GENE, POLYURGIC
      see 2255

      GENE, STERILITY
      see 2674

      GENE, SUPPLEMENTARY
      see 2720

      GENE, SWITCH
      see 2727

      GENEALOGICAL TABLE
      see 2098

1062 GENEALOGY
      f généalogie *f*
      e genealogía *f*
      i genealogia *f*
      d Genealogie *f*

1063 GENE ARRANGEMENT
      f disposition *f* des gènes
      e ordenación *f* génica
      i disposizione *f* dei geni
      d Anordnung *f* der Gene

1064   GENE BALANCE
    f   balance *f* génique
    e   equilibrio *m* genético
    i   equilibrio *m* genico
    d   Genbalance *f*

1065   GENE CENTER
    f   génocentre *m*
    e   genocentro *m*
    i   genocentro *m*
    d   Genzentrum *n*

1066   GENE CHROMATIN
    f   chromatine *f* du gène
    e   cromatina *f* génica
    i   cromatina *f* genica
    d   Genchromatin *n*

1067   GENE COMPLEX
    f   complexe *m* de gènes
    e   complejo *m* de genes
    i   complesso *m* di geni
    d   Genkomplex *m*

1068   GENE-CYTOPLASM ISOLA-
      TION
    f   séparation *f* du gène du
       cytoplasme
    e   aislamiento *m*
       genocitoplásmico
    i   isolamento *m* del gene
       dal citoplasma
    d   Gen-Cytoplasma-Isola-
       tion *f*

1069   GENE DOSAGE
    f   dosage *m* des gènes;
       dosage *m* génique
    e   dosis *f* génica
    i   dosatura *f* dei geni
    d   Gendosis *f*

1070   GENE DRIFT
    f   - -
    e   - -
    i   - -
    d   Gendrift *f*

1071   GENE FLOW
    f   dispersion *f* des gènes
    e   dispersión *f* génica
    i   dispersione *f* genica

    d   Gen-"flow" *m*

1072   GENE FREQUENCY
    f   fréquence *f* génique
    e   frecuencia *f* genética
    i   frequenza *f* genetica
    d   Genhäufigkeit *f*

1073   GENE INFILTRATION
    f   infiltration *f* de gènes
    e   infiltración *f* genética
    i   infiltrazione *f* genica
    d   Geninfiltration *f*

1074   GENE INTERACTION
    f   interaction *f* des gènes
    e   interacción *f* de los genes
    i   interazione *f* fra geni
    d   Wechselwirkung *f* zwischen
       Genen

1075   GENE MUTATION
    f   mutation *f* de gènes
    e   mutación *f* de genes
    i   mutazione *f* di geni
    d   Genmutation *f*

1076   GENE NEST
    f   nid *m* de gènes
    e   nido *m* génico
    i   nido *m* genico
    d   Gennest *n*

1077   GENE POOL
    f   effectif *m* de gènes
    e   acumulación *f* de genes
    i   effectivo *m* disponibile di
       geni
    d   Gen-"pool" *m*

1078   GENE PRODUCT TO GENE
      EQUILIBRIA HYPOTHESIS
    f   hypothèse *f* de l'équilibre
       entre le gène et son produit
    e   hipótesis *f* del equilibrio
       entre el gen y su producto
    i   ipotesi *f* dell'equilibrio
       fra il gene e suo prodotto
    d   Hypothese *f* vom Gleich-
       gewicht zwischen Gen und
       Genprodukt

1079 GENERATION
 f génération *f*
 e generación *f*
 i generazione *f*
 d Generation *f*

1080 GENERATION INTERVAL
 f intervalle *m* entre les
    générations
 e intervalo *m* entre las
    generaciones
 i intervallo *m* fra le
    generazioni
 d Generationenintervall *n*;
    Abstand *m* zwischen den
    Generationen

1081 GENERATIVE
 f génératif
 e generativo
 i generativo
 d generativ

1082 GENERATIVE NUCLEUS
 f noyau *m* génératif
 e núcleo *m* generativo
 i nucleo *m* generativo
 d generativer Kern *m*

1083 GENE REDUPLICATION
 f réduplication *f* des gènes
 e reduplicación *f* de genes
 i seconda duplicazione *f* dei
    geni
 d Genreduplikation *f*

1084 GENE REPLICA HYPOTHESIS
 f hypothèse *f* sur la répli-
    cation du gène
 e hipótesis *f* de la genocopia
 i ipotesi *f* della duplica-
    zione dei geni
 d Gen-Replica-Hypothese *f*

1085 GENERIC
 f générique
 e genérico
 i generico
 d generisch (Θ);
    Gattungs- (in Zusammen-
    setzungen)

1086 GENERIC HYBRID

 f hybride *m* générique
 e híbrido *m* genérico
 i ibrido *m* generico
 d Gattungsbastard *m*

1087 GENERITYPE
 f généritype *m*
 e generotipo *m*
 i tipo del genere *m*
 d Gattungstyp *m*

 GENES, BLOCK OF
 see 390

 GENES, DUPLICATE
 see 883

 GENES, ISOLATION
 see 1582

 GENES, MIMIC
 see 1802

 GENES, PAIR OF
 see 2023

 GENES, TRIPLICATE
 see 2861

 GENES, TWIN
 see 2873

1088 GENE STARVATION HY-
 POTHESIS
 f théorie *f* de la carence
    des gènes
 e teoría *f* de la inanición
    de genes
 i teoria *f* dell'inanizione
    dei geni
 d gene-starvation-Hypothese *f*

1089 GENE STRING
 f chromatide *f*;
    cordon *m* de gènes
 e filamento *m* de los genes
 i filamento *m* dei geni
 d Gen-"string" *m*

1090 GENE TAGGED CHROMO-
 SOMES

f chromosomes *mpl* couverts
de gènes
e cromosomas *mpl* cubiertos
de genes
i cromosomi *mpl* coperti
di geni
d genetisch markierte
Chromosomen *npl*

1091 GENETIC
f génétique
e genético
i genetico
d genetisch

**1092 GENETIC ASSORTATIVE
MATING**
f accouplement *m* assorta-
tif
e apareamiento *m* de
semejanza
i accoppiamento *m* di tipo
assortito
d assortative Paarung *f*

1093 GENETIC BACKGROUND
f milieu *m* génétique
e ambiente *m* genético
i ambiente *m* genetico
d genotypisches Milieu *n;*
Genmilieu *n*

1094 GENETIC COADAPTATION
f coadaptation *f* génétique
e coadaptación *f* genética
i coadattamento *m* genetico
d genetische Coadaptation *f*

1095 GENETIC DEATH
f mort *f* génétique
e muerte *f* genética
i morte *f* genetica
d genetischer Tod *m*

**1096 GENETIC EQUILIBRIUM;
POPULATION EQUILIBRIUM**
f équilibre *m* génétique
e equilibrio *m* génico;
equilibrio *m* genético
i equilibrio *m* genetico
d genetisches Gleichge-
wicht *n*

1097 GENETIC LOAD
f charge *m* génétique
e carga *f* genética
i potenza *f* genetica
d - -

1098 GENETICS
f génétique *f*
e genética *f*
i genetica *f*
d Genetik *f*

1099 GENETIC VALUE
f valeur *f* héréditaire;
valeur *f* de reproduction
e valor hereditario *m*
i valore *m* ereditario
d - -

1100 GENETIC VARIATION
f variation *f* génotypique
e variación *f* genotípica
i variazione *f* genotipica
d genetische Variation *f*

1101 GENETOTROPHIC
f génétotrophique
e genetotrófico
i genetotrofico
d genetotroph (#)

1102 GENE TRANSFER
f transfert *m* de gènes
e transferencia *f* de genes
i trasferimento *m* di gene
d Gentransfer *m*

1103 GENIC
f génique
e génico
i genico
d genisch;
Gen-(in Zusammensetzun-
gen)

1104 GENIC BALANCE
f équilibre *m* génique
e equilibrio *m* génico
i equilibrio *m* genico
d Genbalance *f*

105 GENIC DOSAGE
  f dosage *m* génique
  e dosis *f* génica
  i dosatura *f* genica
  d Gendosis *f*

106 GENIC STERILITY
  f stérilité génique *f*
  e esterilidad *f* génica
  i sterilità *f* genica
  d genische Sterilität *f*

107 GENOCLINE
  f génocline *m*
  e genoclino *m*
  i genocline *m*
  d Genocline *m*

108 GENOCOPY
  f génocopie *f*
  e genocopia *f*
  i genocopia *f*
  d Genokline *m*

109 GENODISPERSION
  f dispersion *f* génique
  e genodispersión *f*
  i dispersione *f* genica
  d Genodispersion *f*

110 GENOID
  f génoïde *m*
  e genoide *m*
  i genoide *m*
  d Genoid *n*

111 GENOM;
  GENOME
  f génome *m*;
    garniture *f* chromosomique
  e genomio *m*;
    genomia *f*
  i genoma *m*;
    genomio *m*
  d Genom *n*

112 GENOME ANALYSIS
  f analyse *f* génomique
  e análisis *f* genómica
  i analisi *f* genomica
  d Genomanalyse *f*

1113 GENOME MUTATION
  f mutation *f* de génome
  e mutación *f* de genomio;
    mutación *f* genómica
  i - -
  d Genommutation *f*

1114 GENOMERE
  f génomère *m*
  e genómero *m*
  i genomero *m*
  d Genomere *f*

1115 GENOMIC
  f génomique
  e genómico
  i genomico
  d genomisch;
    Genom- (in Zusammen-
    setzungen)

1116 GENONEMA
  (pl GENONEMATA)
  f génonème *m*
  e genonema *m*
  i genonema *m*
  d Genonema *n*

1117 GENONOMY
  f génonomie *f*
  e genonomía *f*
  i genonomia *f*
  d - -

1118 GENOSOME
  f génosome *m*
  e genósoma *m*
  i genosoma *m*
  d Genosom *n*

1119 GENOTYPE
  f génotype *m*
  e genótipo *m*
  i genotipo *m*
  d Genotyp *m*

  GENOTYPE, RESIDUAL
  see 2460

1120 GENOTYPE-ENVIRONMENT
  INTERACTION

f interaction *f* entre hérédité et environnement
e interacción *f* entre la
herencia y el medio
i interazione *f* fra eredità
ed ambiente
d Genotyp-Umwelt-Wechsel-
wirkung *f*

1121 GENOTYPIC;
GENOTYPICAL
f génotypique
e genotípico
i genotipico
d genotypisch

1122 GENOTYPICAL;
GENOTYPIC
f génotypique
e genotípico
i genotipico
d genotypisch

1123 GENOVARIATION
f génovariation *f*
e mutación *f* de gene
i mutazione *f* di gene
d Genovariation *f* (Ө);
Genmutation *f*

1124 GENUS (pl GENERA)
f genre *m*
e género *m*
i genere *m*
d Genus *n*;
Gattung *f*

1125 GEOGRAPHICAL RACE
f race *f* géographique
e raza *f* geográfica
i razza *f* geografica
d geographische Rasse *f*

1126 GERMINAL
f germinal
e germinal
i germinale
d Keim- (in Zusammen-
setzungen)

1127 GERMINAL SELECTION
f sélection *f* germinale

e selección *f* germinal
i selezione *f* germinale
d Germinalselektion *f*

1128 GERMINAL SPOT
f tache *f* germinale
e mancha *f* germinal
i - -
d Keimfleck *m*

1129 GERMINAL VESICLE
f vésicule *f* germinative
e vesícula *f* germinal
i vescica *f* germinativa
d Keimblase *f*

1130 GERM PLASM;
GERM PLASMA
f germen *m*;
plasma *m* germinal
e plasma *m* germinal
i plasma *m* germinale
d Keimplasma *n*

1131 GERM TRACK
f tractus *m* germinal
e ruta *f* germinal
i catena *f* germinale
d Keimbahn *f*

1132 GIANT CHROMOSOME
f chromosome *m* géant
e cromosoma *m* gigante
i cromosoma *m* gigantico
d Riesenchromosom *n*

1133 GOLGI APPARATUS
f appareil *m* de Golgi
e aparato *m* de Golgi
i apparato *m* di Golgi
d Golgi-Apparat *m*

1134 GOLGIOGENESIS
f golgiogénèse *f*
e golgiogénesis *f*
i golgiogenesi *f*
d Golgiogenese *f*

1135 GOLGIOKINESIS
f golgiocinèse *f*
e golgiocinesis *f*

i golgiocinesi *f*
d Golgiokinese *f*

1136 GOLGIOLYSIS
f golgiolyse *f*
e golgiolisis *f*
i golgiolisi *f*
d Golgiolysis *f*

1137 GONAD
f gonade *f*
e gónada *f*
i gonade *f*
d Gonade *f*;
Keimdrüse *f*;
Geschlechtsdrüse *f*

1138 GONADAL;
GONADIAL;
GONADIC
f gonadique
e gonádico
i gonadiale;
gonadico
d Keimdrüsen-;
Gonaden-;
Geschlechtsdrüsen- (in
Zusammensetzungen)

1139 GONE
f gonie *f*
e gonio *m*
i gonio *m*
d Gone *f*

1140 GONIDIUM (pl GONIDIA)
f gonidie *f*
e gonidio *m*;
gonidia *f*
i gonidio *m*
d Gonidie *f*

1141 GONOCYTE
f gonocyte *m*
e gonócito *m*
i gonocita *m*
d Gonocyte *f*

1142 GONOGAMETE
f gonogamète *m*
e gonogameto *m*
i gonogameto *m*

d Gonogamet *m*

1143 GONOGENESIS
f gonogénèse *f*
e gonogénesis *f*
i gonogenesi *f*
d Gonogenese *f*

1144 GONOMERE
f gonomère *m*
e gonómero *m*
i gonomero *m*
d Gonomere *f*

1145 GONOMERY
f gonomérie *f*
e gonomeridio *m*
i gonomeria *f*
d Gonomerie *f*

1146 GONOMONOARRHENIC
f gonomonoarrhénique
e - -
i - -
d gonomonoarrhenisch

1147 GONOMONOTHELIDIC
f gonomonothélique
e - -
i - -
d gonomonothelidisch

1148 GONOPHAGE
f gonophage *m*
e gonofago *m*
i gonofaco *m*
d Gonophage *m*

1149 GONOSOME
f gonosome *m*
e gonosoma *m*
i gonosoma *m*
d Gonosom *n* (Θ);
Geschlechtschromosom *n*

1150 GONOTOCONT;
GONOTOKONT
f gonotoconte *m*
e gonotoconte *m*;
gonotoconto *m*;
gonoconto *m*

i gonotoconte
d Gonotokont *m*

1151 GRADE;
      GRADE ANIMAL
      f degré *m*;
         grade *m*;
         animal *m* amélioré
      e grado *m*;
         grade *m*
      i grado *m*;
         grade *m*
      d Grade *m*

1152 GRADING
      f classement *m*
      e clasificación *f*
      i classificazione *f*
      d Einstufung *f*

1153 GRADING UP
      f croisement *m* d'absorption;
         croisement *m* d'implanta-
         tion;
         croisement *m* continu;
         croisement *m* de substitution
      e cruzamiento *m* de
         absorción
      i incrocio *m* sostituivo
         d'assorbimento
      d Verdrängungskreuzung *f*

1154 GRAFT HYBRID
      f hybride *m* de greffe
      e injerto *m* híbrido
      i - -
      d Pfropfbastard *m*

      GRANULES, TERMINAL
      see 2787

1155 GYNANDROID
      f gynandroïde
      e ginandroide
      i ginandroide
      d gynandroid

1156 GYNANDROMORPH
      f gynandromorphe
      e ginandromorfo
      i ginandromorfo
      d gynandromorph

1157 GYNANDROMORPHISM
      f gynandromorphisme *m*
      e ginandromorfismo *m*
      i ginandromorfismo *m*
      d Gynandromorphismus *m*

1158 GYNAUTOSOME
      f gynautosome *m*
      e ginautosoma *m*
      i ginautosoma *m*
      d Gynoautosom *n*

1159 GYNODIOECIOUS
      f gynodioïque
      e ginodioico
      i ginodioico
      d gynodiözisch;
         gynodioezisch

1160 GYNODIOECY
      f gynodioécie *f*
      e ginodioecia *f*
      i ginodioicismo *m*;
         ginodioecia *f*
      d Gynodiözie *f*;
         Gynodioezie *f*

1161 GYNOECIOUS
      f gynoïque
      e ginoeico
      i ginoico
      d gynözisch;
         gynoezisch

1162 GYNOECY
      f gynoécie *f*
      e ginoecia *f*
      i ginoicismo *m*;
         ginoecia *f*
      d Gynözie *f*;
         Gynoezie *f*

1163 GYNOFACTOR
      f gynofacteur *m*
      e ginofactor *m*
      i fattore *m* genico
      d Gynofaktor *m* (#)

1164 GYNOGENESIS
      f gynogénèse *f*
      e ginogénesis *f*

i ginogenesi *f*
d Gynogenese *f*

1165 GYNOGENETIC
f gynogénétique
e ginogenético
i ginogenetico
d gynogenetisch

1166 GYNOGENIC
f gynogénique
e ginogénico
i ginogenico
d gynogen

1167 GYNOGENY
f gynogénie *f*
e ginogenia *f*
i ginogenia *f*
d Gynogenie *f* (#)

1168 GYNOMONOECIOUS
f gynomonoïque
e ginomonoico
i ginomonoico
d gynomonözisch;
   gynomonoezisch

1169 GYNOMONOECY
f gynomonoécie *f*

e ginomonoecia *f*
i ginomonoicismo *m*;
   ginomonoecia *f*
d Gynomonözie *f*;
   Gynomonoezie *f*

1170 GYNOSPERM
f gynosperme *m*
e ginospermia *f*
i ginosperma *f*
d - -

1171 GYNOSPORE
f gynospore *m*
e ginoespora *f*
i ginospora *f*
d Gynospore *f*

1172 GYNOSPOROGENESIS
f gynosporogénèse *f*
e ginoesporagénesis *f*
i ginosporagenesi *f*
d Gynosporogenese *f*

1173 GYRE
f rotation *f*
e rotación
i rotazione *f*
d Chromosomenwindung *f*

H

1174   HALF-BLOODED
 f demi-sang *m*
 e media-sangre *f*
 i mezzo sangue *m*
 d Halbblut-
  (in Zusammensetzungen)

1175   HALF-BREED
 f métis *m;*
  métisse *f*
 e mestizo *m;*
  mestiza *f*
 i meticcio *m;*
  meticcia *f*
 d Mischling *m*

1176   HALF-CHIASMA
 f demi-chiasma *m*
 e semiquiasma *m*
 i mezzo chiasma *m*
 d Halbchiasma *n*

1177   HALF CHROMATID BREAK
 f rupture *f* des demi-chro-
  matides
 e ruptura *f* medio-cromatidio
 i rottura *f* dei mezzi
  cromatidi
 d Halbchromatidenbruch *m*

1178   HALF CHROMATID FRAG-
  MENTATION
 f fragmentation *f* des demi-
  chromatides
 e fragmentación *f* medio-
  cromatidio
 i frammentazione *f* dei
  mezzi cromatidi
 d Halbchromatidenfragmentation *f*

1179   HALF DISJUNCTION
 f semi-disjonction *f*
 e semidisyunción *f*
 i disgiunzione *f* equazionale;

 i semi-disgiunzione *f*
 d Halbdisjunktion *f*

1180   HALF-MUTANT
 f demi-mutant *m*
 e semi-mutante *m*
 i mezzo-mutante *m;*
  semi-mutante *m*
 d Halbmutante *f*

1181   HALF-RACE
 f demi-race *f*
 e media raza *f*
 i mezza razza *f*
 d Halbrasse *f*

1182   HALF-SIB: HALF-SISTER,
  HALF-BROTHER
 f demi-soeur *f;*
  demi-frère *m*
 e hermanastra *f;*
  hermanastro *m*
 i sorellastra *f;*
  fratellastro *m*
 d Halbgeschwister: Halb-
  schwester *f;*
  Halbbruder *m*

1183   HALF-SIB FAMILY;
  HALF-SISTER FAMILY
 f famille *f* de demi-frères et
  demi-soeurs
 e familia *f* de hermanastros
  y hermanastras
 i famiglia *f* di fratellastri e
  sorellastre
 d Halbgeschwisterfamilie *f*

1184   HALF-SIDER
 f - -
 e semilateral *m*
 i - -
 d Halbseitenzwitter *m*

1185   HALF STARVATION
 f demi-carence *f*

   e   semi-inanición *f*
   i   semi-inanizione *f*
   d   Half-starvation *f*

186   HAPLOBIONT
   f   haplobionte *m*
   e   haplobionte *m*
   i   aplobionte *m*
   d   Haplobiont *m*

187   HAPLOBIONTIC
   f   haplobiontique
   e   haplobióntico
   i   aplobiontico
   d   haplobiontisch

188   HAPLOCARYOTYPE
   f   haplocaryotype *m*
   e   haplocariótipo *m*
   i   aplocariotipo *m*
   d   Haplokaryotyp *m*

189   HAPLOCHLAMYDEUS
      CHIMAERA
   f   chimère *f* haplochlamy-
      dée
   e   quimera *f*
      haploclamídea
   i   chimera *f* aploclamidea
   d   haplochlamyde (oder
      haplochlamydeische)
      Chimäre *f*

190   HAPLODIPLOID (adj.)
      HAPLODIPLOID
   f   haplodiploïde
      haplodiploïde *m*
   e   haplodiploide
      haplodiploide *m*
   i   aplodiploide
      aplodiploide *m*
   d   haplodiploid
      Haplodiploide *f*

191   HAPLODIPLOIDY
   f   haplodiploïdie *f*
   e   haplodiploidía *f*
   i   aplodiploidismo *m*
   d   Haplodiploidie *f*

192   HAPLODIPLONT

   f   haplodiplonte *m*
   e   haplodiplonte *m*
   i   aplodiplonte *m*
   d   Haplodiplont *m*

1193   HAPLOGENOTYPIC
   f   haplogénotypique
   e   haplogenotípico
   i   aplogenotipico
   d   haplogenotypisch

1194   HAPLOHETEROECY
   f   haplohétéroécie *f*
   e   haploheteroecia *f*
   i   aploeteroicismo *m*
   d   Haploheterözie *f*

1195   HAPLOID (adj.)
      HAPLOID
   f   haploïde
      haploïde *m*
   e   haploide
      haploide *m*
   i   aploide
      aploide *m*
   d   haploid
      Haploide *f*

1196   HAPLOID-INSUFFICIENT;
      HAPLO-INSUFFICIENT
   f   haploïde semi-dominant
   e   haploide-insuficiente
   i   aplo-insufficiente
   d   haplo-insuffizient

1197   HAPLOID-SUFFICIENT;
      HAPLO-SUFFICIENT
   f   haploïde dominant
   e   haploide-suficiente
   i   aplo-sufficiente
   d   haplo-suffizient

1198   HAPLOIDY
   f   haploïdie *f*
   e   haploidía *f*
   i   aploidismo *m*
   d   Haploidie *f*

1199   HAPLOME
   f   haplome *m*
   e   haploma *m*

i   aploma *m*
d   Haplom *n*

1201   HAPLOMITOSIS
f   haplomitose *f*
e   haplomitosis *f*
i   aplomitosi *f*
d   Haplomitose *f*

1202   HAPLOMIXIS
f   haplomixie *f*
e   haplomixis *f*
i   aplomissi *f*
d   Haplomixis *f*

1203   HAPLONT
f   haplonte *m*
e   haplonte *m*
i   aplonte *m*
d   Haplont *m*

1204   HAPLONTIC
f   haplontique
e   haplóntico
i   aplontico
d   haplontisch

1205   HAPLOPHASE
f   haplophase *f*
e   haplofase *f*
i   aplofase *f*
d   Haplophase *f*

1206   HAPLOPOLYPLOID (adj.)
        HAPLOPOLYPLOID
f   haplopolyploïde
    haplopolyploïde *m*
e   haplopoliploide
    haplopoliploide *m*
i   aplopoliploide
    aplopoliploide *m*
d   haplopolyploid
    Haplopolyploide *f*

1207   HAPLOPOLYPLOIDY
f   haplopolyploïdie *f*
e   haplopoliploidía *f*
i   aplopoliploidismo *m*
d   Haplopolyploidie *f*

1208   HAPLOSIS
f   haplose *f*
e   haplosis *f*
i   aplosi *f*
d   Haplosis *f*

1209   HAPLOSOMY
f   haplosomie *f*
e   haplosomia *f*
i   aplosomia *f*
d   Haplosomie *f*

1210   HAPLOZYGOUS
f   haplozygote
e   haplocigótico
i   aplozigotico
d   haplozygotisch (#)

1211   HEMIALLOPLOID (adj.)
        HEMIALLOPLOID
f   hémialloploïde
    hémialloploïde *m*
e   hemialoploide
    hemialoploide *m*
i   emialloploide
    emialloploide *m*
d   hemialloploid
    Hemialloploide *f*

1212   HEMIALLOPLOIDY
f   hémialloploïdie *f*
e   hemialoploidía *f*
i   emialloploidismo *m*
d   Hemialloploidie *f*

1213   HEMIAUTOPLOID (adj.)
        HEMIAUTOPLOID
f   hémiautoploïde
    hémiautoploïde *m*
e   hemiautoploide
    hemiautoploide *m*
i   emiautoploide
    emiautoploide *m*
d   hemiautoploid
    Hemiautoploide *f*

1214   HEMIAUTOPLOIDY
f   hémiautoploïdie *f*

　　　e　hemiautoploidîa *f*
　　　i　emiautoploidismo *m*
　　　d　Hemiautoploidie *f*

1215　HEMICHROMATIDIC
　　　f　hémichromatidique
　　　e　hemicromatídico
　　　i　emicromatidico
　　　d　halbchromatidisch;
　　　　　Halbchromatiden- (in
　　　　　Zusammensetzungen)

1216　HEMIDIERESIS
　　　f　hémidiérèse *f*
　　　e　hemidiéresis *f*
　　　i　emidieresi *f*
　　　d　Hemidiärese *f*

1217　HEMIHAPLOID (adj.)
　　　HEMIHAPLOID
　　　f　hémihaploïde
　　　　　hémihaploïde *m*
　　　e　hemihaploide
　　　　　hemihaploide *m*
　　　i　emiaploide
　　　　　emiaploide *m*
　　　d　hemihaploid
　　　　　Hemihaploide *f*

1218　HEMIHAPLOIDY
　　　f　hémihaploïde *f*
　　　e　hemihaploidîa *f*
　　　i　emiaploidismo *m*
　　　d　Hemihaploidie *f*

1219　HEMIHOLODIPLOID (adj.)
　　　HEMIHOLODIPLOID
　　　f　hémiholodiploïde
　　　　　hémiholodiploïde *m*
　　　e　hemiholodiploide
　　　　　hemiholodiploide *m*
　　　i　emiolodiploide
　　　　　emiolodiploide *m*
　　　d　hemiholodiploid
　　　　　Hemiholodiploide *f*

1220　HEMIHOLODIPLOIDY
　　　f　hémiholodiploïdie *f*
　　　e　hemiholodiploidîa *f*
　　　i　emiolodiploidismo *m*
　　　d　Hemiholodiploidie *f*

1221　HEMIKARYON

　　　f　hémicarion *m*
　　　e　hemicario *m*
　　　i　emicarion *m*
　　　d　Hemikaryon *n*

1222　HEMIKINESIS
　　　f　hémicinèse *f*
　　　e　hemicinesis *f*
　　　i　emicinesi *f*
　　　d　Hemikinese *f*

1223　HEMIORTHOKINESIS
　　　f　hémiorthocinèse *f*
　　　e　hemiortocinesis *f*
　　　i　emiortocinesi *f*
　　　d　Hemiorthokinese *f*

1224　HEMIXIS
　　　f　hémixie *f*
　　　e　hemixis *f*
　　　i　emissi *f*
　　　d　Hemixis *f*

1225　HEMIZYGOUS
　　　f　hémizygotique
　　　e　hemicigótico
　　　i　emizigotico
　　　d　hemizygot

1226　HEPTAPLOID (adj.)
　　　HEPTAPLOID
　　　f　heptaploïde
　　　　　heptaploïde *m*
　　　e　heptaploide
　　　　　heptaploide *m*
　　　i　eptaploide
　　　　　eptaploide *m*
　　　d　heptaploid
　　　　　Heptaploide *f*

1227　HEPTAPLOIDY
　　　f　heptaploïdie *f*
　　　e　heptaploidîa *f*
　　　i　eptaploidismo *m*
　　　d　Heptaploidie *f*

1228　HERCOGAMY;
　　　HERKOGAMY
　　　f　hercogamie *f*
　　　e　hercogamia *f*
　　　i　ercogamia *f*
　　　d　Herkogamie *f*

1229   HEREDITABILITY;
        HERITABILITY
        f   héritabilité *f*
        e   heredabilidad *f*
        i   ereditarietà *f;*
            ereditabilità
        d   Erblichkeit *f;*
            Vererbbarkeit *f*

1230   HEREDITARY
        f   héréditaire
        e   hereditario
        i   ereditario
        d   erblich

1231   HEREDITY;
        INHERITANCE
        f   hérédité *f*
        e   herencia *f*
        i   eredità *f*
        d   Vererbung *f*

        HEREDITY, RESIDUAL
        see 2461

1232   HERITABILITY;
        HEREDITABILITY
        f   héritabilité *f*
        e   heredabilidad *f*
        i   ereditarietà *f;*
            ereditabilità *f*
        d   Erblichkeit *f;*
            Vererbbarkeit *f*

1233   HERITAGE
        f   héritage *m*
        e   herencia *f*
        i   eredità *f*
        d   Erbe *n*

1234   HERKOGAMY;
        HERCOGAMY
        f   hercogamie *f*
        e   hercogamia *f*
        i   ercogamia *f*
        d   Herkogamie *f*

1235   HERMAPHRODITE (adj.)
        HERMAPHRODITE
        f   hermaphrodite
            hermaphrodite *m*

        e   hermafrodito
            hermafrodito
        i   ermafrodito
            ermafrodito *m*
        d   hermaphrodit;
            zwittrig
            Hermaphrodit *m*

1236   HERMAPHRODITISM
        f   hermaphrodisme *m*
        e   hermafroditismo *m*
        i   ermafroditismo *m*
        d   Hermaphroditismus *m*

1237   HETERAUXESIS
        f   hétérauxèse *f*
        e   heterauxesis *f*
        i   eterauxesi *f*
        d   Heterauxese *f* (Θ)

1238   HETEROALLELE
        f   hétéroallèle *m*
        e   heteroalelo *m*
        i   eteroallelo *m*
        d   Heteroallel *n*

1239   HETEROALLELIC
        f   hétéroallélique
        e   heteroalélico
        i   eteroallelico
        d   heteroallel

1240   HETEROBRACHIAL
        f   hétérobrachial
        e   heterobraquial
        i   eterobrachiale
        d   heterobrachial

1241   HETEROCARYON;
        HETEROKARYON
        f   hétérocaryon *m*
        e   heterocarionte *m*
        i   eterocarion *m*
        d   Heterokaryon *n*

1242   HETEROCARYOSIS;
        HETEROKARYOSIS
        f   hétérocaryose *f*
        e   heterocariosis *f*
        i   eterocariosi *f*
        d   Heterokaryosis *f*

1243 HETEROCARYOTIC;
      HETEROKARYOTIC
      f hétérocaryotique
      e heterocarióticο
      i eterocariotico
      d heterokaryotisch

1244 HETEROCENTRIC
      f hétérocentrique
      e heterocéntrico
      i eterocentrico
      d heterozentrisch

1245 HETEROCENTRICITY
      f hétérocentralité *f*
      e heterocentricidad *f*
      i eterocentricità *f*
      d Heterozentrizität *f*

1246 HETEROCHONDRIC
      f hétérochondrique
      e heterocóndrico
      i eterocondrico
      d heterochondrisch (≠)

1247 HETEROCHROMATIC
      f hétérochromatique
      e heterocromático
      i eterocromatico
      d heterochromatisch

1248 HETEROCHROMATIN
      f hétérochromatine *f*
      e heterocromatina *f*
      i eterocromatina *f*
      d Heterochromatin *n*

1249 HETEROCHROMATINOSOME
      f hétérochromatinosome *m*
      e heterocromatinosoma *m*
      i eterocromatinosoma *m*
      d Heterochromatinosom *n*

1250 HETEROCHROMATISM
      f hétérochromatisme *m*
      e heterocromatismo *m*
      i eterocromatismo *m*
      d Heterochromatie *f*

1251 HETEROCHROMATIZATION
      f hétérochromatinisation *f*

e heterocromatinización *f*
i eterocromatinisazione *f*
d Heterochromatisierung *f*

1252 HETEROCHROMOMERE
      f hétérochromomère *m*
      e heterocromómero *m*
      i eterocromomero *m*
      d Heterochromomer *n*

1253 HETEROCHROMOSOME
      f hétérochromosome *m*
      e heterocromosoma *m*
      i eterocromosoma *m*
      d Heterochromosom *n*

1254 HETERODICHOGAMOUS
      f hétérodichogame
      e heterodicógamo
      i eterodicogamo
      d heterodichogam

1255 HETERODICHOGAMY
      f hétérodichogamie *f*
      e heterodicogamia *f*
      i eterodicogamia *f*
      d Heterodichogamie *f*

1256 HETEROECIOUS
      f hétéroécique
      e heterecio;
         heteroecio
      i eteroico
      d heteroecisch;
         heteroezisch;
         heterözisch

1257 HETEROECY
      f hétéroécie *f*
      e heterecia *f;*
         heteroecia *f*
      i eteroicismo *m*
      d Heteroecie *f;*
         Heteroezie *f;*
         Heterözie *f*

1258 HETEROFERTILIZATION
      f hétérofertilisation *f;*
         hétérofécondation *f*
      e heterofecundación *f*
      i eterofecondazione *f*
      d Heterofertilisation *f*

1259 HETEROGAMEON
  f hétérogaméon *m*
  e heterogameón *m*
  i eterogameon *m*
  d Heterogameon *n*

1260 HETEROGAMETE
  f hétérogamète *m*
  e heterogameto *m*
  i eterogamete *m*
  d Heterogamet *m*

1261 HETEROGAMETIC
  f hétérogamétique
  e heterogamético
  i eterogametico
  d heterogametisch

1262 HETEROGAMOUS
  f hétérogame
  e heterogamo
  i eterogamico
  d heterogam

1263 HETEROGAMY
  f hétérogamie *f*
  e heterogamia *f*
  i eterogamia *f*
  d Heterogamie *f*

1264 HETEROGENESIS
  f hétérogénèse *f*
  e heterogénesis *f*
  i eterogenesi *f*
  d Heterogenese *f* (Θ);
    Generationswechsel *m*

1265 HETEROGENETIC
  f hétérogénétique
  e heterogenético
  i eterogenetico
  d heterogenetisch

1266 HETEROGENIC
  f hétérogénique
  e heterogénico
  i eterogenico
  d heterogenisch

1267 HETEROGENOMATIC
  f hétérogénomatique
  e heterogenomático

  i eterogenomatico
  d heterogenomatisch

1268 HETEROGENOTE
  f hétérogénote *m*
  e heterogenote *m*
  i eterogenote *m*
  d Heterogenote *f*

1269 HETEROGENOTIC
  f hétérogénotique
  e heterogenótico
  i eterogenotico
  d heterogenotisch

1270 HETEROGENY
  f hétérogénie *f*
  e heterogenia *f*
  i eterogenia *f*
  d Heterogenie *f*

1271 HETEROGONIC
  f hétérogonique
  e heterógono
  i eterogonico
  d heterogonisch

1272 HETEROGONY
  f hétérogonie *f*
  e heterogonía *f*
  i eterogonia *f*
  d Heterogonie *f*

1273 HETEROGYNIC;
  HETEROGYNOUS
  f hétérogyne
  e heterogino
  i eterogino
  d heterogyn

1274 HETEROGYNISM;
  HETEROGYNY
  f hétérogynisme *m*
  e heteroginismo *m*;
    heteroginia *f*
  i eteroginismo *m*;
    eteroginia *f*
  d Heterogynie *f*

1275 HETEROGYNOUS;
  HETEROGYNIC

   f hétérogyne
   e heterogino
   i eterogino
   d heterogyn

276 HETEROHAPLOID (adj.)
     HETEROHAPLOID
   f hétérohaploïde
      hétérohaploïde *m*
   e heterohaploide
      heterohaploide *m*
   i eteroaploide
      eteroaploide *m*
   d heterohaploid
      Heterohaploide *f*

277 HETEROHAPLOIDY
   f hétérohaploïdie *f*
   e heterohaploidïa *f*
   i eteroaploidismo *m*
   d Heterohaploidie *f*

278 HETEROKARYON;
     HETEROCARYON
   f hétérocaryon *m*
   e heterocarionte *m*
   i eterocarion *m*
   d Heterokaryon *n*

279 HETEROKARYOSIS;
     HETEROCARYOSIS
   f hétérocaryose *f*
   e heterocariosis *f*
   i eterocariosi *f*
   d Heterokaryosis *f*

280 HETEROKARYOTIC;
     HETEROCARYOTIC
   f hétérocaryotique
   e heterocariótico
   i eterocariotico
   d heterokaryotisch

281 HETEROKINESIS
   f hétérocinèse *f;*
      hétérokinèse *f*
   e heterocinesis *f*
   i eterocinesi *f*
   d Heterokinesis *f;*
      Heterokinese *f*

1282 HETEROLECITHAL
   f hétérolécithe
   e heterolecito
   i eterolecito
   d heterolezithal

1283 HETEROLOGOUS
   f hétérologue
   e heterólogo
   i eterologo
   d heterolog

1284 HETEROLOGY
   f hétérologie *f*
   e heterología *f*
   i eterologia *f*
   d Heterologie *f*

1285 HETEROMIXIS
   f hétéromixie *f*
   e heteromixis *f*
   i eteromissi *f*
   d Heteromixis *f*

1286 HETEROMORPHIC
   f hétéromorphe
   e heteromórfico
   i eteromorfo
   d heteromorph

1287 HETEROMORPHISM
   f hétéromorphisme *m;*
      hétéromorphie *f*
   e heteromorfia *f;*
      heteromorfismo *m*
   i eteromorfismo *m*
   d Heteromorphie *f*

1288 HETEROMORPHOSIS
   f hétéromorphose *f*
   e heteromorfosis
   i eteromorfosi *f*
   d Heteromorphosis *f;*
      Heteromorphose *f*

1289 HETEROMORPHOUS
   f hétéromorphe
   e heteromorfo
   i eteromorfo
   d heteromorph

1290   HETEROPHENOGAMY
        f hétérophénogamie *f*
        e heterofenogamia *f*
        i eterofenogamia *f*
        d Heterophänogamie *f*

1291   HETEROPHYTIC
        f hétérophytique
        e heterofítico
        i eterofitico
        d heterophytisch

1292   HETEROPLASMONY
        f hétéroplasmonie *f*
        e heteroplasmonia *f*
        i eteroplasmonia *f*
        d Heteroplasmonie *f*

1293   HETEROPLOID (adj.)
        HETEROPLOID
        f hétéroploïde
           hétéroploïde *m*
        e heteroploide
           heteroploide *m*
        i eteroploide
           eteroploide *m*
        d heteroploid
           Heteroploide *f*

1294   HETEROPLOIDY
        f hétéroploïdie *f*
        e heteroploidïa *f*
        i eteroploidismo *m*
        d Heteroploidie *f*

1295   HETEROPOLAR
        f hétéro-polaire (☉)
        e heteropolar
        i eteropolare
        d heteropolar

1296   HETEROPYCNOSIS;
        HETEROPYKNOSIS
        f hétéropycnose *f*
        e heteropicnosis *f*
        i eteropicnosi *f*
        d Heteropyknosis *f*;
           Heteropyknose *f*

1297   HETEROPYCNOTIC;
        HETEROPYKNOTIC

        f hétéropycnotique
        e heteropicnótico
        i eteropicnotico
        d heteropyknotisch

1298   HETEROPYKNOSIS;
        HETEROPYCNOSIS
        f hétéropycnose *f*
        e heteropicnosis *f*
        i eteropicnosi *f*
        d Heteropyknosis *f*;
           Heteropyknose *f*

1299   HETEROPYKNOTIC;
        HETEROPYCNOTIC
        f hétéropycnotique
        e heteropicnótico
        i eteropicnotico
        d heteropyknotisch

1300   HETEROSIS;
        HYBRID VIGOUR
        f hétérosis *f*;
           vigueur hybride *f*
        e heterosis *f*;
           vigor híbrido *m*
        i eterosi
        d Heterosis *f*

        HETEROSIS, SINGLE LOCUS
        see 2603

1301   HETEROSOMAL
        f hétérosomal
        e heterosomal
        i eterosomo
        d heterosomal

1302   HETEROSTASIS
        f hétérostase *f*
        e heterostasis *f*
        i eterostasi *f*
        d Heterostasis *f*

1303   HETEROSTYLIC
        f hétérostylé
        e heterostílico
        i eterostilico
        d heterostyl;
           verschiedengriff(e)lig

1304 HETEROSTYLY
  f hétérostylie *f*
  e heterostilia *f*
  i eterostilia *f*
  d Heterostylie *f*

1305 HETEROSYNAPSIS
  f hétérosynapse *f*
  e heterosinapsis *f*
  i eterosinapsi *f*
  d Heterosynapsis *f*

1306 HETEROSYNDESIS
  f hétérosyndèse *f*
  e heterosíndesis *f*
  i eterosindesi *f*
  d Heterosyndese *f*

1307 HETEROTETRAPLOID (adj.)
  HETEROTETRAPLOID
  f hétérotétraploïde
    hétérotétraploïde *m*
  e heterotetraploide
    heterotetraploide *m*
  i eterotetraploide
    eterotetraploide *m*
  d heterotetraploid
    Heterotetraploide *f*

1308 HETEROTETRAPLOIDY
  f hétérotétraploïdie *f*
  e heterotetraploidía *f*
  i eterotetraploidismo *m*
  d Heterotetraploidie *f*

1309 HETEROTHALLIC
  f hétérothallique
  e heterotálico
  i eterotallico
  d heterothallisch

  HETEROTHALLIC, PARTIAL
  see 2082

1310 HETEROTHALLISM;
  HETEROTHALLY
  f hétérothallisme *m*
  e heterotalismo *m*
  i eterotallismo *m*
  d Heterothallie *f*

1311 HETEROTHALLY;
  HETEROTHALLISM
  f hétérothallisme *m*
  e heterotalismo *m*
  i eterotallismo *m*
  d Heterothallie *f*

1312 HETEROTIC
  f hétérotique
  e heterótico
  i eterotico
  d Heterosis- (in Zusammen-
    setzungen)

1313 HETEROTRANSFORMATION
  f hétérotransformation *f*
  e heterotransformación *f*
  i eterotrasformazione *f*
  d Heterotransformation *f*

1314 HETEROTROPIC CHROMOSOME
  f chromosome *m* hétérotrope
  e cromosoma *m* heterotrópico
  i cromosoma *m* eterotropico
  d heterotropisches Chromo-
    som *n*

1315 HETEROTYPE DIVISION;
  HETEROTYPIC DIVISION
  f division *f* hétérotypique
  e división *f* heterotípica
  i divisione *f* eterotipica
  d heterotypische Teilung *f*

1316 HETEROTYPIC DIVISION;
  HETEROTYPE DIVISION
  f division *f* hétérotypique
  e división *f* heterotípica
  i divisione *f* eterotipica
  d heterotypische Teilung *f*

1317 HETEROZYGOSIS;
  HETEROZYGOSITY
  f hétérozygotie *f*
  e heterocigosis *f*
  i eterozigosi *f*
  d Heterozygotie *f*

  HETEROZYGOSIS, ENFORCED
  see 928

1318   HETEROZYGOTE
        f   hétérozygote *m*
        e   heterocigoto *m*
        i   eterozigote *m*
        d   Heterozygote *f*

        HETEROZYGOTE, RESIDUAL
        See 2462

        HETEROZYGOTE, STRUC-
        TURAL
        see 2684

1319   HETEROZYGOUS
        f   hétérozygote
        e   heterocigótico
        i   eterozigotico
        d   heterozygot

1320   HEXAD
        f   hexade *f*
        e   hexada *f*
        i   esade *f*
        d   Hexade *f*

1321   HEXAPLOID (adj.)
        HEXAPLOID
        f   hexaploïde
            hexaploïde *m*
        e   hexaploide
            hexaploide *m*
        i   esaploide
            esaploide *m*
        d   hexaploid
            Hexaploide *f*

1322   HEXAPLOIDY
        f   hexaploïdie *f*
        e   hexaploidía *f*
        i   esaploidismo *m*
        d   Hexaploidie *f*

1323   HEXASOMIC
        f   hexasomique
        e   hexasómico
        i   esasomico
        d   hexasom

1324   H F C MIXTURE
        f   mélange *m* HFC;
            culture *f* HFC

        e   mezcla *f* CAF
        i   miscuglio *m* HFC
        d   - -

1325   HIGH FREQUENCY COLI-
        CINOGENY TRANSFER
        (HFC)
        f   transfert *m* colicinogénique
            à haute fréquence (H.F.C.)
        e   transferencia *f* colicino-
            génica de alta frecuencia
        i   trasferimento *m* colicinó-
            genico ad alta frequenza
        d   Hochfrequenz-
            Colicinogen-transfer *m*

1326   HISTOGENESE;
        HISTOGENY
        f   histogénèse *f*
        e   histogénesis *f*
        i   istogenesi *f*
        d   Histogenese *f*

1327   HISTOGENY;
        HISTOGENESE
        f   histogénèse *f*
        e   histogénesis *f*
        i   istogenesi *f*
        d   Histogenese *f*

1328   HOLANDRIC
        f   holandrique
        e   holándrico
        i   olandrico
        d   holandrisch

1329   HOLLOW SPINDLE
        f   fuseau creux *m*
        e   huso concavo *m*
        i   fuso cavo *m* (Ⓞ)
        d   Hohlspindel *f*

1330   HOLOBLASTIC CLEAVAGE
        f   clivage *m* holoblastique;
            scission *f* holoblastique
        e   segmentación *f* holoblástic
        i   segmentazione *f* oloblastic
        d   holoblastische Furchung *f*

1331   HOLOGAMOUS
        f   hologame

e hológamo;
   hologámico
i ologamico
d hologam

1332 HOLOGAMY
f hologamie *f*
e hologamia *f*
i ologamia *f*
d Hologamie *f*

1333 HOLOGYNIC
f hologynique
e hologínico
i ologinico
d hologyn

1334 HOLOKINETIC
f holocinétique
e holocinético
i olocinetico
d holokinetisch

1335 HOLOMORPHOSIS
f holomorphose *f*
e holomorfosis *f*
i olomorfosi *f*
d Holomorphosis *f* (#);
   Holomorphose *f* (#)

1336 HOLOTOPY
f holotopisme *m*
e holotopismo *m*
i olotopismo *m*
d Holotopie *f* (#)

1337 HOLOTYPE
f holotype *m*
e holótipo *m*
i olotipo *m*
d Holotyp *m* (#)

1338 HOMEOKINESIS;
   HOMOEOKINESIS
f homéocinèse *f*
e homeocinesis *f*
i omeocinesi *f*
d Homöokinesis *f*;
   Homöokinese *f*

1339 HOMEOLOGOUS (adj.);
   HOMOEOLOGOUS (adj.)

HOMEOLOGUE;
HOMOEOLOGUE
f homéologue
   homéologue *f*
e homeólogo
   homeólogo *m*
i omeologo
   omeologo *m*
d homöolog;
   homoiolog
   Homoiologon *n*

1340 HOMEOMORPH
f homéomorphe
e homeomorfo
i omeomorfo
d homoiomorph

1341 HOMEOSIS;
   HOMOEOSIS;
   HOMOOSIS
f homéose *f*
e homeosis *f*
i omeosi *f*
d Homöosis *f*

1342 HOMEOSTASIS;
   HOMOEOSTASIS
f homéostasie *f*
e homeostasis *f*
i omeostasi *f*
d Homöostasis *f*

1343 HOMEOSTAT;
   HOMOEOSTAT;
   HOMEOSTATE;
   HOMOEOSTATE
f homéostat *m*
e homeostato *m*
i omeostato *m*
d Homöostat *m*

1344 HOMEOSYNAPSIS;
   HOMOSYNAPSIS
f homosynapse *f*
e homeosinapsis *f*
i omosinapsi *f*
d Homöosynapsis *f*

1345 HOMEOTIC
f homéotique

e homeótico
i omeotico
d homöotisch

1346 HOMEOTYPE
f homéotype *m*
e homeótipo *m*
i omeotipo *m*
d Homöotyp *m*

1347 HOMEOTYPIC
f homéotypique
e homeotípico
i omeotipico
d homöötypisch

1348 HOMEOTYPIC DIVISION;
EQUATION DIVISION
f division *f* homéotypique;
division *f* équationelle
e división *f* homeotípica;
división *f* ecuacional
i divisione *f* omeotipica;
divisione *f* equazionale
d homoiotypische Teilung *f*;
homöotypische Teilung *f*;
Äquationsteilung *f*

1349 HOMOALLELE
f homoallèle *m*
e homoalelo *m*
i omoallelo *m*
d Homoallel *n*

1350 HOMOALLELIC
f homoallélique
e homoalélico
i omoallelico
d homoallel

1351 HOMOBRACHIAL
f homobrachial
e homobraquial
i omobrachiale
d homobrachial

1352 HOMOCARYON;
HOMOKARYON
f homocaryon *m*
e homocarionte *m*
i omocarion *m*
d Homokaryon *n*

1353 HOMOCARYOSIS;
HOMOKARYOSIS
f homocaryose *f*
e homocariosis *f*
i omocariosi *f*
d Homokaryosis *f*

1354 HOMOCENTRIC
f homocentrique
e homocéntrico
i omocentrico
d homozentrisch

1355 HOMOCENTRICITY
f homocentralité *f*
e homocentricidad *f*
i omocentricità *f*
d Homozentrizität *f*

1356 HOMOCHONDRIC
f homochondrique
e homocóndrico
i omocondrico
d homochondrisch

1357 HOMOCHRONOUS
f homochrone
e homócrono
i omocrono
d homochron

1358 HOMODYNAMIC
f homodyname
e homodinámico
i omodinamico
d homodynam

1359 HOMOEOKINESIS
HOMEOKINESIS
f homéocinèse *f*
e homeocinesis *f*
i omeocinesi *f*
d Homöokinesis *f*;
Homöokinese *f*

HOMOCOLOGY
see 1379

1360 HOMOEOLOGOUS (adj.);
HOMEOLOGOUS (adj.)
HOMEOLOGUE;
HOMOELOGUE

f homéologue
   homéologue *f*
e homeólogo
   homeólogo *m*
i omeologo
   omeologo *m*
d homöolog;
   homoiolog
   Homoiologon *n*

1361 HOMOEOSIS;
      HOMEOSIS;
      HOMOOSIS
f homéose *f*
e homeosis *f*
i omeosi *f*
d Homöosis *f*

1362 HOMOEOSTASIS;
      HOMEOSTASIS
f homéostasie *f*
e homeostasis *f*
i omeostasi *f*
d Homöostasis *f*

1363 HOMOEOSTAT;
      HOMEOSTAT;
      HOMOEOSTATE;
      HOMEOSTATE
f homéostat *m*
e homeostato *m*
i omeostato *m*
d Homöostat *m*

1364 HOMOGAMETE
f homogamète *m*
e homogámeto *m*
i omogameto *m*
d Homogamet *m*

1365 HOMOGAMETIC
f homogamétique
e homogamético
i omogametico
d homogametisch

366 HOMOGAMOUS
f homogame
e homógamo
i omogamo
d homogam

1367 HOMOGAMY
f homogamie *f*
e homogamia *f*
i omogamia *f*
d Homogamie *f*

1368 HOMOGENETIC
f homogénétique
e homogenético
i omogenetico
d homogenetisch

1369 HOMOGENIC
f homogénique
e homogénico
i omogenico
d homogenisch (*#*)

1370 HOMOGENOTE
f homogénote *m*
e homogenote *m*
i omogenote *m*
d Homogenote *f*

1371 HOMOGENOTIC
f homogénotique
e homogenótico
i omogenotico
d homogenotisch

1372 HOMOGENY
f homogénie *f*
e homogenía *f*
i omogenia *f*
d Homogenie *f*

1373 HOMO-HETEROMIXIS
f homohétéromixie *f*
e homoheteromixis *f*
i omo-eteromissi *f*
d Homoheteromixis *f*

1374 HOMOKARYON;
      HOMOCARYON
f homocaryon *m*
e homocarionte *m*
i omocarion *m*
d Homokaryon *n*

1375 HOMOKARYOSIS;
      HOMOCARYOSIS

f homocaryose *f*
e homocaryosis *f*
i omocariosi *f*
d Homokaryosis *f*

1376 HOMOLECITHAL
f homolécithe
e homolecito
i omolecitico
d homolezithal

1377 HOMOLOGOUS
f homologue
e homólogo
i omologo
d homolog

1378 HOMOLOGUE
f homologue *f*
e homólogo *m*
i omologo *m*
d Homologe *m,f,n*

1379 HOMOLOGY;
f homologie *f*
e homología *f*
i omologia *f*
d Homologie *f*

HOMOLOGY, RESIDUAL
see 2463

1380 HOMOLYSOGENIC
f homolysogénique
e homolysogénico
i omolisogenico
d homolysogen

1381 HOMOMERAL;
HOMOMEROUS;
HOMOMERIC
f homomérique
e homómero;
homomérico
i omomero
d homomer

1382 HOMOMERY
f homomérie *f*
e homomeria *f*
i omomeria *f*

d Homomerie *f*

1383 HOMOMIXIS
f homomixie *f*
e homomixis *f*
i omomissi *f*
d Homomixis *f*

1384 HOMOMORPHIC
f homomorphe
e homomórfico
i omomorfo
d homomorph

1385 HOMOMORPHOUS
f homomorphe
e homomorfo
i omomorfo
d homomorph

1386 HOMOMORPHY
f homomorphie *f*
e homomorfia *f*
i omomorfia *f*;
omomorfismo *m*
d Homomorphie *f*

1387 HOMOOSIS;
HOMOEOSIS;
HOMEOSIS
f homéose *f*
e homeosis *f*
i omeosi *f*
d Homöosis *f*

1388 HOMOPHYTIC
f homophytique
e homofítico
i omofitico
d homophytisch

1389 HOMOPLASTIC
f homoplastique
e homoplástico
i omoplastico
d homoplastisch

1390 HOMOPLASY
f homoplasie *f*
e homoplasía *f*

i   omoplasia *f*
d   Homoplasie *f*

1391 HOMOPLOID (adj.)
      HOMOPLOID
      f   homoploïde
          homoploïde *m*
      e   homoploide
          homoploide *m*
      i   omoploide
          omoploide *m*
      d   homoploid
          Homoploide *f*

1392 HOMOPLOIDY
      f   homoploïdie *f*
      e   homoploidía *f*
      i   omoploidismo *m*
      d   Homoploidie *f*

1393 HOMOPOLAR
      f   homopolaire
      e   homopolar
      i   omopolare
      d   homopolar

1394 HOMOSIS
      f   homosis *f*
      e   homosis *f*
      i   omosi *f*
      d   - -

1395 HOMOSOMAL
      f   homosomal
      e   homosomal
      i   omosomale
      d   homosomal

1396 HOMOSTASIS
      f   homostase *f*
      e   homostasis *f*
      i   omostasi *f*
      d   - -

1397 HOMOSYNAPSIS;
      HOMEOSYNAPSIS
      f   homosynapse *f*
      e   homosinapsis *f*
      i   omosinapsi *f*
      d   Homosynapsis *f*

1398 HOMOSYNDESIS

f   homosyndèse *f*
e   homosíndesis *f*
i   omosindesi *f*
d   Homosyndese *f*

1399 HOMOTHALLIC
      f   homothallique
      e   homotálico
      i   omotallico
      d   homothallisch

1400 HOMOTHALLISM
      f   homothallisme *m*
      e   homotalismo *m*
      i   omotallismo *m*
      d   Homothallie *f*

1401 HOMOTRANSFORMATION
      f   homotransformation *f*
      e   homotransformación *f*
      i   omotrasformazione *f*
      d   Homotransformation *f*

1402 HOMOTYPE DIVISION;
      HOMOTYPIC DIVISION
      f   division *f* homotypique
      e   división *f* homotípica
      i   divisione *f* omotipica
      d   homotype Zellteilung *f*

1403 HOMOZYGOSIS;
      HOMOZYGOSITY
      f   homozygotie;
          homozygose *f*
      e   homocigosis *f*
      i   omozigosi *f*
      d   Homozygotie *f*

1404 HOMOZYGOTE
      f   homozygote *m*
      e   homocigoto *m*
      i   omozigote *m*
      d   Homozygote *f*

      HOMOZYGOTE, CROSSOVER
      see 646

1405 HOMOZYGOTISATION
      f   homozygotisation *f*
      e   homocigotización *f*
      i   omozigotizzazione *f*
      d   Homozygotisierung *f*

1406 HOMOZYGOUS
    f homozygote
    e homocigoto
    i omozigotico
    d homozygot

1407 HOROTELIC
    f horotélique
    e - -
    i - -
    d horotelisch

1408 HYBRID (adj.)
    HYBRID
    f hybride
       hybride *f*
    e híbrido
       híbrido *m*
    i ibrido
       ibrido *m*
    d hybrid
       Hybride *f;*
       Bastard *m*

HYBRID, DERIVATIVE
see 730

HYBRID, GENERIC
see 1086

HYBRID, NUMERICAL
see 1968

1409 HYBRID FLOCK;
    HYBRID SWARM
    f population *f* hybride
    e población *f* híbrida
    i popolazione *f* ibrida
    d Bastardpopulation *f;*
       Hybridenschwarm *m*

1410 HYBRID INCAPACITATION;
    HYBRID STERILITY
    f stérilité *f* hybride
    e incapacitación *f* híbrida
    i impedenza *f* ibridativa
    d Bastardsterilität *f*

1411 HYBRIDISATION
    f hybridation *f*
    e hibridación *f*
    i ibridazione *f*

    d Hybridisation *f;*
       Bastardierung *f;*
       Kreuzung *f*

HYBRIDISATION, CONGRUENT
see 607

HYBRIDISATION, CYCLIC
see 669

HYBRIDISATION, INCONGRUEN
see 1477

HYBRIDISATION, INTER-
SPECIFIC
see 1537

1412 HYBRIDISM;
    HYBRIDITY
    f hybridisme *m;*
       hybridité *f*
    e hibridismo *m*
    i ibridismo *m*
    d Hybridität *f;*
       Bastardnatur *f*

1413 HYBRID LETHALITY
    f létalité *f* de l'hybride
    e letalidad *f* del híbrido
    i letalità *f* dell'ibrido
    d Bastard-Letalität *f*

1414 HYBRIDOGENOUS PSEUDO-
    PARTHENOGENESIS
    f pseudo-parthénogénèse
       hybridogène *f*
    e (p)seudo-partenogénesis
       hibridógena *f*
    i pseudo-partenogenesi da
       ibridazione *f*
    d hybridogene Pseudo-
       parthenogenese *f*

HYBRID SWARM
see 1409

HYBRID STERILITY
see 1410

1415 HYBRID VIGOUR;
    HETEROSIS

f vigueur *f* hybride;
  hétérosis *f*
e vigor *m* híbrido;
  heterosis *f*
i vigore *m* degli ibridi
  eterosi *f*
d Heterosis *f*

416   HYBRID ZONE
       f zone *f* hybride
       e zona *f* híbrida
       i zona *f* ibrida
       d Hybridenzone *f*

417   HYPALLELOMORPH
       f hypallélomorphe
       e hipalelomorfo
       i ipo-allelomorfo
       d hypallelomorph

418   HYPERCHIMAERA
       f hyperchimère *f*
       e hiperquimera *f*
       i iperchimera *f*
       d Hyperchimäre *f*

419   HYPERCHROMASY
       f hyperchromasie *f*
       e hipercromasia *f*
       i ipercromasia *f*
       d Hyperchromasie *f*

420   HYPERCHROMATOSIS
       f hyperchromatose *f*
       e hipercromatosis *f*
       i ipercromatosi *f*
       d Hyperchromatose *f*

421   HYPERCYESIS
       f hypercyèse *f*
       e hiperciesis *f*
       i iperciesi *f*
       d Hypercyese *f* (#)

422   HYPERDIPLOID (adj.)
       HYPERDIPLOID
       f hyperdiploïde
         hyperdiploïde *m*
       e hiperdiploide
         hiperdiploide *m*
       i iperdiploide

   iperdiploide *m*
d hyperdiploid
  Hyperdiploide *f*

1423   HYPERDIPLOIDY
       f hyperdiploïdie *f*
       e hiperdiploidía *f*
       i iperdiploidismo *m*
       d Hyperdiploidie *f*

1424   HYPERHAPLOID (adj.)
       HYPERHAPLOID
       f hyperhaploïde
         hyperhaploïde *m*
       e hiperhaploide
         hiperhaploide *m*
       i iperaploide
         iperaploide *m*
       d hyperhaploid
         Hyperhaploide *f*

1425   HYPERHAPLOIDY
       f hyperhaploïdie *f*
       e hiperhaploidía *f*
       i iperaploidismo *m*
       d Hyperhaploidie *f*

1426   HYPERHETEROBRACHIAL
       f hyperhétérobrachial
       e hipereterobraquial
       i ipereterobrachiale
       d hyperheterobrachial

1427   HYPERMORPHIC
       f hypermorphe
       e hipermórfico
       i ipermorfo
       d hypermorph

1428   HYPERMORPHOSIS
       f hypermorphose *f*
       e hipermórfosis *f*
       i ipermorfosi *f*
       d Hypermorphosis *f*

1429   HYPERPLOID (adj.)
       HYPERPLOID
       f hyperploïde
         hyperploïde *m*
       e hiperploide

108

hiperploide *m*
i iperploide
iperploide *m*
d hyperploid
Hyperploide *f*

1430 HYPERPLOIDY
f hyperploïdie
e hiperploidía *f*
i iperploidismo *m*
d Hyperploidie *f*

1431 HYPERPOLYPLOID (adj.)
HYPERPOLYPLOID
f hyperpolyploïde
hyperpolyploïde *m*
e hiperpoliploide
hiperpoliploide *m*
i iperpoliploide
iperpoliploide *m*
d hyperpolyploid
Hyperpolyploide *f*

1432 HYPERPOLYPLOIDY
f hyperpolyploïdie *f*
e hiperpoliploidía *f*
i iperpoliploidismo *m*
d Hyperpolyploidie *f*

1433 HYPERSYNDESIS
f hypersyndèse *f*
e hipersíndesis *f*
i ipersindesi *f*
d Hypersyndese *f*

1434 HYPERTELY
f hypertélie *f*
e hipertelia *f*
i ipertelia *f*
d Hypertelie *f*

1435 HYPHAL AVERSION
f incompatibilité entre
hyphes *f*
e repulsión *f* hifal
i - -
d Hyphenaversion *f*

1436 HYPOCHROMATICITY
f hypochromaticité *f*
e hypocromaticidad *f*

i ipocromaticità *f*
d Hypochromatizität *f*

1437 HYPOGENESIS
f hypogénèse *f;*
hypogénésie *f*
e hipogénesis *f*
i ipogenesi *f*
d Hypogenese *f* (#)

1438 HYPOGENETIC
f hypogénétique
e hipogenético
i ipogenetico
d hypogenetisch (#)

1439 HYPOHAPLOID (adj.)
HYPOHAPLOID
f hypohaploïde
hypohaploïde *m*
e hipohaploide
hipohaploide *m*
i ipoaploide
ipoaploide *m*
d hypohaploid
Hypohaploide *f*

1440 HYPOHAPLOIDY
f hypohaploïdie *f*
e hipohaploidía *f*
i ipoaploidismo *m*
d Hypohaploidie *f*

1441 HYPOMORPH
f hypomorphe
e hipomorfo
i ipomorfo
d hypomorph

1442 HYPOMORPHIC
f hypomorphe
e hipomórfico
i ipomorfo
d hypomorph

1443 HYPOPLOID (adj.)
HYPOPLOID
f hypoploïde
hypoploïde *m*
e hipoploide
hipoploide *m*

i   ipoploide
    ipoploide *m*
d   hypoploid
    Hypoploide *f*

1444   HYPOPLOIDY
f   hypoploïdie *f*
e   hipoploidía *f*
i   ipoploidismo *m*
d   Hypoploidie *f*

1445   HYPOPOLYPLOID (adj.)
       HYPOPOLYPLOID
f   hypopolyploïde
    hypopolyploïde *m*
e   hipopoliploide
    hipopoliploide *m*
i   ipopoliploide
    ipopoliploide *m*
d   hypopolyploid
    Hypopolyploide *f*

1446   HYPOPOLYPLOIDY
f   hypopolyploïdie *f*
e   hipopoliploidía *f*
i   ipopoliploidismo *m*
d   Hypopolyploidie *f*

1447   HYPOSTASIS
f   hypostase *f*
e   hipostasia *f;*
    hipostasis *f*
i   ipostasi *f*
d   Hypostasis *f*

1448   HYPOSTATIC
f   hypostatique
e   hipostático
i   ipostatico
d   hypostatisch

1449   HYPOSYNDESIS
f   hyposyndèse *f*
e   hiposíndesis *f*
i   iposindesi *f*
d   Hyposyndese *f*

HYPOTHESIS, ALVEOLAR
see 118

HYPOTHESIS, ONE BAND
ONE GENE see 1974

HYPOTHESIS, ONE GENE ONE
ENZYME
see 1977

1450   HYPOTYPE
f   hypotype *m*
e   hipótipo *m*
i   ipotipo *m*
d   Hypotyp *m* (#)

1451   HYSTERESIS
f   hystérésis *f*
e   histeresis *f*
i   isteresi *f*
d   Hysteresis *f*

I

1452   ID
    f   ide *m*
    e   id *m*
    i   ide *m*
    d   Id *n*

1453   IDENTICAL
    f   identique
    e   idéntico
    i   identico
    d   identisch

1454   IDENTICAL TWINS;
     MONOZYGOTIC TWINS;
     UNIOVULAR TWINS
    f   jumeaux *mpl* uniovulaires;
       jumeaux *mpl* univitellins;
       jumeaux *mpl* vrais
    e   gemellos *mpl* idénticos
    i   gemelli *mpl* identici;
       gemelli *mpl* monozigotici;
       gemelli *mpl* uniovulari
    d   eineiige Zwillinge *mpl*

1455   IDIOCHROMATIN
    f   idiochromatine *f*
    e   idiocromatina *f*
    i   idiocromatina *f*
    d   Idiochromatin *n*

1456   IDIOCHROMIDIA
    f   idiochromidie *f*
    e   idiocromidio *m*
    i   idiocromidio *m*
    d   Idiochromidie *f*

1457   IDIOCHROMOSOME
    f   idiochromosome *m*
    e   idiocromosoma *m*
    i   idiocromosoma *m*
    d   Idiochromosom *n*

1458   IDIOGRAM
    f   idiogramme *m*
    e   idiograma *m*
    i   idiograma *m*
    d   Idiogramm *n*

1459   IDIOMUTATION
    f   idiomutation *f*
    e   idiomutación *f*
    i   idiomutazione *f*
    d   Idiomutation *f*

1460   IDIOPLASM
    f   idioplasma *m*
    e   idioplasma *m*
    i   idioplasma *m*
    d   Idioplasma *n*

1461   IDIOSOME;
     IDIOZOME
    f   idiosome *m*;
       idiozome *m*
    e   idiosoma *m*
    i   idiosoma *m*
    d   Idiosom *n*

1462   IDIOTYPE
    f   idiotype *m*
    e   idiótipo *m*
    i   idiotipo *m*
    d   Idiotyp *m*

1463   IDIOVARIATION
    f   idiovariation *f*
    e   idiovariación *f*
    i   idiovariazione *f*
    d   Idiovariation *f*

1464   IDIOZOME;
     IDIOSOME
    f   idiozome *m*;
       idiosome *m*
    e   idiosoma *m*
    i   idiosoma *m*
    d   Idiosom *n*

1465   ILLEGITIMATE
    f   illégitime
    e   ilegítimo
    i   illegittimo
    d   illegitim

1466   IMMIGRATION COEFFICIENT
       f   coefficient *m* d'immigration
       e   coeficiente *m* de inmigra-
           ción
       i   coefficiente *m* d'immigra-
           zione
       d   Einwanderungskoeffizient *m*

1467   IMMIGRATION PRESSURE
       f   pression *f* d'immigration
       e   presión *f* de inmigración
       i   pressione *f* d'immigra-
           zione
       d   Wanderungsdruck *m*

1468   INBREEDING
       f   consanguinité *f;*
           inbreeding *m*
       e   cruzamiento *m* en
           consanguinidad
           consanguinidad *f;*
       i   consanguinità *f*
       d   Inzucht *f*

1469   INBREEDING COEFFICIENT
       f   coefficient *m* de consangui-
           nité
       e   coeficiente *m* de
           consanguinidad
       i   coefficiente *m* di
           consanguinità
       d   Inzuchtkoeffizient *m*

1470   INBREEDING DEPRESSION
       f   dégénérescence *f*
           consanguine
       e   depresión *f* consanguínea
       i   degenerescenza *f* con-
           sanguinea
       d   Inzuchtdepression *f*

1471   INCOMPATIBILITY
       f   incompatibilité *f*
       e   incompatibilidad *f*
       i   incompatibilità *f*
       d   Inkompatibilität *f*

1472   INCOMPATIBILITY FACTOR
       f   facteur *m* d'incompatibilité
       e   factor *m* de incompatibilidad
       i   fattore *m* d'incompatibilità
       d   Inkompatibilitätsfaktor *m*

1473   INCOMPATIBILITY GENES
       f   gènes *mpl* d'incompatibi-
           lité
       e   genes *mpl* de
           incompatibilidad
       i   geni *mpl* d'incompatibilità
       d   Inkompatibilitätsgene *npl;*
           Unverträglichkeitsgenen *npl*

1474   INCOMPATIBLE
       f   incompatible
       e   incompatible
       i   incompatibile
       d   inkompatibel

1475   INCOMPLETE DOMINANCE;
       PARTIAL DOMINANCE
       f   dominance *f* incomplète;
           dominance *f* partielle
       e   dominancia *f* incompleta;
           dominancia *f* parcial
       i   dominanza *f* incompleta;
           dominanza *f* parziale
       d   unvollständige Dominanz *f*

1476   INCOMPLETE PENETRANCE
       f   pénétrance *f* incomplète
       e   penetración *f* incompleta
       i   penetranza *f* incompleta
       d   unvollständige Penetranz *f*

1477   INCONGRUENT CROSSING;
       INCONGRUENT HYBRIDISATION
       f   croisement *m* incongru
       e   cruzamiento *m* incongruente;
           hibridación *f* incongruente
       i   incrocio *m* incongruente;
           ibridazione *f* incongruente
       d   inkongruente Kreuzung *f*

1478   INDEPENDENT ASSORTIMENT
       f   assortiment *m* indépendant
       e   distribución *f* independiente
       i   - -
       d   unabhängige Genverteilung *f*

1479   INDEPENDENT CHARACTER
       f   caractère *m* indépendant
       e   carácter *m* independiente
       i   carattere *m* indipendente
       d   unabhängiges Merkmal *n*

1480 INDEPENDENT VARIABLE
f variable *f* indépendante
e variable *f* independiente
i variabile *f* indipendente
d unabhängige Veränderliche *f*

1481 INDEX OF FREE CROSSING
OVER
f indice *m* de crossing-over
libre
e índice *m* de libre
entrecruzamiento
i indice *m* di libero
sovrincrocio
d Index *m* des freien Crossing-
over

1482 INDICATOR BACTERIUM
f bactérie *f* indicatrice
e bacteria *f* indicadora
i batterio *m* indicatore
d Indikatorbakterium *n*

1483 INDIVIDUAL
f individu *m*
e individuo *m*
i individuo *m*
d Individuum *n*

1484 INDIVIDUAL SELECTION;
MASS SELECTION
f selection *f* individuelle;
selection *f* massale
e selección *f* individual;
selección *f* masal
i selezione *f* individuale;
selezione *f* massale
d Individualselektion *f* (#);
Massenselektion

1485 INDUCED
f induit
e inducido
i indotto
d induziert

1486 INDUCTION
f induction *f*
e inducción *f*
i induzione *f*
d Induktion *f*

1487 INERT
f inerte
e inerte
i inerte
d inert

1488 INERTIA
f inertie *f*
e inercia *f*
i inerzia *f*
d Inertie *f* (#)

INFECTION, ABORTIVE
see 7

INFECTION, REDUCTIVE
see 2428

1489 INHERITANCE;
HEREDITY
f hérédité *f*;
héritage *m*;
patrimoine *m* héréditaire
e herencia *f*;
patrimonio *m* hereditario
i eredità *f*;
ereditarietà *f*
d Vererbung *f*

INHERITANCE, ALTERNATIV
see 116

INHERITANCE, BLENDING
see 386

INHERITANCE, COLLATERA
see 569

INHERITANCE, CRISS-CROSS
see 634

INHERITANCE, CYTOPLAS-
MIC
see 696

INHERITANCE, INTERMEDI-
ATE
see 1527

INHERITANCE, MATERNAL
see 1718

INHERITANCE, MATROCLINAL
see 1730

1493 INHERITANCE, NON-MEN-
DELIAN
see 1922

INHERITANCE, PARTICULATE
see 2089

INHERITANCE, PATRO-
CLINAL see 2092

INHERITANCE, PLASMATIC
see 2151

1495 INHERITANCE, PLASTID
see 2164

INHERITANCE, QUALITATIVE
see 2383

INHERITANCE, QUANTITATIVE
see 2386

1496 INHERITANCE, SEXUAL
see 2580

INHERITANCE, UNILATERAL
see 2893

1490 INHERITANCE OF ʹACQUIRED
CHARACTERS
f   hérédité *f* des caractères
    acquis
e   herencia *f* de los carác-
    teres adquiridos
i   eredità *f* di caratteri
    acquisti
d   Vererbung *f* erworbener
    Eigenschaften

1491 INHERITED CHARACTER
f   caractère *m* héréditaire
e   carácter *m* hereditario
i   carattere *m* ereditario
d   erbliches Merkmal *n*

1492 INHIBITING GENE
f   gène inhibiteur *m*
e   gen *m* inhibidor
i   gene *m* inibitore
d   Inhibitor *m*;

Inhibitorgen *n*

1493 INHIBITION
f   inhibition *f*
e   inhibición *f*
i   inibizione *f*
d   Hemmung *f*

1494 INHIBITION PLAQUE
f   plage *f* d'inhibition
e   placa *f* de inhibición
i   placca *f* d'inibizione;
    posto *m* d'inibizione
d   - -

1495 INHIBITOR
f   inhibiteur *m*
e   inhibidor *m*
i   inibitore *m*
d   Inhibitor *m*

INHIBITOR, PREPROPHASE
see 2291

1496 INHIBITORY
f   inhibiteur
e   inhibitorio
i   inibitore
d   hemmend;
    Hemm-, Hemmungs-
    (in Zusammensetzungen)

1497 INITIAL
f   initial
e   inicial
i   iniziale
d   anfänglich;
    Initial- (in Zusammen-
    setzungen)

1498 INSERTION
f   insertion *f*
e   inserción *f*
i   inserzione *f*
d   Insertion *f*

1499 INSERTIONAL TRANS-
LOCATION
f   translocation *f* insertion-
    nelle
e   translocación *f* de
    inserción

i traslocazione *f* inserzionale
d insertionale Translokation *f*

intensificanti
d Intensitätsfaktoren *mpl*;
Intensivierungsgene *npl*

1500 INSERTION BREAKAGE
f cassure *f* d'insertion
e ruptura *f* de inserción
i rottura *f* inserzionale
d Insertionsbruch *m*

1507 INTERACTING MUTATIONS
f - -
e mutaciones recíprocas *fpl*
i mutazioni interattive *fpl*
d - -

1501 INSERTION REGION
f région *f* d'insertion
e zona *f* de inserción
i regione *f* d'inserimento
d Insertionsstelle *f*

1508 INTERACTION
f interaction *f*
e interacción *f*
i interazione *f*
d Wechselwirkung *f*

1502 INTEGRATED FACTOR
f facteur *f* intégré
e factor *m* integrado
i fattore *m* integrato
d integrierter Faktor *m* (*#*)

1509 INTERACTION THEORY
f théorie *f* de l'interaction
e teoría *f* de la interacción
i teoria *f* dell'interazione
d Wechselwirkungstheorie *f*

1503 INTEGRATED STATE
f état *m* intégré
e estado *m* integrado
i stato *m* integrato
d - -

1510 INTERARM PAIRING
f appariement *m* interbra-
chial
e apareamiento *m* inter-
braquial
i appaiamento *m* inter-
brachiale
d Interarm-Paarung *f*

1504 INTEGRATION
f intégration *f*
e integración *f*
i integrazione *f*
d Integration *f*

1505 INTEMPERATE PHAGE;
VIRULENT PHAGE
f phage *m* intempéré;
phage *m* virulent
e fago *m* intemperado;
fago *m* virulento
i fago *m* intemperato;
fago *m* virulento
d nichttemperierter Phage *m;*
nichttemperenter Phage *m;*
virulenter Phage *m*

1511 INTERBAND
f interbande *f*
e entrebanda *f*
i interbanda *f*
d Zwischenscheibe *f*

1512 INTERBIVALENT CON-
NECTIONS
f connections *fpl* interbivalentes
e conexiones *fpl*
interbivalentes
i connessioni *fpl*
interbivalenti
d Interbivalent-Konnektionen
fpl

1506 INTENSIFYING FACTORS
f facteurs *mpl* d'intensi-
fication
e factores *mpl* de
intensificación
i fattori *mpl*

1513 INTERCENTRIC REGION;
INTERCENTROMERIC REGIO
f région *f* intercentrique
e zona *f* intercéntrica

i   regione *f* intercentrica
d   Intercentromer-Region *f*

1514   INTERCHANGE
f   interchange *m*;
translocation *f* réciproque
e   intercambio *m*
i   interscambio *m*
d   Austausch *m*;
reziproke Translokation *f*

1515   INTERCHANGE TRISOMIC
f   trisomique *m* par interchange
e   trisómico *m* de
intercambio
i   trisomico *m* d'interscambio
d   "interchange"-Trisome *f* (#)

1516   INTERCHROMIDIA
f   interchromidie *f*
e   intercromidio *m*
i   intercromidio *m*
d   Interchromidie *f*

1517   INTERCHROMOMERE
f   interchromomère *n*
e   intercromómero *m*
i   intercromomero *m*
d   Interchromomere *f*

1518   INTERCHROMOSOMAL;
INTERCHROMOSOMIC
f   interchromosomique
e   intercromosómico
i   intracromosomico
d   interchromosomisch (#);
Inter-Chromosomen-
(in Zusammensetzungen)

1519   INTERFERENCE
f   interférence *f*
e   interferencia *f*
i   interferenza *f*
d   Interferenz *f*

1520   INTERFERENCE DISTANCE
f   distance *f* d'interférence
e   distancia *f* de interferencia
i   distanza *f* d'interferenza;

distanza *f* interferenziale
d   Interferenzabstand *m*

1521   INTERGENIC
f   intergénique
e   intergénico
i   intergenico
d   intergenisch

1522   INTERGRADATION
f   intergradation *f*
e   intergradación *f*
i   intergradazione *f*
d   Intergradation *f*

1523   INTERKINESIS
f   intercinèse *f*
e   intercinesis *f*
i   intercinesi *f*
d   Interkinese *f*

1524   INTERLOCKING
f   interlocking
e   intercierre *m*
i   interlocking *m*
d   Interlocking *n*

1525   INTERMEDIARY
f   intermédiaire *m*
e   intermediario *m*
i   intermediario *m*
d   Vermittler *m*

1526   INTERMEDIATE
f   intermédiaire
e   intermedio
i   intermedio
d   intermediär;
Mittel-, Zwischen-
(in Zusammensetzungen)

1527   INTERMEDIATE INHERITANCE
f   hérédité *f* intermédiaire
e   herencia *f* intermediaria
i   eredità *f* intermedia
d   intermediäre Vererbung *f*

1528   INTERMITOSIS
f   intermitose *f*
e   intermitosis *f*

i intermitosi *f*
d Intermitose *f;*
   Intermitosis *f*

1529 INTERNAL
f interne
e interno
i interno
d internal

1530 INTERNAL SPIRAL
f spirale *f* interne
e espiral *f* interna
i spirale *f* interna
d Internalspirale *f*

1531 INTERPHASE
f interphase *f*
e interfase *f*
i interfase *f*
d Interphase *f*

INTERPHASE, POSTSYNDETIC
see 2272

1532 INTERSECTING SPINDLES
f fuseaux *mpl* chevauchants
e husos *mpl* intersectores
i fusi *mpl* accavallanti
d sich überkreuzende
   Spindeln *f pl*

1533 INTERSEX
f intersexué *m;*
   intersexe *m*
e intersexo *m;*
   intersexuado *m*
i intersesso *m*
d Intersex *m*

1534 INTERSEXUAL
f intersexué
e intersexuado
i intersessuale
d intersexuell

1535 INTERSEXUALITY
f intersexualité *f*
e intersexualidad *f*
i intersessualità *f*
d Intersexualität *f*

1536 INTERSPECIFIC
f interspécifique
e interespecífico;
   intraespecífico
i interspecifico
d interspezifisch

1537 INTERSPECIFIC CROSSING;
INTERSPECIFIC HYBRIDISATI〈
f croisement *m* interspécifique
e hibridación *f* intraespecí-
   fica
i ibridazione *f* interspeci-
   fica
d Artkreuzung *f*

1538 INTERSTERILITY
f interstérilité *f*
e interesterilidad *f*
i intersterilità *f*
d Intersterilität *f*

1539 INTERSTITIAL
f intersticiel
e intersticial
i interstiziale
d interstitiell;
   Interstitial-
   (in Zusammensetzungen)

1540 INTERZONAL
f interzone
e interzonal
i interzonale
d interzonal;
   Interzonal-
   (in Zusammensetzungen)

1541 INTRABREEDING POPULATI〈
f population *f* consanguine
e población *f* consanguínea
i popolazione *f* incrociata
d Inzuchtpopulation *f*

1542 INTRACHROMOSOMAL;
INTRACHROMOSOMIC
f intrachromosomique
e intracromosómico
i intracromosomico
d intrachromosomal

1543 INTRACLASS CORRELATION
f corrélation *f* intraclasse
e correlación *f* intraclase
i correlazione *f* fra classi
d Intraklassenkorrelation *f* (#)

1544 INTRAGENIC
f intragénique
e intragénico
i intragenico
d intragenisch

1545 INTRAHAPLOID (adj.)
INTRAHAPLOID
f intrahaploïde
intrahaploïde *m*
e intrahaploide
intrahaploide *m*
i intraaploide
intraaploide *m*
d intrahaploid
Intrahaploide *f*

1546 INTRAHAPLOIDY
f intrahaploïdie *f*
e intrahaploidía *f*
i intraaploidismo *m*
d Intrahaploidie *f*

1547 INTRARADIAL
f intra-radial
e intraradial
i intra-radiale
d intraradial

1548 INTROGRESSION
f introgression *f*
e introgresión *f*
i introgressione *f*
d Introgression *f*

1549 INTROGRESSIVE
f introgressif
e introgresivo
i introgressivo
d introgressiv

1550 INVERSION
f inversion *f*
e inversión *f*
i inversione *f*
d Inversion *f*

INVERSION, ACENTRIC
see 17

1551 INVERSION HETEROZYGOTE
f hétérozygote *f* pour une
inversion
e heterocigote *m* de inversión
i eterozigote *m* da inversione
d Inversionsheterozygote *f*

1552 INVERSION MORPHISM
f morphisme *m* d'inversion
e morfismo *m* de inversión
i morfismo *m* da inversione
d Inversionsmorphismus *m*

1553 INVERSION-CHIASMA
f chiasma *m* d'inversion
e quiasma *m* de inversión
i chiasma *m* da inversione
d Inversionschiasmata *mpl*

1554 INVIABILITY
f inviabilité *f*
e inviabilidad *f*
i invitalità *f;*
avitalità *f*
d Lebensunfähigkeit *f*

1555 INVIABLE
f inviable
e inviable
i invitale *;*
avitale
d lebensunfähig

1556 INVISIBLE GENE
f gène *m* occulte
e gen *m* invisible
i gene *m* invisibile
d latentes Gen *n* (#)

1557 IRREGULARITY DELAY
f délai *m* irrégulier
e retraso *m* irregular
i indugio *m* irregolare
d irreguläre Verzögerung *f*

1558 IRREPARABLE MUTANT
f mutant *m* irréparable
e mutante *m* irreparable

i   mutante *m* irreparabile
d   irreparable Mutante *f*

1559   ISOALLELE;
       ISO-ALLELE
       f   isoallèle *m*
       e   isoalelo *m*
       i   isoalelo *m*
       d   Isoallel *n*

1560   ISOALLELY;
       ISO-ALLELY
       f   isoallélisme *m*
       e   isoalelismo *m*
       i   isoallelismo *m*
       d   Isoallelie *f*

1561   ISOAUTOPOLYPLOID (adj.)
       ISOAUTOPOLYPLOID
       f   isoautopolyploïde
           isoautopolyploïde *m*
       e   isoautopoliploide
           isoautopoliploide *m*
       i   isoautopoliploide
           isoautopoliploide *m*
       d   isoautopolyploid
           Isoautopolyploide *f*

1562   ISOAUTOPOLYPLOIDY
       f   isoautopolyploïdie *f*
       e   isoautopoliploidía *f*
       i   isoautopoliploidismo *m*
       d   Isoautopolyploidie *f*

1563   ISOBRACHIAL
       f   isobrachial
       e   isobraquial
       i   isobrachiale
       d   isobrachial

1564   ISOCHROMATID
       f   isochromatide *f*
       e   isocromatidio *m*
       i   isocromatide *m*
       d   Isochromatide *f*

1565   ISOCHROMATID BREAK
       f   rupture *f* isochromatidique
       e   ruptura *m* de isocromatidio
       i   rottura *f* isocromatidica
       d   Isochromatidenbruch *m*

1566   ISOCHROMATIDIC
       f   isochromatidique
       e   isocromatídico
       i   isocromatidico
       d   isochromatidisch

1567   ISOCHROMOSOME
       f   isochromosome *m*
       e   isocromosoma *m*
       i   isocromosoma *m*
       d   Isochromosom *n*

1568   ISOCYTOSIS
       f   isocytose *f*
       e   isocitosis *f*
       i   isocitosi *f*
       d   Isocytose *f* (#)

1569   ISODICENTRIC
       f   isodicentrique
       e   isodicéntrico
       i   isodicentrico
       d   isodizentrisch

1570   ISOFRAGMENT
       f   isofragment *m*
       e   isofragmento *m*
       i   isoframmento *m*
       d   Isofragment *n*

1571   ISOGAMETE
       f   isogamète *m*
       e   isogameto *m*
       i   isogamete *m*
       d   Isogamet *m*

1572   ISOGAMOUS
       f   isogame
       e   isogamo
       i   isogamo
       d   isogam

1573   ISOGAMY
       f   isogamie *f*
       e   isogamia *f*
       i   isogamia *f*
       d   Isogamie *f*

1574   ISOGENETIC
       f   isogénétique
       e   isogenético

i isogenetico
d isogenetisch (#)

1575 ISOGENIC;
ISOGENOUS
f isogénique
e isogénico
i isogenico
d isogen;
isogenisch

1576 ISOGENOMATIC;
ISOGENOMIC
f isogénomique
e isogenómico
i isogenomico
d isogenomatisch

1577 ISOGENOUS;
ISOGENIC
f isogénique
e isogénico
i isogenico
d isogen;
isogenisch

1578 ISOGENY
f isogénie f
e isogenia f
i isogenia f
d Isogenie f

1579 ISOKARYOSIS
f isocaryose f
e isokariosis f
i isocariosis f
d Isokaryose f

1580 ISOLATE
f isolat m
e - -
i isolate m
d Isolat n;
Fortpflanzungsgemein-
schaft f

1581 ISOLATION
f isolation f
e aislamiento m
i isolamento m
d Isolation f;
Isolierung f

1582 ISOLATION GENES
f gènes mpl d'isolation
e genes mpl de aislamiento
i geni mpl d'isolamento
d Isolationsgene npl

1583 ISOLATION INDEX
f indice d'isolation m
e índice de aislamiento m
i indice d'isolamento m
d Isolationsindex m

1584 ISOLECITHAL
f isolécithe
e isolecito
i isolecito
d isolezithal

1585 ISOMAR
f isomar m (Θ)
e isomar m
i isomar m (Θ)
d Isomare f

1586 ISOMORPHIC;
ISOMORPHOUS
f isomorphe
e isomórfico
i isomorfico
d isomorph

1587 ISOMORPHISM
f isomorphisme
e isomorfismo
i isomorfismo m
d Isomorphie f

1588 ISOMORPHOUS;
ISOMORPHIC
f isomorphe
e isomórfico
i isomorfico
d isomorph

1589 ISOPHANE;
ISOPHENE;
ISOMAR;
PHENOCONTOUR
f isophane m (Θ);
isomar m (Θ)
e isófano m;
isófeno m

i isofano *m* ;
isofeno *m*;
isomar *m* (#)
d Isophäne *f* (#);
Isomare *f*

1590 ISOPHENIC
f isophénique
e isofénico
i isofenico
d isophän

1591 ISOPHENOGAMY
f isophénogamie *f*
e isofenogamia *f*
i isofenogamia *f*
d Isophänogamie *f*

1592 ISOPHENOUS
f isophène
e isófeno
i isofenoso
d isophän

1593 ISOPLOID (adj.)
ISOPLOID
f isoploïde
isoploïde *m*
e isoploide
isoploide *m*
i isoploide
isoploide *m*
d isoploid
Isoploide *f*

1594 ISOPLOIDY
f isoploïdie *f*
e isoploidía *f*
i isoploidismo *m*
d Isoploidie *f*

1595 ISOPOLYPLOID (adj.)
ISOPOLYPLOID
f isopolyploïde
isopolyploïde *m*
e isopoliploide
isopoliploide *m*
i isopoliploide
isopoliploide *m*
d isopolyploid
Isopolyploide *f*

1596 ISOPOLYPLOIDY
f isopolyploïdie *f*
e isopoliploidía *f*
i isopoliploidismo *m*
d Isopolyploidie *f*

1597 ISOPYCNOSIS
f isopycnose *f*
e isopicnosis *f*
i isopicnosi *f*
d Isopyknose *f*

1598 ISOSYNDESIS
f isosyndèse *f*
e isosíndesis *f*
i isosindesi *f*
d Isosyndese *f*

1599 ISOSYNDETIC
f isosyndétique
e isosindético
i isosindetico
d isosyndetisch

1600 ISOTRANSFORMATION
f isotransformation *f*
e isotransformación *f*
i isotrasformazione *f*
d Isotransformation *f*

1601 ISOTRISOMY
f isotrisomie *f*
e isotrisomia *f*
i isotrisomia *f*
d Isotrisomie *f*

1602 ISOTYPY
f isotypie *f*
e isotipia *f*
i isotipia *f*
d Isotypie *f*

1603 ISOZYGOTIC
f isozygotique
e isocigótico;
isozigótico
i isozigotico
d isozygotisch

1604 ISOZYGOTY;
ISOZYGOSITY
f isozygotie *f*
e isocigotía *f*

    i isozigotia *f*
    d Isozygotie *f*

1605   ITERATIVE MUTATION
    f mutation itérative *f*
    e mutación iterativa *f*
    i mutazione iterativa *f*
    d iterative Mutation *f*

K

KACOGENESIS
see 415

1606 KAKOGENIC
f cacogénique
e cacogenético
i cacogenetico
d kakogen

1607 KAPPA-FACTOR
f facteur *m* Kappa
e factor *m* Kappa
i fattore-cappa *m*
d Kappa-Faktor *m*

1608 KARYASTER
f karyaster *m;*
caryaster *m*
e cariaster *m*
i cariaster *m*
d - -

1609 KARYOCLASTIC
f caryoclasique
e carioclásico
i carioclastico
d karyoklastisch

1610 KARYOGAMY
f caryogamie *f*
e cariogamia *f*
i cariogamia *f*
d Karyogamie *f*

1611 KARYOGENE
f caryogène *m*
e cariogen *m*
i cariogene *m*
d Karyogen *n*

1612 KARYOGRAM
f caryogramme *m*
e cariograma *m*
i cariogramma *m*
d Karyogramm *n*

1613 KARYOID
f caryoïde *m*
e carioide *m*
i carioide *m*
d Karyoid *n*

1614 KARYOKINESIS
f caryocinèse *f*
e cariocinesis *f*
i cariocinesi *f*
d Karyokinese *f*

1615 KARYOKINETIC
f caryocinétique
e cariocinético
i cariocinetico
d karyokinetisch

1616 KARYOLOGIC
f caryologique
e cariológico
i cariologico
d karyologisch

1617 KARYOLOGY
f caryologie *f*
e cariología *f*
i cariologia *f*
d Karyologie *f*

1618 KARYOLYMPH
f caryolymphe *f*
e cariolinfa *f*
i cariolinfa *f*
d Karyolymphe *f*

1619 KARYOLYSIS
f caryolyse *f*
e cariólisis *f*
i cariolisi *f*
d Karyolysis *f*

1620 KARYOMERE
f caryomère *m*
e cariómero *m*
i cariomero *m*
d Karyomer *n;*
Karyomere *f*

1621 KARYOMEROKINESIS
 f caryomérocinèse *f*
 e cariomerocínesis *f*
 i cariomerocinesi *f*
 d Karyomerokinese *f*

1622 KARYOMITOSIS
 f caryomitose *f*
 e cariomitosis *f*
 i cariomitosi *f*
 d Karyomitose *f*

1623 KARYONUCLEOLUS
 f caryonucléole *m*
 e carionucléolo *m*
 i carionucleolo *m*
 d - -

1624 KARYOPLASMOGAMY
 f caryoplasmogamie *f*
 e carioplasmogamia *f*
 i carioplasmogamia *f*
 d Karyoplasmogamie *f*

1625 KARYORHEXIS
 f caryorrhexis *f*
 e cariorrexis *m*
 i cariorressi *f*
 d Karyorrhexis *f*

1626 KARYOSOME
 f caryosome *m*
 e cariosoma *m*
 i cariosoma *m*
 d Karyosom *n*

1627 KARYOSPHERE
 f caryosphère *f*
 e cariosfera *f*
 i cariosfera *f*
 d Karyosphäre *f*

1628 KARYOTHECA
 f caryothèque *f*
 e carioteca *f*
 i carioteca *f*
 d Karyotheka *f*;
  Kernmembran *f*

1629 KARYOTIN
 f caryotine *f*

 e cariotina *f*
 i cariotina *f*
 d Karyotin *n*

1630 KARYOTYPE
 f caryotype *m*
 e cariotipo *m*
 i cariotipo *m*
 d Karyotyp *m*

1631 KATACHROMASIS
 f catachromasie *f*
 e catacromasis *f*
 i catacromasi *f*
 d Katachromasis *f*

1632 KATAPHASE
 f cataphase *f*
 e catafase *f*
 i catafase *f*
 d Kataphase *f*

1633 KEY GENE
 f gène-clé *m*
 e oligogén *m*
 i - -
 d Hauptgen *n*

1634 KEY MUTATION
 f mutation-clé *f*
 e - -
 i - -
 d Schlüsselmutation *f*

1635 KILLER
 f killer *m*
 e killer *m*
 i killer *m*
 d Killer *m*

1636 KIN
 f race *f*;
  parent *m*
 e raza *f*;
  pariente *m*
 i razza *f*;
  parente *m*
 d Sippe *f*;
  Verwandtschaft *f*

1637 KINETIC BODY
 f corpuscule *m* cinétique
 e cuerpo *m* cinético
 i corpuscolo *m* cinetico
 d kinetische Körperchen *npl*

1638 KINETIC CONSTRICTION
 f constriction *f* cinétique
 e constricción *f* cinética
 i costrizione *f* cinetica
 d Primäreinschnürung *f*

1639 KINETIC NUCLEI
 f noyaux *mpl* cinétiques
 e núcleos *mpl* cinéticos
 i nuclei *mpl* cinetici
 d kinetische Kerne *mpl*

1640 KINETOCHORE
 f kinétochore *m*
 e cinetócoro *m*
 i cinetocoro *m*
 d Kinetochor *n*

 KINETOCHORE, NAKED
 see 1905

 KINETOCHORE, SECONDARY
 see 2505

1641 KINETOGENE
 f cinétogène *m*
 e cinetogén *m*
 i cinetogeno *m*
 d Kinetogen *n*

1642 KINETOGENESIS
 f cinétogénèse *f*
 e cinetogénesis *f*
 i cinetogenesi *f*
 d Kinetogenese *f*

1643 KINETOMERE
 f cinétomère *m*
 e cinetómero *m*
 i cinetomero *m*
 d Kinetomer *n*

1644 KINETONEMA
 f kinétonéma *m*
 e cinetonema *f*
 i cinetonema *m*
 d Kinetonema *n*

1645 KINETONUCLEUS
 f cinétonucléus *m*
 e cinetonúcleo *m*
 i cinetonucleo *m*
 d Kinetonukleus *m*;
 Kinetonucleus *m*

1646 KINETOPLAST
 f cinétoplaste *m*
 e cinetoplasto *m*
 i cinetoplasto *m*
 d Kinetoplast *m*

1647 KINOPLASM
 f cinoplasme *m*
 e cinoplasma *m*
 i cinoplasma *m*
 d Kinoplasma *n*

1648 KINOSOME
 f cinosome *m*
 e cinosoma *m*
 i cinosoma *m*
 d Kinosom *n*

1649 KLON;
 CLON;
 CLONE
 f clône *m*
 e clon *m*
 i clone *m*
 d Klon *m*

1650 KNOB
 f protubérance *f*
 e nudo *m*
 i nodo *m* (Θ)
 d Knob *m*;
 Knopf *m*

L

1651 LABILE
  f labile
  e lábil
  i labile
  d labil

1652 LABILITY
  f labilité *f*
  e labilidad *f*
  i labilità *f*
  d Labilität *f*

1653 LACUNA
  f lacune *f*
  e lacunula *f*
  i lacuna *f*
  d Lakune *f*;
     Lücke *f*

1654 LAG
  f retard *m*
  e retardo *m*
  i ritardo *m*
  d Verzögerung *f*

  LAG, MUTATIONAL
  see 1896

1655 LAGGARD (adj.)
  LAGGARD
  f retardataire
     retardataire *m*
  e retardatario
     retardatario *m*
  i ritardatario
     ritardatario *m*
  d langsam
     Nachzügler *m*;
     verzögertes Chromosom *n*

1656 LAGGING;
  DELAYED
  f retardé
  e retrasado
  i ritardato
  d verzögert

1657 LAMARCKISM
  f lamarckisme *m*
  e lamarckismo *m*
  i lamarckismo *m*
  d Lamarckismus *m*

1658 LAMPBRUSH CHROMATID
  f chromatide *f* plumeuse;
     chromatide *f* en lampbrush
  e cromatidio *m* plumoso
  i cromatide *f* piumosa
  d Lampenbürstenchromatide *f*

1659 LAMPBRUSH CHROMOSOME
  f chromosome *m* plumeux;
     chromosome *m* en lampbrush
  e cromosoma *m* plumoso
  i cromosoma *m* piumoso
  d Lampenbürstenchromosom *n*

1660 LATENT
  f latent
  e latente
  i latente
  d latent

1661 LATERAL
  f latéral
  e lateral
  i laterale
  d lateral;
     seitlich

1662 LATERAL CHIASMA
  f chiasma *m* latéral
  e quiasma *m* lateral
  i chiasma *m* laterale
  d Lateralchiasma *n*

1663 LATTICE DESIGN
  f forme *f* de treillis
  e diseño *m* de latice
  i disegno *m* di graticolato
  d Gitteranlage *f*

1664 LECTOTYPE

f lectotype *m*
e lectótipo *m*
i lectotipo *m*
d Lektotyp *m*

1665 LEPTONEMA
f leptonème *m*
e leptonema *m*
i leptonema *m*
d Leptonema *n*

1666 LEPTOTENE
f leptotène
e leptoteno
i leptotene
d Leptotän *n*

1667 LETHAL
f léthal;
létal
e letal
i letale
d letal

LETHAL, CONDITIONED
see 603

LETHAL FACTOR, GAMETIC
see 1027

1668 LETHALITY
f léthalité *f*;
létalité *f*
e letalidad *f*
i letalità *f*
d Letalität *f*

LETHALITY, HYBRID
see 1413

LETHALS, BALANCED
see 325

1669 LIMITED
f limité
e limitado
i limitato
d limitiert;
begrenzt

1670 LINE

f lignée *f*
e línea *f*
i linea *f*
d Linie *f*

1671 LINEAL ORDER
f ordre *m* linéaire
e orden *m* lineal
i ordine *m* lineare
d lineare Anordnung *f*
(von Genen u. dgl.)

1672 LINEAR CHROMOSOME
f chromosome *m* linéaire
e cromosoma *m* lineal
i cromosoma *m* lineare
d lineares Chromosom *n*

1673 LINEAR REGRESSION
f régression *f* linéaire
e regresión *f* lineal
i regressione *f* lineare
d lineare Regression *f*

1674 LINE OF BREEDING
f lignée *f* généalogique
e línea *f* genealógica
i linea *f* genealogica
d Zuchtlinie *f*

1675 LINE OF DESCENT
f filiation *f* généalogique
e filiación *f* genealógica
i linea *f* di discendenza
d Abstammungslinie *f*

1676 LINE OF INBREEDING
f lignée *f* consanguine;
filiation *f* consanguine
e línea *f* consanguínea;
filiación *f* consanguínea
i linea *f* consanguinea
d Inzuchtlinie *f*

1677 LINKAGE
f linkage *m*;
liaison *f*
e ligamiento *m*
i linkage *m*;
associazione *f*
d Kopp(e)lung *f*

LINKAGE, COMPLETE
see 593

LINKAGE, FALSE
see 984

1678 LOAD
f charge *f* (génétique)
e carga *f* genética
i potenza *f* (genetica)
d – –

1679 LOCALIZATION
f localisation *f*
e localización *f*
i localizzazione *f*
d Lokalisation *f*

1680 LOCALIZED
f localisé
e localizado
i localizzato
d lokalisiert

1681 LOCK AND KEY THEORY
f théorie *f* de la clé et de
la serrure
e teoría *f* de cerradura y
llave
i teoria *f* della chiave e
della serratura
d Schloss- und Schlüssel-
Theorie *f*

1682 LOCUS, *pl* LOCI (≠)
f locus *m*, *pl* loci (≠)
e locus *m*, *pl* loci (≠)
i locus *m*, *pl* loci (≠);
luogo *m*
d Locus *m*, *pl* Loci

1683 LOCUS SPECIFIC EFFECT
f effet *m* spécifique sur un
locus
e efecto *m* especffico en un
locus
i effetto *m* specifico in un
locus
d locusspezifische Wirkung *f*

1684 LOOP UNIVALENT
f univalent *m* en boucle

e lazo *m* univalente
i anello *m* univalente (Θ)
d geschlossenes Univalent *n*

1685 LOOSELY PAIRED
f uni lâchement
e apareamiento *m* flojo
i – –
d unvollständig gepaart

1686 LOSS MUTATION
f mutation *f* de perte
e – –
i mutazione *f* di perdita (Θ)
d Verlustmutation *f*

1687 LOW FREQUENCY TRANSFER
(LFC)
f transfert *m* à basse fréquence (LFC)
e transferencia *f* de baja
frecuencia (CBF)
i trasferimento *m* a bassa
frequenza (LFC)
d – –

1688 LUXURIANCE
f luxuriance *f*
e exuberancia *f*
i lussureggiamento *m*
d Luxurieren *n*

1689 LYSOGENIC
f lysogénique
e lisogénico;
lisigénico;
lisógeno;
lisígeno
i lisogenico
d lysogen

1690 LYSOGENISATION
f lysogénisation *f*
e lisogenisación *f*
i lisogenizzazione *f*
d Lysogenisierung *f*

1691 LYSOGENY
f lysogénie *f*
e – –
i lisogenia *f*
d Lysogenie *f*

1692   LYSOTYPE
     f lysotype *m*
     e lisótipo *m*
     i lisotipo *m*
     d Lysotyp *m*

M

1693   MACROEVOLUTION
f   macroévolution *f*
e   macroevolución *f*
i   macroevoluzione *f*
d   Makroevolution *f*

1694   MACROGAMETE
f   macrogamète *m*
e   macrogámeto *m*
i   macrogamete *m*
d   Makrogamet *m*

1695   MACROMUTATION
f   macromutation *f*
e   macromutación *f*
i   macromutazione *f*
d   Makromutation *f*;
Grossmutation *f*

1696   MACRONUCLEAR REGENER-
ATION
f   régénération *f* macronu-
cléaire
e   regeneración *f* macronuclear
i   regenerazione *f*
macronucleare
d   Makronukleusregenera-
tion *f*

1697   MACRONUCLEUS
f   macronucléus *m*;
macronoyau *m*
e   macronúcleo *m*
i   macronucleo *m*
d   Makronukleus *f*

1698   MACROPHYLOGENESIS
f   macrophylogénèse *f*
e   macrofilogénesis *f*
i   macrofilogenesi *f*
d   Makrophylogenese *f*

1699   MACROSPORE
f   macrospore *f*
e   macrospora *f*

i   macrospora *f*
d   Makrospore *f*

1700   MACROSPOROGENESIS
f   macrosporogénèse *f*
e   macrosporogénesis *f*
i   macrosporogenesi *f*
d   Makrosporogenese *f*

1701   MAJOR
f   majeur
e   mayor
i   maggiore
d   Gross- (in Zusammen-
setzungen wie Grossspirale);
Haupt- (in Zusammen-
setzungen wie Hauptgen)

1702   MAJOR GENE;
OLIGOGENE
f   oligogène *m*;
gène *m* majeur
e   oligogen *m*;
gen *m* mayor
i   gene *m* maggiore;
oligogene *m*
d   Oligogen *n*;
Hauptgen *n*

1703   MAJOR MUTATION
f   mutation *f* majeure
e   mutación *f* mayor
i   mutazione *f* maggiore
d   Grossmutation *f*

1704   MAJOR SPIRAL
f   spirale *f* majeure
e   espiral *f* mayor
i   spirale *f* maggiore
d   Grossspirale *f*

1705   MALE LINE OF BREEDING
f   lignée *f* mâle;
lignée *f* paternelle;
filiation *f* mâle
e   línea *f* masculina;

línea *f* paterna;
filiación *f* paterna
i linea *f* maschile;
linea *f* paterna
d väterliche Linie *f*;
männliche Linie *f*

1706 MANIFESTATION
f manifestation *f*
e manifestación *f*
i manifestazione *f*
d Manifestation *f*

1707 MANIFOLD EFFECTS
f effets *mpl* multiples
e efectos *mpl* múltiples
i effetti *mpl* molteplici
d manifold effects *mpl*

1708 MANTLE FIBRES
f fibres *fpl* du fuseau
e fibras *fpl* del manto
i mantello *m* del fuso
d Mantelfasern *fpl*

1709 MAP
f carte *f*
e mapa *f*
i carta *f*
d Chromosomenkarte *f*

1710 MAP DISTANCE
f distance *f* sur la carte;
distance *f* graphique
e distancia *f* de mapa
i distanza *f* su mappa
d map distance *f*;
Genabstand *m*

1711 MAP UNIT
f unité *f*;
unité *f* graphique
e unidad *f* de mapa
i unità *f* di mappa;
d map unit *f*;
Chromosomenkarteneinheit *f*

1712 MARKER
f marqueur *m*
e marcador *m*
i marcatore *m*
d Markierungsgen *n*

MARKERS, UNSELECTED
see 2901

1713 MASS . . . .
f massal
e masal
i in massa;
massale
d Massen-
(in Zusammensetzungen)

1714 MASS MUTATION
f mutation *f* massale
e mutación *f* masal
i mutazione *f* in massa
d Massenmutation *f*

1715 MASS SELECTION;
INDIVIDUAL SELECTION
f sélection *f* massale;
sélection *f* individuelle
e selección *f* masal;
selección *f* individual
i selezione *f* massale;
selezione *f* individuale
d Massenselektion *f*;
Individualselektion *f*

1716 MATE KILLER
f killer *m* du partenaire (Θ)
e killer *m* del compañero
i killer *m* del compagno (Θ)
d mate killer *m*

1717 MATERNAL
f maternel
e materno
i materno
d mütterlich

1718 MATERNAL INHERITANCE
f hérédité *f* maternelle
e herencia *f* materna
i eredità *f* materna
d mütterliche Vererbung *f*

1719 MATING
f a. copulation *f*;
coït *m*
b. accouplement *m*
e a. cópula *f*;
coito *m*

b. apareamiento *m*
i accoppiamento *m;*
copulazione *f*
d a. Begattung *f;*
Kopulation *f;*
Coitus *m*
b. Paarung *f*

1720 MATING (USA),
syn. of BREEDING
f élevage *m*
e crianza *f*
i allevamento;
accoppiato *m*
d Züchtung *f*

MATING, ASSORTATIVE
see 244

MATING, CORRECTIVE
see 625

MATING, DISASSORTATIVE
see 830

MATING, GENETIC AS-
SORTATIVE
see 1092

1721 MATING BROTH
f bouillon *m* de croisement;
bouillon *m* de conjugaison
e caldo *m* de apareamiento
i brodo *m* d'accoppiamento
d - -

1722 MATING GROUP
f groupe *m* de croisement
e grupo *m* de apareamiento
i gruppo d'accoppiamento *m*
d mating group *f;*
Fortpflanzungsgruppe *f*

1723 MATING SYSTEM
f méthode *f* d'accouplement;
système *m* d'accouplement
e método *m* de apareamiento;
sistema *m* de cubrición
i metodo *m* d'accoppiamento
d Fortpflanzungssystem *n*

1724 MATING THEORY
f théorie *f* de croisement
e teoría *f* de apareamiento
i teoria *f* d'accoppiamento
d mating theory *f;*
Kopulationstheorie (der
Bakteriophagenrekombi-
nation) *f*

1725 MATING TYPE
f type *m* d'appariement
e tipo *m* de apareamiento
i tipo *m* d'accoppiamento
d Paarungstyp *m*

MATING UNLIKE TO UNLIKE
see 830

1726 MATRICLINOUS;
MATROCLINOUS
f matrocline
e matroclino
i matrilineare;
matroclino
d matroklin

1727 MATRIX
f matrice *f;*
matrix *f;*
calymma *f;*
kalymma *f*
e matriz *f*
i matrice *f*
d Matrix *f;*
Kalymma *n*

1728 MATRIX BRIDGE
f pont *m* de matrice
e puente *m* de matriz
i ponte *m* di matrice
d Matrixbrücke *f*

1729 MATRIX STICKINESS
f viscosité *f* de la matrice
e viscosidad *f* de la matriz
i viscosità *f* della matrice
d matrix stickiness *f;*
Matrixverklebung *f*

1730 MATROCLINAL INHERITANCE
f hérédité *f* matroclinale;
hérédité *f* matrocline

e herencia *f* matroclinal
i eredità *f* matrilineare;
   eredità *f* matroclina
d matrokline Vererbung *f*

1731 MATROCLINY
      f matroclinie *f*
      e matroclinia *f*
      i matroclinia *f*
      d Matroklinie *f*;
         Metroklinie *f*

1732 MATURATION
      f maturation *f*
      e maduración *f*
      i maturazione *f*
      d Reifung *f*

1733 MATURATION DIVISION
      f division *f* de maturation
      e división *f* de maduración
      i divisione *f* maturativa
      d Reifungsteilung *f*

1734 MAXIMUM LIKEHOOD
      METHOD
      f méthode *f* du "maximum
         likehood"
      e método *m* del "maximum-
         likehood" (Θ)
      i metodo *m* del "Maximum-
         likehood"
      d "Maximum likehood"-
         Methode *f*;
         Methode *f* des Möglich-
         keitsmaximums

1735 MEDIOCENTRIC
      f médiocentrique
      e mediocéntrico
      i mediocentrico
      d mediozentrisch (#)

1736 MEGA-EVOLUTION
      f mégaévolution *f*
      e megaevolución *f*
      i megaevoluzione *f*
      d Mega-Evolution *f*

1737 MEGASPORE
      f mégaspore *f*

e megaspora *f*
i megaspora *f*
d Megaspore *f*;
   Makrospore *f*

1738 MEGASPOROCYTE
      f mégasporocyte *m*
      e megasporocito *m*
      i megasporocito *m*
      d Megasporocyte *f* (Θ);
         Embryosackmutterzelle *f*

1739 MEGASPOROGENESIS
      f mégasporogénèse *f*
      e megasporogénesis *f*
      i megasporogenesi *f*
      d Megasporogenese *f*;
         Makrosporogenese *f*

1740 MEGAHETEROCHROMATIC
      f mégahétérochromatique
      e megaheterocromático
      i megaeterocromatico
      d megaheterochromatisch

1741 MEIOCYTE
      f méiocyte *m*
      e meiocito *m*
      i meiocito *m*
      d Meiocyte *f*;
         Meiozyte *f*

1742 MEIOMERY
      f méiomérie *f*
      e meiomería *f*;
         meyomería *f*
      i meiomeria *f*
      d Meiomerie *f*

1743 MEIOSIS
      f méiose *f*
      e meiosis *f*;
         meyosis *f*
      i meiosi *f*
      d Meiose *f*;
         Meiosis *f*

      MEIOSIS, C-
      see 548

      MEIOSIS, ONE DIVISION
      see 1976

1744   MEIOSOME
       f méiosome *m*
       e meiosoma *m*
       i meiosoma *m*
       d Meiosom *n* ;
          Mikrosom *n*;
          Sphärosom *n*

1745   MEIOSPORE
       f méiospore *f*
       e meioespora *f*
       i meiospora *f*
       d Meiospore *f*

1746   MEIOTIC
       f méiotique
       e meiótico
       i meiotico
       d meiotisch

1747   MEIOTROPHIC
       f méiotrophique
       e meiótrofo
       i meiotrofico
       d meiotroph (≠)

       MEMBRANE, NUCLEAR
       see 1934

1748   MENDELIAN
       f mendélien
       e mendeliano
       i mendeliano
       d Mendelsche

1749   MENDELIAN RATIO
       f proportion *f* mendélienne
       e proporción *f* mendeliana
       i proporzione *f* mendeliana
       d Aufspaltungsverhältnis *n*

1750   MENDELISM
       f mendélisme *m*
       e mendelismo *m*
       i mendelismo *m*
       d Mendelismus *m*

1751   MENDEL'S LAWS
       f lois *f pl* de Mendel
       e leyes *f pl* de Mendel
       i leggi *f pl* di Mendel
       d Mendelsche Regeln *f pl*;
          Mendelsche Gesetze *n pl*

1752   MENTOR METHOD
       f méthode *f* mentor
       e método *m* mentor
       i metodo *m* mentore
       d Mentormethode *f*

1753   MERICLINAL
       f mériclinal
       e mericlinal
       i mericlinale
       d meriklinal

1754   MERICLINAL CHIMAERA
       f chimère *f* mériclinale
       e quimera *f* mericlinal
       i chimera *f* mericlinale
       d Meriklinalchimäre *f*

1755   MERISTEM
       f meristème *m*
       e meristema *m*
       i meristema *m*
       d Meristem *n*

1756   MEROBLASTIC CLEAVAGE
       f clivage *m* méroblastique ;
          scission *f* méroblastique
       e escisión *f* meroblástica
       i scissione *f* meroblastica
       d meroblastische Furchung *f*

1757   MEROGAMY
       f mérogamie *f*
       e merogamia *f*
       i merogamia *f*
       d Merogamie *f*

1758   MEROGONE
       f mérogon
       e merogonio *m*
       i merogonio *m*
       d Merogon *n*

1759   MEROGONY
       f mérogonie *f*
       e merogonía *f*
       i merogonia *f*
       d Merogonie *f*

1760   MEROKINESIS
       f mérocinèse *f*

e merocinesis *f*
i merocinesi *f*
d Merokinese *f*

1761 MEROMIXIS
f méromixie *f*
e meromixis *f*;
meromixia *f*
i meromixi *f*;
meromissi *f*
d Meromixis *f*

1762 MEROSTATHMOKINESIS
f merostatmocinèse *f*
e merostatmocinesis *f*
i merostatmocinesi *f*
d Merostathmokinese *f*

1763 MEROZYGOTE
f mérozygote *m*
e merocigoto *m*
i merozigote *m*
d Merozygote *f*

1764 MESOMITIC
f mésomitique
e mesomítico
i mesomitico
d mesomitisch (#)

1765 MESOMITOSIS
f mésomitose *f*
e mesomitosis *f*
i mesomitosi *f*
d Mesomitose *f*

1766 METABOLIC NUCLEUS
f noyau *m* métabolique;
noyau *m* en activité
e núcleo *m* metabólico
i nucleo *m* metabolico
d Arbeitskern *m*

1767 METACENTRIC
f métacentrique
e metacéntrico
i metacentrico
d metazentrisch

1768 METAGENESIS
f métagénèse *f*

e metagénesis *f*
i metagenesi *f*
d Metagenesis (Ⓞ);
Generationswechsel *m*

1769 METAGYNY
f métagynie *f*
e metaginia *f*
i metaginia *f*
d Metagynie *f*

1770 METAKINESIS
f métakinèse *f*
e metacinesis *f*
i metacinesi *f*
d Metakinese *f*

1771 METAMITOSIS
f métamitose *f*
e metamitosis *f*
i metamitosi *f*
d Metamitose *f*

1772 METANDRY
f métandrie *f*
e metandria *f*
i metandria *f*
d Metandrie *f*

1773 METAPHASE
f métaphase *f*
e metafase *f*
i metafase *f*
d Metaphase *f*

1774 METAPHASE PAIRING INDE.
f indice *m* d'appariement
métaphasique
e índice *m* de apareamiento
metafásico
i indice *m* d'appaiamento
metafasico
d Metaphasepaarungsindex *n*

1775 METAPHASIC
f métaphasique
e metafásico
i metafasico
d metaphasisch;
Metaphase- (in Zusammen
setzungen)

1776   METAPLASIA
        f   métaplasie *f*
        e   metaplasia *f*
        i   metaplasia *f*
        d   Metaplasie *f*

1777   METAREDUPLICATION
        f   métaréduplication *f*
        e   metareduplicación *f*
        i   metareduplicazione *f*
        d   Metareduplikation *f*

1778   METASYNDESIS
        f   métasyndèse *f*
        e   metasíndesis *f*
        i   metasindesi *f*
        d   Metasyndese *f*

1779   METATACTIC
        f   métatactique
        e   metatáctico
        i   metatattico
        d   metataktisch (#)

1780   METAXENIA
        f   métaxénie *f*
        e   metaxenia *f*
        i   metaxenia *f*
        d   Metaxenie *f*

1781   METROMORPHOUS
        f   métromorphe
        e   metrómorfo
        i   metromorfo
        d   metromorph

1782   MICROCENTRUM
        f   microcentre *m*
        e   microcentro *m*
        i   microcentro *m*
        d   Mikrozentrum *n*

1783   MICROCHROMOSOME
        f   microchromosome *m*
        e   microcromosoma *f*
        i   microcromosoma *m*
        d   Mikrochromosom *n*

1784   MICROCLUSTER
        f   microagglomération *f*
        e   microrracimo *m*

        i   microagglomerazione *f*
        d   Microcluster *n*

1785   MICRO-EVOLUTION;
        MICROEVOLUTION
        f   microévolution *f*
        e   microevolución *f*
        i   microevoluzione *f*
        d   Mikroevolution *f*

1786   MICROGAMETE
        f   microgamète *m*
        e   microgameto *m*
        i   microgameto *m*
        d   Mikrogamet

1787   MICROGENE
        f   microgène *m*
        e   microgén *m*
        i   microgene *m*
        d   Mikrogen *n*

1788   MICROHETEROCHROMATIC
        f   microhétérochromatique
        e   microheterocromático
        i   microeterocromatico
        d   mikroheterochromatisch

1789   MICROMUTATION
        f   micromutation *f*
        e   micromutación *f*
        i   micromutazione *f*
        d   Mikromutation *f*

1790   MICRONUCLEUS
        f   micronucléus *m*;
            micronoyau *m*
        e   micronúcleo *m*
        i   micronucleo *m*
        d   Mikronukleus

1791   MICROPHYLOGENESE
        f   microphylogénèse *f*
        e   microfilogénesis *f*
        i   microfilogenesi *f*
        d   Mikrophylogenese *f*

1792   MICROPYRENIC
        f   micropyrénique
        e   micropirénico
        i   micropirenico
        d   mikropyrenisch

1793 MICROSOME
    f microsome *m*
    e microsoma *m*
    i microsoma *m*
    d Mikrosom *n*

1794 MICROSPECIES
    f micro-espèce *f*
    e microespecie *f*
    i microspecie *f*
    d Mikrospecies *f*;
       Kleinart *f*

1795 MICROSPHERE
    f microsphère *f*
    e microsfera *f*
    i microsfera *f*
    d Mikrosphäre *f*

1796 MICROSPORE
    f microspore *f*
    e microspora *f*
    i microspora *f*
    d Mikrospore *f*

1797 MICROSPOROCYTE
    f microsporocyte *m*
    e microesporócito *m*
    i microsporocito *m*
    d Mikrosporocyte *f*;
       Pollenmutterzelle *f*

1798 MICROSUBSPECIES
    f micro-sous-espèces *f*
    e microsubespecie *f*
    i microsubspecie *f*
    d Mikrosubspecies *f*

1799 MICTON
    f micton *m*
    e micton *m*
    i micton *m* (#)
    d Mikton *n* (#)

1800 MIGRATION
    f migration *f*
    e migración *f*
    i migrazione *f*
    d Migration *f*;
       Wanderung *f*

1801 MIMICRY

    f mimétisme *m*
    e mimetismo *m*
    i mimetismo *m*
    d Mimikry *f*

1802 MIMICS;
MIMIC GENES
    f gènes *mpl* mimétiques
    e genes *mpl* miméticos
    i geni *mpl* mimetici
    d mimics *npl*;
       mimetische Gene *npl* (#)

1803 MINOR
    f mineur
    e menor
    i minore
    d Klein- (in Zusammen-
       setzungen wie Kleinspirale

1804 MINOR GENE
    f gène mineur *m*
    e gen menor *m*
    i gene minore *m*
    d minor gene *n*
       a. Polygen *n*
       b. Modifikationsgen *n*

1805 MISDIVISION
    f division *f* transversale ;
       misdivision *f*
    e misdivisión *f*
    i misdivisione *f*;
       divisione *f* ortogonale
    d Missteilung *f*

    MISDIVISION , A-
    see 124

    MISDIVISION, CENTROMERE
    see 447

1806 MIS-MEIOSIS
    f misméiose *f*
    e mismeiosis *f*
    i mismeiosis *f*
    d Missmeiose *f*

1807 MITOCLASIC
    f mitoclasique
    e mitoclásico
    i mitoclastico
    d mitoklastisch (#)

1808   MITOGENETIC
       f   mitogénétique
       e   mitogenético
       i   mitogenetico
       d   mitogenetisch

       MITOSE
       see 677

1809   MITOSIS
       f   mitose f
       e   mitosis f
       i   mitosi f
       d   Mitose f;
           Mitosis f

       MITOSIS, ABORTIVE
       see 8

       MITOSIS, C-
       see 549

       MITOSIS, REDUCTIONAL
       see 2427

1810   MITOSOME
       f   mitosome m
       e   mitosoma m
       i   mitosoma m
       d   Mitosom n

1811   MITOSTATIC
       f   mitostatique
       e   mitostático
       i   mitostatico
       d   mitostatisch

1812   MITOTIC
       f   mitotique
       e   mitótico
       i   mitotico
       d   mitotisch;
           Mitose- (in Zusammen-
           setzungen)

1813   MITOTIC INDEX
       f   indice m mitotique
       e   índice mitótico
       i   indice mitotico m
       d   Mitoseindex m

1814   MITOTIC INHIBITION

       f   inhibition f mitotique
       e   inhibición f mitótica
       i   inibizione f mitotica
       d   Mitosehemmung f

1815   MITOTIC POISON
       f   poison m mitotique
       e   tóxico m mitótico
       i   veleno m mitotico
       d   Mitosegift n

       MIXING, PHENOTYPIC
       see 2130

1816   MIXOCHROMOSOME
       f   mixochromosome m
       e   mixocromosoma m
       i   mixocromosoma m
       d   Mixochromosom n

1817   MIXOPLOID (adj.)
       MIXOPLOID
       f   mixoploïde
           mixoploïde m
       e   mixoploide
           mixoploide m
       i   mixoploide
           mixoploide m
       d   mixoploid
           Mixoploide f

1818   MIXOPLOIDY
       f   mixoploïdie f
       e   mixoploidía f
       i   mixoploidismo m
       d   Mixoploidie f

1819   MOCK DOMINANCE
       f   pseudo-dominance f
       e   (p)seudodominancia f
       i   pseudodominanza f
       d   mock dominance f;
           Pseudodominanz f

1820   MODAL NUMBER
       f   nombre m modal
       e   número m modal
       i   numero m modale
       d   Moduszahl f (#)

1821   MODIFICATION
       f   modification f

e modificación *f*
i modificazione *f*
d Modifikation *f*

MODIFICATION, ADAPTIVE
see 33

1822 MODIFIER;
MODIFYING FACTOR
f modificateur *m*
e modificador *m*
i modificatore *m*
d Modifikationsgen *n*

MODIFIER, CROSSING OVER
see 642

1823 MODIFIER COMPLEX
f complexe *m* modificateur
e complejo *m* modificador
i complesso *m* modificatore
d modifier complex *m*;
Modifikationsgenkomplex *m*

1824 MODIFYING FACTOR;
MODIFIER
f modificateur *m*
e modificador *m*
i modificatore *m*
d Modifikationsgen *n*

1825 MODULATION
f modulation *f*
e modulación *f*
i modulazione *f*
d Modulation *f*

1826 MODULATOR
f modulateur *m*
e modulador *m*
i modulatore *m*
d Modulator *m*

1827 MOLECULAR
f moléculaire
e molecular
i molecolare
d molekular;
Molekular- (in Zusammen-
setzungen)

1828 MOLECULAR SPIRAL
f spirale *f* moléculaire
e espiral *f* molecular
i spirale *f* molecolare
d Molekularspirale *f*

1829 MONAD
f monade *f*
e mónade *f*
i monade *f*
d Monade *f*

1830 MONASTER
f monaster *m*
e monáster *m*
i monaster *m*
d Monaster *n*

1831 MONID
f monide *f*
e mónida *f*
i - -
d Monide *f*

1832 MONOBRACHIAL
f monobrachial
e monobraquial
i monobrachiale
d monobrachial;
einarmig;
einschenkelig

1833 MONOCENTRIC
f monocentrique
e monocéntrico
i monocentrico
d monozentrisch

1834 MONOCHROMOSOME
f monochromosome *m*
e monocromosoma *m*
i monocromosoma *m*
d Monochromosom *n*

1835 MONOECIOUS
f monoïque
e monoico
i monoico
d monoecisch;
monoezisch;
monözisch

1836   MONOECY
       f monoécie *f*
       e monoecia *f*
       i monoicismo *m*
       d Monoecie *f*;
          Monoezie *f*;
          Monözie *f*

1837   MONOFACTORIAL
       f monofactoriel
       e monofactorial
       i monofattoriale
       d unifaktoriell;
          monofaktoriell (!)

1838   MONOGENE
       f monogène *m*
       e monogén *m*
       i monogene *m*
       d Monogen *n*

1839   MONOGENESIS
       f monogénèse *f*
       e monogénesis *f*
       i monogenesi *f*
       d Monogenese *f*

1840   MONOGENIC
       f monogénique
       e monogénico
       i monogenico
       d monogen

1841   MONOGENOMATIC;
       MONOGENOMIC
       f monogénomique
       e monogenómico
       i monogenomico
       d monogenomatisch

1842   MONOGENOMIC;
       MONOGENOMATIC
       f monogénomique
       e monogenómico
       i monogenomico
       d monogenomatisch

1843   MONOGENY
       f monogénie *f*
       e monogenia *f*
       i monogenia *f*
       d Monogenie *f*

1844   MONOGONY
       f monogonie *f*
       e monogonia *f*
       i monogonia *f*
       d Monogonie *f*

1845   MONOHAPLOID (adj.)
       MONOHAPLOID
       f monohaploïde
          monohaploïde *m*
       e monohaploide
          monohaploide *m*
       i monoaploide
          monoaploide *m*
       d monohaploid
          Monohaploide *f*

1846   MONOHAPLOIDY
       f monohaploïdie *f*
       e monohaploidïa *f*
       i monoaploidismo *m*
       d Monohaploidie *f*

1847   MONOHYBRID (adj.)
       MONOHYBRID
       f monohybride
          monohybride *m*
       e monohíbrido
          monohíbrido *m*
       i monoibrido
          monoibrido *m*
       d monohybrid
          Monohybride *f*

1848   MONOHYBRIDISM
       f monohybridisme *m*
       e monohibridismo *m*
       i monoibridismo *m*
       d Monohybridie *f*

1849   MONOLEPSIS
       f monolepsie *f*
       e monolepsis *f*
       i monolepsi *f*
       d Monolepsis *f*

1850   MONOMERICAL
       f monomérique
       e monomérico
       i monomerico
       d monomer

1851   MONOMERY
        f   monomérie *f*
        e   monomería *f*
        i   monomeria *f*
        d   Monomerie *f*

1852   MONOPHYLETIC
        f   monophylétique
        e   monofilético
        i   monofiletico
        d   monophyletisch

1853   MONOPLOID (adj.)
        MONOPLOID
        f   monoploïde
            monoploïde *m*
        e   monoploide
            monoploide *m*
        i   monoploide
            monoploide *m*
        d   monoploid
            Monoploide *f*

1854   MONOPLOIDY
        f   monoploïdie *f*
        e   monoploidía *f*
        i   monoploidismo *m*
        d   Monoploidie *f*

1855   MONOPLONT
        f   monoplonte *m*
        e   monoplonte *m*
        i   monoplonte *m*
        d   Monoplont *m*;
            Haplont *m*

1856   MONO-REPRODUCTION
        f   mono-reproduction *f*
        e   mono-reproducción *f*
        i   mono-riproduzione *f*
        d   --

1857   MONOSOME
        f   monosome *m*
        e   monosoma *m*
        i   monosoma *m*
        d   Monosom *n*

1858   MONOSOMIC
        f   monosomique
        e   monosómico

        i   monosomico
        d   monosom

1859   MONOTHALLIC
        f   monothallique
        e   monotálico
        i   monotallico
        d   monothallisch

1860   MONOTOPY
        f   monotopie *m*
        e   monotopismo *m*
        i   monotopismo *m*
        d   Monotopie *f* (#)

1861   MONOZYGOTIC
        f   monozygotique;
            monozygote
        e   monocigótico
        i   monozigotico
        d   monozygotisch;
            monozygot

1862   MONOZYGOTIC TWINS;
        IDENTICAL TWINS;
        UNIOVULAR TWINS
        f   jumeaux *mpl* uniovulaires;
            jumeaux *mpl* univitellins;
            jumeaux *mpl* monozygotes
        e   gemellos *mpl* idénticos;
            gemellos *mpl* uniovulares;
            gemellos *mpl* univitelinos
        i   gemelli *mpl* identici;
            gemelli *mpl* monozigotici;
            gemelli *mpl* uniovulari
        d   eineiige Zwillinge *mpl*

1863   MORGAN
        f   morgan *m*
        e   morgan *m*
        i   morgan *m*
        d   Morgan-Einheit *f*

1864   MORPH
        f   morphe *m*
        e   morfo *m*
        i   morfo *m*
        d   morph *n*

1865   MORPHISM
        f   morphisme *m*

e morfismo *m*
i morfismo *m*
d Morphismus *m*

1866 MORPHOGENESIS
f morphogénèse *f*
e morfogénesis *f*
i morfogenesi *f*
d Morphogenese *f*

1867 MORPHOGENETIC
f morphogénétique
e morfogenético;
   morfogénico
i morfogenetico
d morphogenetisch

1868 MORPHOSE
f morphose *f*
e morfosis *f*
i morfosi *f*
d Morphose *f*

1869 MOSAIC
f mosaïque *f*
e mosaico *m*
i mosaico *m*
d Mosaikform *f*;
   Mosaik *n*

1870 MOTTLED
f panaché
e variegado
i variegato
d mottled;
   panaschiert;
   variegat

1871 MOTTLING
f panachure *f*
e variegación *f*
i variegazione *f*
d mottling *n*

1872 MOVEMENT INDEX
f indice *n* du mouvement
   chiasmatique
e índice *m* del movimiento
   quiasmático
i indice *m* del movimento
   chiasmatico
d Chiasmabewegungsindex *m*

1873 MULTI-BREAK REAR-
      RANGEMENT
f remaniement *m* dans le cas
   de ruptures multiples
e reacomodo *m* de rupturas
   múltiples
i ricomposizione *f* di rotture
   multiple
d Multibruch-Rearrangement *n*

1874 MULTIGENE
f multigène *m*
e multigén *m*
i multigene *m*
d Multigen *n*

1875 MULTIPLE
f multiple
e múltiple
i multiplo
d mehrfach;
   Mehrfach- (in Zusammensetzungen);
   multipel

1876 MULTIPLE ALLELES
f allèles *mpl* multiples
e alelos *mpl* múltiples
i multipli alleli *mpl*
d multiple Allele *npl*

1877 MULTIPLE CENTROMERE
f centromère *m* multiple
e centrómero *m* múltiple
i centromero *m* molteplice
d multiples Centromer *n*

1878 MULTIPLE CHIASMA
f chiasma *m* multiple
e quiasma *m* múltiple
i chiasma *m* multiplo
d multiples Chiasma *n*;
   Mehrfachchiasma *n*

1879 MULTIPLE CORRELATION
f corrélation *f* multiple
e correlación *f* múltiple
i correlazione *f* molteplice
d multiple Korrelation *f*;
   Mehrfachkorrelation *f*

1880   MULTIPLE DRUG RE-
       SISTANCE TRANSFER
       f   transfert *m* de la résis-
           tance multiple
       e   transferencia *f* de
           resistencia a multidroga (O)
       i   trasferimento *m* della
           resistenza multipla
       d   - -

1881   MULTIPLE-POINT TEST
       CROSS
       f   test-cross *m* à points
           multiples
       e   cruza *f* probadora de
           puntos múltiples
       i   incrocio test *m* ;
           incrocio *m* di controllo
       d   n-Punkt-Versuch *m*

1882   MULTIPLE RE-
       ACTIVATION (MR)
       f   réactivation *f* multiple (RM)
       e   reactivación *f* múltiple
           (RM)
       i   riattivazione *f* molteplice
           (RM)
       d   Mehrfachreaktivierung *f*

1883   MULTIPLE REGRESSION
       f   régression *f* multiple
       e   regresión *f* múltiple
       i   regressione *f* molteplice
       d   multiple Regression *f*;
           Mehrfachregression *f*

1884   MULTIPLE SEX CHROMO-
       SOMES
       f   chromosomes *mpl* sexuels
           multiples
       e   cromosomas *mpl* sexuales
           múltiples
       i   cromosomi *mpl* sessuali
           molteplici
       d   multiple Geschlechtschro-
           mosomen *npl*

1885   MULTIVALENT
       f   multivalent
       e   multivalente
       i   multivalente
       d   Multivalent *n*

1886   MUTABLE
       f   mutable
       e   mutable
       i   mutabile
       d   mutabel

1887   MUTAFACIENT
       f   mutateur
       e   mutafaciente
       i   mutafaciente
       d   mutafacient

1888   MUTAGENE
       f   mutagène
       e   mutagén
       i   mutageno
       d   Mutagen *n*

1889   MUTAGENESIS
       f   mutagénèse *f*
       e   mutagénesis *f*
       i   mutagenesi *f*
       d   Mutagenese *f*

1890   MUTAGENE SPECIFICITY
       f   spécifité *f* mutagénique
       e   especificidad *f* mutagénica
       i   specificità *f* mutagena
       d   Mutagenspezifität *f*

1891   MUTAGENE STABILITY
       f   stabilité *f* mutagénique
       e   estabilidad *f* mutagénica
       i   stabilità *f* mutagena
       d   Mutagenstabilität *f*

1892   MUTAGENIC
       f   mutagénique
       e   mutagénico
       i   mutagenico
       d   mutagen

1893   MUTANT
       f   mutant *m*
       e   mutante *m*
       i   mutante *m*
       d   Mutante *f*

       MUTANT, IRREPARABLE
       see 1558

MUTANT, REPARABLE
see 2446

MUTANT, SUPPRESSIBLE
see 2721

MUTANT, TEMPERATURE
see 2782

1894 MUTATION
f mutation *f*
e mutación *f*
i mutazione *f*
d Mutation *f*

MUTATION, CYTOPLASMIC
see 697

MUTATION, GAMETIC
see 1028

MUTATION, ITERATIVE
see 1605

MUTATION, KEY
see 1634

MUTATION, LOSS
see 1686

MUTATION, MAJOR
see 1703

MUTATION, MASS
see 1714

MUTATION, PLASTID
see 2165

MUTATION, REVERSE
see 2476

MUTATION, SIMULTANEOUS
see 2599

MUTATION, SOMATIC
see 2612

MUTATION, SYSTEMIC
see 2748

MUTATION, ZERO POINT
see 2941

1895 MUTATIONAL EQUILIBRIUM
f équilibre *m* mutationnel;
    équilibre *m* réalisé par les
    mutations
e equilibrio *m* mutacional
i equilibrio *m* mutazionale
d Mutationsgleichgewicht *n*

1896 MUTATIONAL LAG;
MUTATION DELAY
f délai *m* de mutacion
e retardo *m* de mutación
i ritardo *m* di mutazione
d Mutationsverzögerung *f*

1897 MUTATION COEFFICIENT
f coefficient *m* de mutation
e coeficiente *m* de mutación
i coefficiente *m* di mutazione
d Mutationskoeffizient *m*

1898 MUTATION DELAY;
MUTATIONAL LAG
f délai de mutation *m*
e retardo de mutación *m*
i ritardo di mutazione *m*
d Mutationsverzögerung *f*

1899 MUTATION PRESSURE
f pression *f* de mutation
e presión *f* de mutación
i pressione *f* di mutazione
d Mutationsdruck *m*

1900 MUTATION RATE
f taux *m* de mutation
e frecuencia *f* de las muta-
    ciones
i frequenza *f* delle muta-
    zioni
d Mutationsrate *f*

1901 MUTATOR GENE
f gène-mutateur *m*
e gen *m* mutador
i gene *m* mutatore
d Mutatorgen *n*

1902   MUTATOR SUBSTANCE
 f substance *f* mutatrice
 e substancia *f* mutadora
 i sostanza *f* mutatrice
 d Mutatorsubstanz *f*

1903   MUTON
 f muton *m*
 e mutón *m*

 i muton *m*
 d Muton *n*

1904   MUTUAL
 f mutuel
 e mutuo
 i mutuo
 d gegenseitig

N

1905 NAKED KINETOCHORE
f cinétochore *m* nu
e cinetócoro *m* desnudo
i - -
d chromatinfreie Kinetochor-
   region *f*

1906 NATURAL SELECTION
f sélection *f* naturelle
e selección *f* natural
i selezione *f* naturale
d natürliche Selektion *f*

1907 "NECKTIE" ASSOCIATION
f association *f* en noeud de
   cravate
e asociación de "corbata"
i associazione *f* di cravatta
d - -

1908 NEMAMERE
f némamère *m*
e nemámero *m*
i nemamero *m*
d Nemamer *n*

1909 NEOCENTROMERE
f néocentromère *m*
e neocentrómero *m*
i neocentromero *m*
d Neocentromer *n*

1910 NEO-DARWINISM
f néo-darwinisme *m*
e neodarwinismo *m*
i neodarwinismo *m*
d Neo-Darwinismus *m*

1911 NEOMORPH (adj.)
   NEOMORPH
f néomorphe
   néomorphe *m*
e neomorfo
   neomorfo *m*
i neomorfo
   neomorfo *m*
d neomorph

1912 NEOTEINIA;
   NEOTENY
f néoténie *f*
e neotenia *f*
i neotenia *f*
d Neotenie *f*

1913 NEO TWO PLANE THEORY
f nouvelle théorie *f* des
   deux plans
e neoteoría *f* de dos planos
i neoteoria *f* dei due piani
d neo two plane theory *f*

1914 NEW REUNION
f nouveau recollement *m*;
   nouvelle réunion *f*
e nueva reunión *f*
i nuova riunione *f*;
   nuovo riattacco *m*
d (neue) Reunion *f*

1915 NEW SPIRAL PROPHASE
f prophase *f* "nouvelles
   spirales"
e profase *f* nueva espiral
i profase *f* nuova spirale
d new spiral prophase *f*

1916 NICKING
f entaille *f*
e - -
i - -
d nicking *n*

1917 NLG CHROMOSOME
f chromosome *m* nucléogène;
   nlg chromosome *m*
e nlg cromosoma *m*
i cromosoma *m* nucleogeno;
   nlg cromosoma
d Nlg-Chromosom *n*

1918 NOMBRE FONDAMENTAL
f nombre *m* fondamental
e número *m* fundamental

i	numero *m* fondamentale
d	nombre fondamental

1919	NON-DISJUNCTION (CHRO-
	MATIDS .... CHROMOSOMES
	....)
	f	non-disjonction *f* (des
		chromosomes....
		des chromatides....)
	e	no disyunción *f*
	i	mancata *f* disgiunzione
		(dei cromatidi o dei
		cromosomi)
	d	Non-Disjunction *f*
		(Chromatiden-....
		Chromosomen-....)

1920	NON HOMOLOGOUS PAIRING
	f	appariement *m* non homo-
		logue
	e	apareamiento *m* no homó-
		logo
	i	appaiamento *m* non omo-
		logo
	d	nichthomologe Paarung *f*

1921	NON-LINEAR REGRESSION
	f	régression *f* non-linéaire
	e	regresión *f* no lineal
	i	regressione *f* non lineare
	d	nichtlineare Regression *f*

1922	NON-MENDELIAN IN-
	HERITANCE
	f	hérédité *f* non-mendélienne
	e	herencia *f* no mendeliana
	i	eredità *f* non mendeliana
	d	nichtmendelnde Vererbung *f*

1923	NON-PAIRING
	f	non-appariement *m*
	e	no apareamiento *m*
	i	non-appaiamento *m*
	d	Nichtpaarung *f*

1924	NON PARENTAL DITYPE
	TETRAD (N.P.D.)
	f	tétrade *f* ditype recombinée;
		tétrade *f* ditype non-
		parentale
	e	- -

i	- -
d	non parental ditype tetrad

1925	NON REDUCTION
	f	non-réduction *f*
	e	no reducción *f*
	i	non-riduzione *f*
	d	Non-reduction *f*;
		Nichtreduktion *f*

1926	NON-SISTER CHROMATIDS
	f	chromatides *f pl* non-
		soeurs
	e	cromatidas *f pl* no
		hermanas
	i	cromatidi *m pl* non
		fratelli
	d	Nicht-Schwesterchroma-
		tiden *f pl*

1927	NOTOMORPH (adj.)
	NOTOMORPH
	f	notomorphe
		notomorphe *m*
	e	notomorfo
		notomorfo *m*
	i	notomorfo
		notomorfo *m*
	d	Notomorphe *f*

1928	NUCLEAR ASSOCIATION
	f	association *f* nucléaire
	e	asociación *f* nuclear
	i	associazione *f* nucleare
	d	Kernassoziation *f*

1929	NUCLEAR CAP
	f	capsule *f* nucléaire
	e	casquete *f* nuclear
	i	capsula *f* nucleare (Θ)
	d	Kernkappe *f*

1930	NUCLEAR DIMORPHISM
	f	dimorphisme *m* nucléaire
	e	dimorfismo *m* nuclear
	i	dimorfismo *m* nucleare
	d	Kerndimorphismus *m*

1931	NUCLEAR DISRUPTION
	f	dislocation *f* nucléaire
	e	desgarradura *f* nuclear

i   dislocamento *m* nucleare
d   nuclear disruption *f*
    Kernbruch *m*

1932   NUCLEAR DIVISION
f   division *f* nucléaire
e   división *f* nuclear
i   divisione *f* nucleare
d   Kernteilung *f*

1933   NUCLEAR FRAGMENTATION
f   fragmentation *f* nucléaire
e   fragmentación *f* nuclear
i   frammentazione *f* nucleare
d   Kernfragmentation *f*

1934   NUCLEAR MEMBRANE
f   membrane *f* nucléaire
e   membrana *f* nuclear
i   membrana *f* nucleare
d   Kernmembran *f*

1935   NUCLEAR PHENOTYPE
f   phénotype *m* nucléaire
e   fenótipo *m* nuclear
i   fenotipo *m* nucleare
d   Kernphänotyp *m*

1936   NUCLEAR SAP
f   suc *m* nucléaire
e   jugo *m* nuclear
i   succo *m* nucleare
d   Kernsaft *m;*
    Karyolymphe *f*

1937   NUCLEAR SEGREGATION
f   ségrégation *f* nucléaire
e   segregación *f* nuclear
i   segregazione *f* nucleare
d   Kernsegregation *f*

NUCLEI, COMPLEMENTARY
see 618

NUCLEI, KINETIC
see 1639

1938   NUCLEIC ACID STARVATION
f   inhibition *f* de la synthèse
    d'acides nucléiques
e   enrarecimiento *m* del ácido
    nucléico

i   carenza *f* d'acido nucleico
d   Hypochromatizität *f*

1939   NUCLEIN
f   nucléine *f*
e   nucleína *f*
i   nucleina *f*
d   Nuklein *n*

1940   NUCLEINATION
f   nucléisation *f*
e   nucleinación *f*
i   nucleinazione *f*
d   Nukleinisierung *f*

1941   NUCLEOCENTROSOME
f   nucléocentrosome *m*
e   nucleocentrosoma *m*
i   nucleocentrosoma *m*
d   Nukleocentrosom *n*

1942   NUCLEOGENIC REGION
f   région *f* nucléogénique
e   región *f* nucleogénica
i   regione *f* nucleogenica
d   nukleogene Region *f;*
    Nukleolareinschnürung *f*

1943   NUCLEOID
f   nucléoïde *m*
e   nucleoide *m*
i   nucleoide *m*
d   Nukleoid *n*

1944   NUCLEOLAR
f   nucléolaire
e   nucleolar
i   nucleolare
d   Nukleolus-;
    Nukleolar- (in Zusammen-
    setzungen)

1945   NUCLEOLAR ASSOCIATED
       CHROMATIN
f   chromatine *f* nucléolaire
    associée
e   cromatina *f* nucleolar
    asociada
i   cromatina *f* nucleolare
    associata
d   nucleolar associated chro-
    matin *n*

1946   NUCLEOLAR CONSTRICTION
        f   constriction f nucléolaire
        e   constricción f nucleolar
        i   costrizione f nucleolare
        d   Nukleolareinschnürung f

1947   NUCLEOLAR FRAGMEN-
        TATION
        f   fragmentation f nucléolaire
        e   fragmentación f nucleolar
        i   frammentazione f nucleo-
            lare
        d   Nukleolusfragmentation f

1948   NUCLEOLAR ORGANIZER;
        NUCLEOLUS ORGANISER
        f   organisateur m nucléolaire
        e   organizador m nucleolar
        i   organizzatore m nucleolare
        d   Nukleolus-Organisator m

1949   NUCLEOLAR TRACK
        f   tractus m nucléolaire
        e   tracto m nucleolar
        i   tratto m nucleolare
        d   Nukleolusschlauch m

1950   NUCLEOLAR ZONE
        f   zone f nucléolaire
        e   zona f nucleolar
        i   zona f nucleolare
        d   Nukleolarzone f

1951   NUCLEOLONEMA
        f   nucléolonème m
        e   nucleolonema m
        i   nucleolonema m
        d   Nukleolonema n

1952   NUCLEOLUS
        f   nucléole m
        e   nucléolo m
        i   nucleolo m
        d   Nukleolus m;
            Nucleolus m

        NUCLEOLUS, SATELLITE
        see 2498

1953   NUCLEOLUS ORGANISER;
        NUCLEOLAR ORGANIZER

        f   organisateur m nucléolaire
        e   organizador m nucleolar
        i   organizzatore m nucleolare
        d   Nukleolus-Organisator m

1954   NUCLEOLUS ORGANIZER
        REGION (N.O.R.)
        f   région f de l'organisateur
            nucléolaire (R.O.N.)
        e   región f del organizador
            nucleolar (R.O.N.)
        i   regione f dell'organizzatore
            nucleolare (R.O.N.)
        d   Nukleolus-Organisator-Re-
            gion (N.O.R.) f

1955   NUCLEOLUS SICKLE-STAGE
        f   paranucléole m
        e   paranucléolo m
        i   paranucleolo m
        d   Paranukleolus m

1956   NUCLEOME
        f   nucléome m
        e   nucleoma m
        i   nucleoma m
        d   Nukleom n

1957   NUCLEOMIXIS
        f   nucléomixie f
        e   nucleomixis f
        i   nucleomixi f;
            nucleomissi f
        d   Nukleomixis f

1958   NUCLEOPLASM
        f   nucléoplasme m
        e   nucleoplasma m
        i   nucleoplasma m
        d   Nukleoplasma n;
            Kernplasma n

1959   NUCLEOPLASMATIC RATIO;
        NUCLEOPLASMIC RATIO
        f   rapport m nucléo-plasma-
            tique
        e   relación f nucleoplasmática
            relación f nucleoplásmica
        i   rapporto m nucleo-plasma-
            tico
        d   Kern-Plasma-Relation f

1960   NUCLEOPROTEIN
        f   nucléoprotéine *f*
        e   nucleoproteína *f*
        i   nucleoproteina *f*
        d   Nukleoprotein *n*

1961   NUCLEOSOME
        f   nucléosome *m*
        e   nucleosoma *m*
        i   nucleosoma *m*
        d   Nukleosom *n*

1962   NUCLEOTID
        f   nucléotide
        e   nucleótida *f*
        i   nucleotide *m*
        d   Nukleotid *n*

1963   NUCLEUS
        f   noyau *m* ;
            nucléus *m*
        e   núcleo *m*
        i   nucleo *m*
        d   Kern *m* ;
            Zellkern *m*;
            Nukleus *m*;
            Nucleus *m*

        NUCLEUS, DEFINITIVE
        see 718

        NUCLEUS, ENERGIC
        see 925

        NUCLEUS, FUSION
        see 1022

        NUCLEUS, GENERATIVE
        see 1082

        NUCLEUS, METABOLIC

        see 1766

        NUCLEUS, RESTING
        see 2464

        NUCLEUS, TROPHIC
        see 2867

1964   NULLIPLEX
        f   nulliplex
        e   nuliplexo
        i   nulliplex
        d   nulliplex

1965   NULLISOMIC
        f   nullisomique
        e   nulisómico
        i   nullisomico
        d   nullisom

1966   NULLISOMY;
        NULLOSOMY
        f   nullisomie *f*
        e   nulisomia *f*
        i   nullisomia *f*
        d   Nullisomie *f*;
            Nullosomie *f*

1967   NUMERICAL
        f   numérique
        e   numérico
        i   numerico
        d   numerisch

1968   NUMERICAL HYBRID
        f   hybride *m* numérique
        e   híbrido *m* numérico
        i   ibrido *m* numerico
        d   numerische Hybride *f*;
            Chromosomenzahlen-
            bastard *m*

O

1969   OFFSPRING
        f descendant *m* ;
           rejeton *m*
        e descendiente *m*
        i discendente *m*
        d Nachkommenschaft *f*

1970   OLD SPIRAL PROPHASE
        f prophase *f* "vieilles spi-
           rales"
        e - -
        i spirale *f* profasica
        d old spiral prophase *f*

1971   OLIGOGENE ;
        MAJOR GENE
        f oligogène *m*;
           gène *m* majeur
        e oligogén *m*;
           gen *m* mayor
        i gene *m* maggiore ;
           oligogene *m*
        d Oligogen *n* ;
           Hauptgen *n*

1972   OLIGOPYRENE
        f oligopyrène
        e oligopirene
        i oligopirene
        d oligopyren

1973   OLISTHETEROZONE
        f olisthétérozone *f*
        e olistheterozona *f*
        i olisteterozona *f*
        d Olistheterozone *f*

1974   ONE BAND ONE GENE
        HYPOTHESIS
        f hypothèse *f* d'une bande
           pour un gène
        e hipótesis *f* de una banda
           por un gene
        i ipotesi *f* di ogni striscia
           per ogni gene
        d Eine-Querscheibe-ein-Gen-
           Hypothese *f*

1975   ONE-BAND TANDEM REPEA
        f répétition *f* d'une bande
           double
        e repetición *f* de una banda
           dobla
        i ripetizione *f* unilaterale
           abbinante
        d One band tandem repeat;
           Duplikation *f* einer Quers

1976   ONE-DIVISION MEIOSIS
        f méiose *f* en une seule
           division
        e meiosis *f* en una división
        i meiosi *f* in una divisione
        d Ein-Schritt-Meiose *f*

1977   ONE-GENE ONE ENZYME
        HYPOTHESIS
        f hypothèse *f* d'un gène pou
           un enzyme
        e hipótesis *f* de un gene por
           una enzima
        i ipotese *f* d'un gene per
           ciascun ensima
        d Ein-Gen-ein-Ferment-
           Hypothese *f*

1978   ONTOGENESIS ;
        ONTOGENY
        f ontogénie *f*
        e ontogenia *f*
        i ontogenesi *f*
        d Ontogenie *f*;
           Ontogenese *f*

1979   ONTOGENETIC ;
        ONTOGENIC
        f ontogénétique;
           ontogénique
        e ontogenético;
           ontogénico
        i ontogenetico ;
           ontogenico
        d ontogenetisch

1980   ONTOGENY;
        ONTOGENESIS
        f   ontogénie *f*
        e   ontogenia *f*
        i   ontogenesi *f*
        d   Ontogenie *f*;
            Ontogenese *f*

1981   OOCENTRE;
        OVOCENTER
        f   ovocentre *m*
        e   ovocentro *m*
        i   oocentro *m*
        d   Oozentrum *n*

1982   OOCYTE;
        OÖCYTE;
        OVOCYTE
        f   oocyte *m*;
            ovocyte *m*
        e   oocito *m*;
            ovocito *m*
        i   oocito *m*;
            ovocito *m*
        d   Oocyte *f*;
            Oozyte *f*

1983   OOGENESIS;
        OÖGENESIS;
        OVOGENESIS
        f   ovogénèse *f*;
            ovogénie *f*
        e   oogénesis *f*;
            ovogénesis *f*
        i   oogenesi *f*;
            ovogenesi *f*
        d   Oogenese *f*;
            Eizellenbildung *f*

1984   OOKINESIS
        f   oocinèse *f*
        e   oocinesis *f*
        i   oocinesi *f*
        d   Ookinesis *f*

1985   OOSOME
        f   oosome *m*
        e   oosoma *m*
        i   oosoma *m*
        d   Oosom *n*

1986   OOSPHERE
        f   oosphère *f*

        e   oosfera *f*
        i   oosfera *f*
        d   Oosphäre *f*

1987   OOTID
        f   ovotide *f*
        e   oótida *f*
        i   ootidio *m*
        d   Ootide *f*

1988   OPERON
        f   opéron *m*
        e   operón *m*
        i   operon *m*
        d   Operon *n*

1989   OPPOSITION FACTOR
        f   facteur *m* d'opposition
        e   factor *m* de oposición
        i   fattore *m* d'opposizione
        d   Oppositionsfaktor *m*

1990   ORGANISER
        f   organisateur *m*
        e   organizador *m*
        i   organizzatore *m*
        d   Organisator *m*

1991   ORGANIZATION CENTRE
        f   centre *m* d'organisation
        e   centro *m* de organización
        i   centro *m* d'organizzazione
        d   Organisationszentrum *n*

1992   ORGANIZATION EFFECT
        f   effet *m* d'organisation
        e   efecto *m* de organización
        i   effetto *m* d'organizzazione
        d   Organisationseffekt *m*

1993   ORIENTATION
        f   orientation *f*
        e   orientación *f*
        i   orientazione *f*
        d   Orientierung *f*

1994   ORTHOAMITOSIS
        f   orthoamitose *f*
        e   ortoamitosis *f*
        i   orto-amitosi *f*
        d   Orthoamitose *f*

1995   ORTHOGENESIS
    f   orthogénèse *f*
    e   ortogénesis *f*
    i   ortogenesi *f*
    d   Orthogenese *f*

1996   ORTHOKINESIS
    f   orthocinèse *f*
    e   ortocinesis
    i   ortocinesi *f*
    d   Orthokinese *f*

1997   ORTHOPLOID (adj.)
    ORTHOPLOID
    f   orthoploïde
       orthoploïde *m*
    e   ortoploide
       ortoploide *m*
    i   ortoploide
       ortoploide *m*
    d   orthoploid
       Orthoploide *f*

1998   ORTHOPLOIDY
    f   orthoploïdie *f*
    e   ortoploidía *f*
    i   ortoploidismo *m*
    d   Orthoploidie *f*

1999   ORTHOSELECTION
    f   orthosélection *f*
    e   ortoselección *f*
    i   ortoselezione *f*
    d   Orthoselektion *f*

2000   ORTHOSPIRAL
    f   orthospiral
    e   ortoespiral
    i   ortospirale
    d   Orthospirale *f*

2001   ORTHOTACTIC
    f   orthotactique
    e   ortotáctico
    i   ortotattico
    d   orthotaktisch (#)

2002   ORTHOTELOMITIC
    f   orthotélomitique
    e   ortotelomítico
    i   ortotelomitico
    d   orthotelomitisch

2003   OUTBREEDING;
    OUTCROSSING
    f   outbreeding *m*
    e   outbreeding *m*
    i   outbreeding *m*;
       esincrocio *m*
    d   Fremdzucht *f*;
       Outbreeding *n*

2004   OUTCROSS
    f   outcross *m*
    e   outcross *m*
    i   outcross *m*
    d   Outcross *n*

2005   OVERDOMINANCE
    f   overdominance *f*;
       superdominance *f*
    e   sobredominancia *f*
    i   superdominanza *f*;
       dominanza *f* esteriore
    d   Überdominanz *f*;
       Superdominanz *f*

2006   OVERNUCLEATED REGION
    f   zone *f* d'hétéropycnose
       positive
    e   zona *f* de heteropicnosis
       positiva
    i   zona *f* d'eteropicnosi
       positiva
    d   positiv heteropyknotischer
       Abschnitt *m*

2007   OVULE
    f   ovule *m*
    e   óvulo *m*
    i   ovulo *m*
    d   Samenanlage *f*

2008   OVUM
    f   oeuf *m*
    e   huevo *m*
    i   uovo *m*
    d   Ovum *n* ;
       Ei *n*

2009   OXYCHROMATIN
    f   oxychromatine *f*
    e   oxicromatina *f*
    i   ossicromatina *f*
    d   Oxychromatin *n*

P

2010   PACHYNEMA
    f   pachynème *m*
    e   paquinema *m*
    i   pachinema *m*
    d   Pachynema *n*

2011   PACHYTENE (adj.)
    PACHYTENE
    f   pachytène
       pachytène *f*
    e   paquiteno
       paquiteno *m*
    i   pachitene
       pachitene *f*
    d   Pachytän *n*

2012   PACKING FACTOR
    f   facteur *m* d'empaquetage
    e   factor *m* de empaque
    i   fattore *m*
       d'impacchettamento
    d   Packungsfaktor *m*

2013   PAEDOGAMY
    f   paedogamie *f*
    e   paidogamia *f*
    i   paedogamia *f*
    d   Paedogamie *f*

2014   PAEDOGENESIS
    f   paedogénèse *f*
    e   paidogénesis *f*
    i   paedogenesi *f*
    d   Paedogenese *f*

2015   PAIR ALLELES
    f   allèles *mpl* couplés;
       pseudo-allèles *mpl*
    e   pares *mpl* alelomórficos
    i   alleli *mpl* appaiati
    d   Pseudoallele *npl*

2016   PAIRING
    f   appariement *m*
    e   apareamiento *m*
    i   appaiamento *m*
    d   Paarung *f*

PAIRING, ECTOPIC
see 902

PAIRING, FALSE
see 985

PAIRING, NON HOMOL-
OGOUS
see 1920

PAIRING, TOUCH AND GO
see 2818

2017   PAIRING BLOCK
    f   bloc *m* d'appariement
    e   bloque *m* aparente
    i   blocco *m* d'appaiamento
    d   Paarungsblock *m*

2018   PAIRING COEFFICIENT
    f   coefficient *m* d'appariement
    e   coeficiente *m* de apareamiento
    i   coefficiente *m* d'appaiamento
    d   Paarungskoeffizient *m*

2019   PAIRING HETEROSIS
    f   hétérosis *f* d'appariement
    e   heterosis *f* aparente
    i   eterosi *f* d'appaiamento
    d   Paarungsheterosis *f*

2020   PAIRING SEGMENT
    f   segment *m* d'appariement
    e   segmento *m* apareante
    i   segmento *m* appaiante
    d   Paarungssegment *n*

2021   PAIRING TYPE
    f   type d'appariement *m*
    e   tipo *m* de apareamiento
    i   tipo *m* d'appaiamento
    d   Paarungstyp *m*

2022   PAIR MATING
    f   croisement *m* de paires
    e   apareamiento *m* de pares

i   appaiamento *m* di paia
d   pair mating *n*

2023   PAIR OF GENES
f   couple *m* de gènes
e   par *m* de genes
i   coppia *f* di geni
d   Genpaar *n*

2024   PALINGENESIS;
PALINGENY
f   palingénèse *f*
e   palingenesia *f*;
palingenia *f*
i   palingenesia *f*;
palingenia *f*
d   Palingenese *f*

2025   PANALLELE
f   panallèle *m*
e   panalelo *m*
i   panallelo *m*
d   Panallel *n*

2026   PANALLOPLOID (adj.)
PANALLOPLOID
f   panalloploïde
panalloploïde *m*
e   panaloploide
panaloploide *m*
i   panalloploide
panalloploide *m*
d   pan-alloploid
Pan-Alloploide *f*

2027   PANALLOPLOIDY
f   panalloploïdie *f*
e   panaloploidia *f*
i   panalloploidismo *m*
d   Pan-Alloploidie *f*

2028   PANAUTOPLOID (adj.)
PANAUTOPLOID
f   panautoploïde
panautoploïde *m*
e   panautoploide
panautoploide *m*
i   panautoploide
panautoploide *m*
d   pan-autoploid
Pan-Autoploide *f*

2029   PANAUTOPLOIDY
f   panautoploïdie *f*
e   panautoploidia *f*
i   panautoploidismo *m*
d   Pan-Autoploidie *f*

2030   PANGEN;
PANGENE
f   pangène *m*
e   pángene *m*
i   pangene *m*
d   Pangen *n*

2031   PANGENESIS
f   pangénèse *f*
e   pangénesis *f*
i   pangenesi *f*
d   Pangenesis *f*

2032   PANGENOSOME
f   pangénosome *m*
e   pangenosoma *m*
i   pangenosoma *m*
d   Pangenosom *n*

2033   PANMICTIC
f   panmictique
e   panmíctico
i   panmittico
d   panmiktisch

2034   PANMIXIA;
PANMIXIE;
PANMIXIS;
RANDOM MATING
f   panmixie *f*
e   panmixis *f*
i   panmissi *f*
d   Panmixie *f*;
Panmixis *f*

2035   PARA-ALLELE
f   para-allèle *m*
e   para-alelo *m*
i   para-allelo *m*
d   Paraallel *n*

2036   PARACENTRIC
f   paracentrique
e   paracéntrico
i   paracentrico
d   parazentrisch

2037 PARACHROMATIN
    f parachromatine *f*
    e paracromatina *f*
    i paracromatina *f*
    d Parachromatin *n*

2038 PARACHUTE X Y -BIVALENT
    f bivalent *m* XY en forme de
       parachute
    e bivalente *m* XY en forma
       de paracaídas
    i bivalente *m* XY in forma
       di paracadute
    d XY-Fallschirm-
       Bivalent *n* (#)

2039 PARAGENEON
    f paragénéon *m*
    e parageneon *m*
    i parageneon *m*
    d Parageneon *n*

2040 PARAGENOPLAST
    f paragénoplaste *m*
    e paragenoplasto *m*
    i paragenoplasto *m*
    d Paragenoplast *m*

2041 PARALLEL DISJUNCTION;
     PARALLEL SEPARATION
    f disjonction *f* parallèle
    e disyunción *f* paralela
    i disgiunzione *f* parallela
    d Paralleldisjunktion *f*

2042 PARALLEL EVOLUTION
    f évolution *f* parallèle
    e evolución *f* paralela
    i evoluzione *f* parallela
    d Parallelevolution *f*

2043 PARALLEL MUTATION
    f mutation *f* parallèle
    e mutación *f* paralela
    i - -
    d Parallelmutation *f*

2044 PARALLEL SEPARATION
    f disjonction *f* parallèle
    e separación *f* paralela
    i disgiunzione *f* parallela
    d Paralleldisjunktion *f*

2045 PARALOCUS
    f paralocus *m*
    e paralocus *m*
    i paralocus *m*
    d Paralocus *m*

2046 PARAMECIN
    f paramécine *f*
    e paramecina *f*
    i paramecina *f*
    d Paramaecin *n*

2047 PARAMEIOSIS
    f paraméiose *f*
    e parameiosis *f*
    i parameiosi *f*
    d Parameiose *f*

2048 PARAMICTIC
    f paramictique
    e paramíctico
    i paramittico
    d paramiktisch

2049 PARAMITOSIS
    f paramitose *f*
    e paramitosis *f*
    i paramitosi *f*
    d Paramitose *f*

2050 PARAMIXIS
    f paramixie *f*
    e paramixis *f*
    i paramissi *f*
    d Paramixie *f*

2051 PARANEMIC
    f paranémique
    e paranémico
    i paranemico
    d paranematisch (#)

2052 PARANEMIC COIL
    f spirale *f* paranémique
    e espiral *f* paranémica
    i spirale *f* paranemica
    d paranematische
       Wicklung *f* (#)

2053 PARANUCLEIN
    f paranucléine *f*
    e paranucleína *f*

i paranucleina *f*
d Paranuclein *n*

2054 PARANUCLEOPLASM
f paranucléoplasme *m*
e paranucleoplasma *m*
i paranucleoplasma *m*
d Paranukleoplasma *n*

2055 PARASELECTIVITY
f parasélectivité *f*
e paraselectividad *f*
i paraselettività *f*
d Paraselektivität *f*

2056 PARASEXUAL
f parasexuel
e parasexual
i parasessuale
d parasexuell

2057 PARASYNAPSIS
f parasynapsis *f*
e parasinapsis *f*
i parasinapsi *f*
d Parasynapsis *f*

2058 PARASYNDESIS
f parasyndèse *f*
e parasíndesis *f*
i parasindesi *f*
d Parasyndese *f*

2059 PARATACTIC
f paratactique
e paratáctico
i paratattico
d parataktisch (#)

2060 PARATROPHIC
f paratrophe
e parátrofo
i paratrofo
d paratroph

2061 PARAVARIATION
f paravariation *f*
e paravariación *f*
i paravariazione *f*
d Paravariation *f*

2062 PARENT (adj.)
PARENT
f géniteur ;
parent
géniteur *m* ;
parent *m*
e genitor
genitor *m*
i genitore
genitore *m*
d elterlich
Elternteil *m* ;
Eltern- (in Zusammen-
setzungen)

2063 PARENTAL
f parent
e parental
i parentale
d elterlich

2064 PARENTAL GENERATION
f génération *f* des parents
e generación *f* paterna
i generazione *f* dei genitori
d Elterngeneration *f* ;
Parentalgeneration *f*

2065 PARENTS
f parents *mpl*
e parientes *mpl;*
genitores *mpl*
i genitori *mpl*
d Eltern *mpl*

2066 PARTHENAPOGAMY
f parthénapogamie *f*
e partenapogamia *f*
i partenapogamia *f*
d Parthenapogamie *f*

2067 PARTHENOCARPY
f parthénocarpie *f*
e partenocarpia *f*
i partenocarpia *f*
d Parthenokarpie *f*

2068 PARTHENOGAMY
f parthénogamie *f*
e partenogamia *f*
i partenogamia *f*
d Parthenogamie *f*

2069 PARTHENOGENESIS
f parthénogénèse *f*
e partenogénesis *f*
i partenogenesi *f*
d Jungfernzeugung *f*;
Parthenogenese *f*

PARTHENOGENESIS,
ARTIFICIAL
see 238

2070 PARTHENOGENETIC
f parthénogénétique
e partenogenético
i partenogenetico
d parthenogenetisch

2071 PARTHENOGENONE
f parthenogénone *m* (⊙)
e partenogenona *f*
i partenogenone *m*
d - -

2072 PARTHENOMIXIS
f parthénomixie *f*
e partenomixis *f*
i partenomissia *f*
d Parthenomixis *f*

2073 PARTHENOSPERM
f parthénosperme *m*
e partenosperma *m*;
partenospermium *m*
i partenosperma *m*
d Parthenospermium *n*

2074 PARTHENOSPORE
f parthénospore *f*
e partenospora *f*
i partenospora *f*
d Parthenospore *f*

2075 PARTHENOTE
f parthénote *m*
e partenote *m*
i partenote *m*
d Parthenote *f*

2076 PARTIAL BREAKAGE
f rupture *f* partielle
e ruptura *f* parcial

i rottura *f* parziale
d Partialbruch *m*

2077 PARTIAL CHIASMA
f chiasma *m* partiel
e quiasma *m* parcial
i chiasma *m* parziale
d Partialchiasma *n*

2078 PARTIAL CORRELATION
f corrélation *f* partielle
e correlación *f* parcial
i correlazione *f* parziale
d partielle Korrelation *f*

2079 PARTIAL CROSSING-OVER
f crossing-over *m* partiel;
enjambement *m* partiel
e sobrecruzamiento *m* parcial;
crossing-over *m* parcial
i crossing-over *m* parziale;
scambio *m* parziale
d Partial-Crossing-over *n*

2080 PARTIAL DOMINANCE;
INCOMPLETE DOMINANCE
f dominance *f* incomplète
e dominancia *f* incompleta
i dominanza *f* incompleta
d unvollständige Dominanz *f*

2081 PARTIAL FERTILIZATION
f fertilisation partielle *f*
e fertilización parcial *f*
i fertilizzazione parziale *f*
d Partialbefruchtung *f*

2082 PARTIAL HETEROTHALLIC
f hétérothallique partiel
e heterotálico parcial
i eterotallico parziale
d partialheterothallisch

2083 PARTIAL POLYPLOID
f polyploïde *m* partiel
e poliploide *m* parcial
i poliploide *m* parziale
d partielle Polyploide *f*

2084 PARTIAL POTENCY;
INCOMPLETE DOMINANCE

f puissance *f* partielle;
dominance *f* incomplète
e potencia *f* parcial;
dominancia *f* incompleta
i potenza *f* parziale;
dominanza *f* incompleta
d unvollständige Dominanz *f*

2085 PARTIAL REGRESSION
f régression *f* partielle
e regresión *f* parcial
i regressione *f* parziale
d partielle Regression *f*

2086 PARTIAL RENEWAL OF BLOOD
f rafraîchissement *m* de sang;
renouvellement *m* partiel du sang
e refrescamiento *m* parcial de sangre
i rinfrescamento *m* di sangue
d Blutauffrischung *f*

2087 PARTIAL REPLICA HYPOTHESIS
f hypothèse *f* des répliques partielles
e hipótesis *f* de reproducción parcial
i ipotesi *f* di replica parziale (Θ)
d Teilreplikahypothese *f*

2088 PARTIAL RESTITUTION
f restitution *f* partielle
e restitución *f* parcial
i restituzione *f* parziale
d Partialrestitution *f*

2089 PARTICULATE INHERITANCE
f hérédité *f* mendélienne
e herencia *f* mendeliana
i eredità *f* mendeliana (Θ)
d Mendelvererbung *f*

2090 PATERNAL
f paternel
e paternal
i paternale;
paterno;
d väterlich

2091 PATH COEFFICIENT
f path-coefficient *m*
e coeficiente *m* de trayectoria
i coefficiente di traiettoria *m*
d Pfadkoeffizient *m*

2092 PATROCLINAL INHERITANCE
f hérédité *f* patroclinale
e herencia *f* patroclinal
i eredità *f* patroclinale
d patrokline Vererbung *f*

**2093 PATRICLINOUS;
PATROCLINAL**
f patrocline
e patroclino
i patroclino
d patroklin

2094 PATROGENESIS
f patrogénèse *f*
e patrogénesis *f*
i patrogenesi *f*
d Patrogenese *f*

2095 PATTERN GENES
f gènes-type *mpl* (#)
e genes modelo *mpl*
i geni modelli *mpl* (#)
d Musterbildungsgene *npl* (#)

2096 PATTERN MODIFIERS
f modificateurs-type *mpl*
e modificadores modelo *mpl*
i modificatori modelli *mpl*
d mustermodifizierende Gene *npl*

2097 PATTERN OF DAMAGE
f type *m* de lésion (Θ)
e modelo *m* de daño
i modello *m* di danno (Θ)
d Schädigungsmuster *n*

PEAK, ADAPTIVE
see 34

**2098 PEDIGREE;
GENEALOGICAL TABLE**
f tableau *m* généalogique;
pedigree *m* (animaux)

e table *f* genealógica;
pedigree *m* (animales)
i tavola *f* genealogica;
pedigree *m*
d Stammbaum *m*;
Stammtafel *f*

2099 PENETRANCE
f pénétrance *f*
e penetración *f*
i penetranza *f*
d Penetranz *f*;
Ausprägungshäufigkeit *f*
der Gene

PENETRANCE, INCOMPLETE
see 1475

2100 PENTAPLOID (adj.)
PENTAPLOID
f pentaploïde
pentaploïde *m*
e pentaploide
pentaploide *m*
i pentaploide
pentaploide *m*
d pentaploid
Pentaploide *f*

2101 PENTAPLOIDY
f pentaploïdie *f*
e pentaploidía *f*
i pentaploidismo *m*
d Pentaploidie *f*

2102 PENTASOMIC
f pentasomique
e pentasómico
i pentasomico
d pentasom

2103 PERCENTAGE OF BLOOD
f pourcentage *m* de sang;
pourcentage *m* sanguin;
part *f* du sang
e porcentaje *m* de sangre;
porcentaje *m* sanguíneo
i percentuale *f* di sangue
d Blutanteil *m*

2104 PERICENTRIC

f péricentrique
e pericéntrico
i pericentrico
d perizentrisch

2105 PERICLINAL
f périclinal
e periclinal
i periclinale
d periklin;
periklinal

2106 PERIPHERIC MOVEMENT
f mouvement *m* périphérique
e movimiento *m* periférico
i movimento *m* periferico
d Peripheriewanderung *f*

2107 PERIPLASMA
f périplasme *m*
e periplasma *m*
i periplasma *m*
d Periplasma *n*

2108 PERIPLAST
f périplaste
e periplasto *m*
i periplasto *m*
d Periplast *m*

2109 PERISSOPLOID (adj.)
PERISSOPLOID
f périssoploïde
périssoploïde *m*
e perisoploide
perisoploide *m*
i perissoploide
perissoploide *m*
d perissoploid
Perissoploide *f*

2110 PERISSOPLOIDY
f périssoploïdie *f*
e perisoploidía *f*
i perissoploidismo *m*
d Perissoploidie *f*

2111 PERISTASIS
f péristase *f*
e peristasis *f*
i peristasi *f*
d Peristase *f*

2112 PERMANENT
    f permanent
    e permanente
    i permanente
    d permanent

2113 PERMISSIBLE DOSIS
    f dose *f* tolérée
    e dosis *f* permisible
    i dose *f* tollerata
    d Toleranzdosis *f*

2114 PERSISTENCE
    f persistance *f*
    e persistencia *f*
    i persistenza *f*
    d Persistenz *f*

2115 PERSISTENT
    f persistant
    e persistente
    i persistente
    d persistent

      PHAGE, INTEMPERATE
      see 1505

      PHAGE, TEMPERATE
      see 2781

2116 PHAGE SPLITTING
    f rupture *f* phagique
    e ruptura *f* de fago
    i rottura *f* del fago
    d Phagenspaltung *f*

2117 PHASE SPECIFICITY;
      PHASIC SPECIFICITY
    f spécificité *f* phasique
    e especificidad *f* fásica
    i specificità *f* fasica
    d Phasenspezifität *f*

2118 PHEN
    f phène *m*
    e fen *m*
    i fene *m*
    d Phän *n*

2119 PHENOCLINE
    f phénocline *m*

    e fenoclino *m*
    i fenoclino *m*
    d Phänokline *f*

      PHENOCONTOUR
      see 1589

2120 PHENOCOPY
    f phénocopie *f*
    e fenocopia *f*
    i fenocopia *f*
    d Phänokopie *f*

2121 PHENOCYTOLOGY
    f phénocytologie *f*
    e fenocitología *f*
    i fenocitologia *f*
    d Phänocytologie *f*

2122 PHENODEVIANT (adj.)
      PHENODEVIANT
    f phénodéviant
        phénodéviant *m*
    e fenodesviante
        fenodesviante *m*
    i fenodeviante
        fenodeviante *m*
    d Phänodeviante *f* (#)

2123 PHENOGENETICS
    f phénogénétique *f*
    e fenogenética *f*
    i fenogenetica *f*
    d Phänogenetik *f*

2124 PHENOME
    f phénome *m*
    e fenomio *m*
    i fenomio *m*
    d Phänom *n*

2125 PHENOMIC
    f phénomique
    e fenómico
    i fenomico
    d phänomatisch

2126 PHENOMIC DELAY
    f retard *m* phénomique
    e retardo *m* fenómico
    i ritardo *m* fenomico
    d phänomatischer Verzug *m*

2127 PHENOTYPE
  f phénotype *m*
  e fenótipo *m*
  i fenotipo *m*
  d Phänotyp *m*

PHENOTYPE, NUCLEAR
see 1935

2128 PHENOTYPIC
  f phénotypique
  e fenotípico
  i fenotipico
  d phänotypisch

2129 PHENOTYPIC CORRELATION
  f corrélation *f* phénotypique
  e correlación *f* fenotípica
  i correlazione *f* fenotipica
  d phänotypische Korrelation *f*

2130 PHENOTYPIC MIXING
  f mélange *m* phénotypique
  e mezcla *f* fenotípica
  i miscuglio *m* fenotipico (Θ)
  d phänotypische Vermischung *f*

PHENOTYPIC STABILITY
see 743

2131 PHOTOREACTIVATION
  f photoréactivation *f*
  e fotoreactivación *f*
  i fotoriattivazione *f*
  d Photoreaktivierung *f*

2132 PHOTOTROPHIC
  f phototrophique
  e fototrófico
  i fototrofico
  d phototroph

2133 PHRAGMOPLAST
  f phragmoplaste *m*
  e fragmoplasto *m*
  i fragmoplasto *m*
  d Phragmoplast *m*

2134 PHRAGMOSOME
  f phragmosome *m*
  e fragmosoma *m*

  i fragmosoma *m*
  d Phragmosom *n*

2135 PHYLETIC
  f phylétique
  e filético
  i filetico
  d phyletisch

2136 PHYLOGENESIS;
PHYLOGENY;
RACE DEVELOPMENT;
RACE HISTORY
  f phylogénèse *f*;
    phylogénie *f*
  e filogenia *f*;
    filogénesis *f*
  i filogenesi *f*;
    filogenia *f*
  d Phylogenie *f*;
    Phylogenese *f*;
    Stammesentwicklung *f*;
    Stammesgeschichte *f*

2137 PHYLOGENETIC
  f phylogénétique
  e filogénico
  i filogenico
  d phylogenetisch

2138 PHYLOGENETICS
  f phylogénétique *f*
  e filogenética *f*
  i filogenetica *f*
  d Phylogenetik *f*

2139 PHYLOGENY;
PHYLOGENESIS;
RACE DEVELOPMENT;
RACE HISTORY
  f phylogénèse *f*;
    phylogénie *f*
  e filogenia *f*;
    filogénesis *f*
  i filogenia *f*;
    filogenesi *f*
  d Phylogenie *f*;
    Phylogenese *f*;
    Stammesentwicklung *f*;
    Stammesgeschichte *f*

2148 PLASMACHROMATIN
 f plasmachromatine *f*
 e plasmacromatina *f*
 i plasmacromatina *f*
 d Plasmachromatin *n*

2140 PHYLUM
 f phylum *m*
 e filum *m*
 i phylum *m* ;
  derivazione *f*
 d Stamm *m* ;
  Phylum *n*

2149 PLASMAGENE
 f plasmagène *m*
 e plasmagén *m*
 i plasmagene *m*
 d Plasmagen *n*

2141 PHYSIOCHROMATIN
 f physiochromatine *f*
 e fisiocromatina *f*
 i fisiocromatina *f*
 d Physiochromatin *n*

2150 PLASMALEMMA
 f a. plasmalemma *m* (absence
  de membrane solide chez
  certaines bactéries)
  b. plasmalemma *m* (surface
  externe du protoplasme)
 e plasmalema *m*
 i plasmalemma *m*
 d Plasmalemma *n*

2142 PHYSIOLOGICAL RACE
 f race *f* physiologique
 e raza *f* fisiológica
 i razza *f* fisiologica
 d physiologische Rasse *f*

2151 PLASMATIC INHERITANCE
 f hérédité *f* plasmatique
 e herencia *f* plasmática
 i eredità *f* plasmatica
 d Plasmavererbung *f* ;
  plasmatische Vererbung *f* ;
  cytoplasmatische Verer-
  bung *f*

2143 PHYTOGENETICS
 f phytogénétique *f*
 e fitogenética *f*
 i fitogenetica
 d Phytogenetik *f* ;
  Pflanzengenetik *f*

2144 PLACODESMOSE
 f placodesmose *f*
 e placo-desmosa *f*
 i placodesmosi *f*
 d Placodesmose *f*

2152 PLASMID
 f plasmide *f*
 e plasmidio *m*
 i plasmidio *m*
 d Plasmid *n*

2145 PLANOGAMETE
 f planogamète *m*
 e planogameta *m*
 i planogameto *m*
 d Planogamet *m*

2153 PLASMOCHROMATIN
 f plasmochromatine *f*
 e plasmocromatina *f*
 i plasmocromatina *f*
 d Plasmochromatin

2146 PLANOSOME
 f planosome *m*
 e planosoma *m*
 i planosoma *m*
 d Planosom *n*

2154 PLASMODESMA,
 pl PLASMODESMATA
 f plasmodesme *m*
 e plasmodesma *m*
 i plasmodesma *m*
 d Plasmodesma *n*

2147 PLAQUE
 f plaque *f*
 e placa *f*
 i placca *f*
 d Plaque *f* ;
  Phagenloch *n*

2155 PLASMODIUM
 f plasmodium *m*

e plasmodio *m*
i plasmodio *m*
d Plasmodium *n*

2156 PLASMOGAMY
f plasmogamie *f*
e plasmogamia *f*
i plasmogamia *f*
d Plasmogamie *f*

2157 PLASMON
f plasmone *m*
e plasmón *m*
i plasmon *m*
d Plasmon *n*

2158 PLASMOSOME
f plasmosome *m*
e plasmosoma *m*
i plasmosoma *m*
d Plasmosom *n*

2159 PLASMOTOMY
f plasmotomie *f*
e plasmotomia *f*
i plasmotomia *f*
d Plasmotomie *f*

2160 PLASMOTYPE
f plasmotype *m*
e plasmótipo *m*
i plasmotipo *m*
d Plasmotyp *m*

2161 PLAST
f plaste *f*
e plasto *m*
i plasto *m*
d Plast *m*

2162 PLASTICITY
f plasticité *f*
e plasticidad *f*
i plasticità *f*
d Plastizität *f*

2163 PLASTID
f plastide *m*
e plastidio *m*
i plastidio *m*
d Plastid *m*

2164 PLASTID INHERITANCE
f hérédité *f* plastidique
e herencia *f* plastídica
i eredità *f* plastidica
d Plastidenvererbung *f*

2165 PLASTID MUTATION
f mutation *f* plastidique
e mutación *f* plastídica
i mutazione *f* plastidica
d Plastidenmutation *f*

2166 PLASTIDOGENIC COMPLEX
f complexe *m* plastidogénique
e complejo *m* plastidógeno
i complesso *m* plastidiogenico
d Plastidogen-Komplex *m*

2167 PLASTIDOME
f plastidome *m*
e plastidomio *m*
i plastidoma *m*
d Plastidom *n*

2168 PLASTIDOTYPE
f plastidotype *m*
e plastidiótipo *m*
i plastidiotipo *m*
d Plastidotyp *m*

2169 PLASTOCONT;
PLASTOKONT
f plastoconte *m*
e plastoconto *m*
i plastoconto *m*
d Plastokont *m*

2170 PLASTODESMA;
pl PLASTODESMATA
f plastodesme *m*
e plastodesma *m*
i plastodesma *m*
d Plastodesma *n*

2171 PLASTOGENE
f plastogène *m*
e plastogén *m*
i plastogene *m*
d Plastogen *n*

2172   PLASTOKONT;
       PLASTOCONT
       f plastoconte *m*
       e plastoconto *m*
       i plastoconto *m*
       d Plastokont *m*

2173   PLASTOME
       f plastome *m*
       e plastoma *m*
       i plastoma *m*
       d Plastom *n*

2174   PLASTOMERE
       f plastomère *m*
       e plastomero *m*
       i plastomero *m*
       d Plastomer *n*

2175   PLASTOSOME
       f plastosome *m*
       e plastosoma *m*
       i plastosoma *m*
       d Plastosom *n*

2176   PLATE
       f plaque *f*
       e placa *f*
       i placca *f*
       d Platte *f*

2177   PLECTONEMIC
       f plectonémique
       e plectonemático;
          plectonémico
       i plectonemico
       d plektonem(at)isch

2178   PLECTONEMIC COIL
       f enroulement *m* plectoné-
          mique
       e arrollamiento *m* plectoné-
          mico;
          arrollamiento *m* plectone-
          mático
       i avvolgimento *m* plectone-
          mico
       d plektonem(at)ische Wick-
          lung *f*

2179   PLEIOTROPIC

       f pléiotropique
       e pleiotrópico
       i pleiotropico
       d pleiotrop

2180   PLEIOTROPISM;
       PLEIOTROPY
       f pléiotropie *f*
       e pleiotropía *f*
       i pleiotropismo *m*
       d Pleiotropie *f*

2181   PLEIOTROPY;
       PLEIOTROPISM
       f pléiotropie *f*
       e pleiotropía *f*
       i pleiotropismo *m*
       d Pleiotropie *f*

2182   PLEUROMITIC
       f pleuromictique
       e pleuromíctico
       i pleuromittico
       d pleuromitisch

2183   PLOIDY
       f ploïdie *f*
       e ploidia *f*
       i ploidia *f*;
          ploidismo *m*
       d Ploidie *f*

2184   PLUS MODIFIER
       f modificateur *m* positif;
          modificateur *m* plus
       e modificador *m* plus
       i modificatore *m* plus (☉);
          arcimodificatore *m*
       d plus modifier *m*

2185   POIKILOPLOID (adj.)
       POIKILOPLOID
       f poikiloploïde
          poikiloploïde *m*
       e poiquiloploide
          poiquiloploide *m*
       i poichiloploide
          poichiloploide *m*
       d poikiloploid
          Poikiloploide *f*

86 POIKILOPLOIDY
 f poikiploïdie *f*
 e poiquiploploidia *f*
 i poichiloploidismo *m*
 d Poikiloploidie *f*

87 POIKILOSYNDESIS
 f poikilosyndèse *f*
 e poiquilosíndesis *f*
 i poichilosindesi *f*
 d Poikilosyndese *f*

88 POINT ERROR
 f erreur *f* ponctuelle
 e error *m* puntual
 i errore *m* puntuale
 d point error *m*

89 POINT MUTATION
 f mutation *f* ponctuelle
 e mutación *f* puntual
 i mutazione *f* puntuale
 d Punktmutation *f*;
  Genmutation *f*

90 POINT STICKINESS
 f adhérence *f* ponctuelle
 e aglutinamiento *m* puntual
 i stickiness *f* puntuale (☉)
 d point stickiness *f*

91 POISSON'S LAW;
POISSON'S DISTRIBUTION;
SERIE DE POISSON
 f loi *f* de Poisson;
  distribution *f* de Poisson;
  série *f* de Poisson;
  équation *f* de Poisson
 e ley *f* de Poisson;
  distribución *f* de Poisson;
  serie *f* de Poisson;
  ecuación *f* de Poisson
 i legge *f* di Poisson;
  distribuzione *f* di Poisson;
  serie *f* di Poisson
 d Poissonsches Gesetz *n*;
  Poisson-Verteilung *f*

92 POLAR
 f polaire
 e polar
 i polare
 d polar

2193 POLAR BODY;
POLOCYTE
 f globule *m* polaire;
  polocyte *m*
 e glóbulo *m* polar;
  polocito *m*
 i polocita *m*;
  globulo *m* polare
 d Polkörperchen *n*;
  Polocyte *f*

2194 POLAR CAP
 f capsule *f* polaire
 e casquete *f* polar
 i capsula *f* polare
 d Polkappe *f*

2195 POLARITY
 f polarité *f*
 e polaridad *f*
 i polarità *f*
 d Polarität *f*

2196 POLARIZATION
 f polarisation *f*
 e polarización *f*
 i polarizzazione *f*
 d Polarisation *f*;
  Polarisierung *f*

2197 POLARIZED REDUCTION;
PREFERENTIAL REDUCTION
 f réduction *f* préférentielle
 e reducción *f* polarizada;
  reducción *f* preferencial
 i riduzione *f* preferenziale
 d gerichtete Reduktion *f*

2198 POLAR PLATE;
POLE PLATE
 f plaque *f* polaire
 e placa *f* polar
 i placca *f* polare
 d Polplatte *f*

2199 POLE
 f pôle *m*
 e polo *m*
 i polo *m*
 d Pol *m*; Zentrum *n*

2200   POLE FIELD
       f région f polaire
       e área f polar
       i campo m polare
       d Polfeld n

2201   POLE PLATE;
       POLAR PLATE
       f plaque polaire f
       e placa polar f
       i placca polare f
       d Polplatte f

2202   POLLEN LETHALS
       f léthaux mpl pour le
          pollène
       e letales mpl por el polen
       i letali mpl per il polline
       d Pollenletalfaktoren mpl

2203   POLLEN STERILITY
       f stérilité f du pollen;
          stérilité f pollinique
       e esterilidad f polínica
       i sterilità f pollinica
       d Pollensterilität f

2204   POLOCYTE;
       POLAR BODY
       f globule m polaire;
          polocyte (m) (peu usité)
       e polocito m;
          glóbulo m polar
       i polocita m;
          globulo m polare
       d Polocyte f;
          Polkörperchen n

2205   POLYALLELE CROSSING
       f croisement m polyallélique
       e cruzamiento m polialelo
       i incrocio m poliallele
       d diallele Kreuzung f

2206   POLYBASIC
       f polybasique
       e polibásico
       i polibasico
       d polybasisch

2207   POLYCARYOTIC;
       POLYKARYOTIC

       f à noyaux mpl multiples
       e policario
       i policariotico
       d polycaryotisch;
          polykaryotisch;
          mehrkernig

2208   POLYCENTRIC
       f polycentrique
       e policéntrico
       i policentrico
       d polyzentrisch

2209   POLYCHONDRIC
       f polychondrique
       e policóndrico
       i policondrico
       d polychondrisch (#)

2210   POLYCHROMOSOME
       f polychromosome m
       e policromosoma m
       i policromosoma m
       d Polychromosom n

2211   POLYCROSS
       f polycroisement m
       e policruzamiento m
       i poli-incrocio m;
          incrocio plurimo m
       d Massenkreuzung f

2212   POLYEMBRYONY
       f polyembryonie f
       e poliembrionía f
       i poliembrionia f
       d Polyembryonie f

2213   POLYENERGID
       f polyénergide m
       e polienérgida f
       i polienergide m
       d polyenergid
          Polyenergide f

2214   POLYERGISTIC
       f polyergistique
       e poliergístico
       i poliergittico
       d polyergistisch

2215 POLYFACTORIAL
  f polyfactoriel
  e polifactorial
  i polifattoriale
  d plurifaktoriell;
    polyfaktoriell (!)

2216 POLYGAMOUS
  f polygame
  e polígamo
  i poligamo
  d polygam

2217 POLYGAMY
  f polygamie *f*
  e poligamia *f*
  i poligamia *f*
  d Polygamie *f*

2218 POLYGENE;
    MINOR GENE
  f polygène *m*;
    gène *m* mineur
  e poligén *m*;
    gen *m* menor
  i poligene *m*;
    gene *m* minore
  d Polygen *n*

2219 POLYGENERIC
  f polygénérique
  e poligenérico
  i poligenerico
  d polygenerisch;
    Mehrgattungs- (in Zusam-
    mensetzungen mit Bastard)

2220 POLYGENIC
  f polygénique
  e poligénico
  i poligenico
  d polygen;
    polygenisch

2221 POLYGENIC COMBINATION
  f combinaison *f* polygénique
  e combinación *f* poligénica
  i combinazione *f* poligenica
  d Polygenkombination *f*

2222 POLYGENOMATIC
  f polygénomatique

  e poligenomático
  i poligenomatico
  d polygenomatisch

2223 POLYGENY
  f polygénie *f*
  e poligenia *f*
  i poligenia *f*
  d Polygenie *f*

2224 POLYHAPLOID (adj.)
    POLYHAPLOID
  f polyhaploïde
    polyhaploïde *m*
  e polihaploide
    polihaploide *m*
  i poliaploide
    poliaploide *m*
  d polyhaploid
    Polyhaploide *f*

2225 POLYHAPLOIDY
  f polyhaploïdie *f*
  e polihaploidía *f*
  i poliaploidismo *m*
  d Polyhaploidie *f*

2226 POLYHYBRID (adj.)
    POLYHYBRID
  f polyhybride
    polyhybride *m*
  e polihíbrido
    polihíbrido *m*
  i poliibrido
    poliibrido *m*
  d polyhybrid
    Polyhybride *f*

2227 POLYHYBRIDISM
  f polyhybridisme *m*
  e polihibridismo *m*
  i poliibridismo *m*
  d Polyhybridismus *m*

2228 POLYKARYOTIC;
    POLYCARYOTIC
  f à noyaux multiples *mpl*
  e policario
  i policariotico
  d polycaryotisch;
    polykaryotisch;
    mehrkernig

2229 POLYLYSOGENIC
    f polylysogénique
    e polilisogénico
    i polilisogenico
    d polylysogen

2230 POLYMERIC;
    POLYMERICAL;
    POLYMEROUS
    (when syn. of polymeric)
    f polymère;
       polymérique
    e polimérico;
       polímero
    i polimerico
    d polymer

2231 POLYMERIC CHROMOSOME
    f chromosome *m* polymère
    e cromosoma *m* polímero
    i cromosomo *m* polimerico
    d Polymerchromosom *n*

2232 POLYMEROUS (when syn. of
    polymeric);
    POLYMERIC;
    POLYMERICAL
    f polymère;
       polymérique
    e polimérico;
       polímero
    i polimerico
    d polymer

2233 POLYMERY
    f polymérie *f*
    e polimería *f*
    i polimerismo *m*
    d Polymerie *f*

2234 POLYMITOSIS
    f polymitose *f*
    e polimitosis *f*
    i polimitosi *f*
    d Polymitose *f*

2235 POLYMORPHISM
    f polymorphisme *m*
    e polimorfismo *m*
    i polimorfismo *m*
    d Polymorphismus *m*

POLYMORPHISM, BALANCE
see 326

POLYMORPHISM, TRANSIEN
see 2833

2236 POLYMORPHOUS
    f polymorphe
    e polimorfo
    i polimorfo
    d polymorph

2237 POLYPHASY
    f polyphasie *f*
    e polifasia *f*
    i polifasia *f*
    d Polyphasie *f*

2238 POLYPHENY
    f polyphénie *f*
    e polifenia *f*
    i polifenia *f*
    d Polyphänie *f*

2239 POLYPHYLETIC
    f polyphylétique
    e polifilético
    i polifiletico
    d polyphyletisch

2240 POLYPLOID (adj.)
    POLYPLOID
    f polyploïde
       polyploïde *m*
    e poliploide
       poliploide *m*
    i poliploide
       poliploide *m*
    d polyploid
       Polyploide *f*

POLYPLOID, DUPLI-
CATIONAL
see 885

POLYPLOID, PARTIAL
see 2083

2241 POLYPLOID COMPLEX
    f complexe *m* de polyploïdes
    e complejo *m* de poliploides

i complesso *m* di poliploidi
d Polyploidenkomplex *m*

2242 POLYPLOIDY
f polyploïdie *f*
e poliploidía *f*
i poliploidismo *m*;
   poliploidia *f*
d Polyploidie *f*

POLYPLOIDY, BALANCED
see 327

POLIPLOIDY, RAMPANT
see 2395

POLIPLOIDY, SECONDARY
see 2506

POLYPLOIDY, STRUC-
TURALLY CHANGED
see 2686

POLYPLOIDY, UNBALANCED
see 2888

2243 POLYPLOTYPE
f polyplotype *m*
e poliplótipo *m*
i poliplotipo *m*
d Polyplotypus *m* (#)

2244 POLYRADIAL
f polyradial
e polirradial
i poli-radiale
d polyradial

2245 POLY-REPRODUCTION
f poly-reproduction *f*
e polireproducción *f*
i poli-riproduzione *f*
d Mehrfachverdopplung *f*

2246 POLYSATELLITIC
f polysatellitique
e polisatelítico
i polisatellitico
d polysatellitisch

2247 POLYSOMIC
f polysomique

e polisómico
i polisomico
d polysom

2248 POLYSOMY
f polysomie *f*
e polisomía *m*
i polisomia *f*
d Polysomie *f*

2249 POLYSPERMIC
f polysperme
e polispérmico
i polispermo
d polysperm

2250 POLYSPERMY
f polyspermie *f*
e polispermia *f*
i polispermia *f*
d Polyspermie *f*

2251 POLYTENE
f polytène (noyau, chromosome)
e politénico
   (núcleo, cromosoma)
i politene
   (da nucleo, cromosoma)
d polytän;
   Polytän- (in Zusammen-
   setzungen)

2252 POLYTENY
f polyténie *f*
e politenia *f*
i politenia *f*
d Polytänie *f*

2253 POLYTOPY
f polytopisme *m*; polytopie *f*
e politopismo *m*
i politopismo *m*
d Polytopie *f*

2254 POLYTYPIC
f polytypique
e politípico
i politipico
d polytypisch

2255 POLYURGIC GENE
f gène polyurgique *m*

e gen poliúrgico *m*
i gene poliurgico *m*
d polyurgisches Gen *n (≠)*

2256 POPULATION
f population *f*
e población *f*
i popolazione *f*
d Population *f*

2257 POPULATION DENSITY
f densité *f* de la population
e densidad *f* de la población
i densità *f* della popolazione
d Populationsdichte *f*

2258 POPULATION EQUILIBRIUM;
GENETIC EQUILIBRIUM
f équilibre *m* génétique
e equilibrio *m* genético;
equilibrio *m* de la población
i equilibrio *m* genetico
d Populationsgleichgewicht *n*

2259 POPULATION GENETICS
f génétique *f* des populations
e genética *f* de las poblaciones
i genetica *f* delle popolazioni
d Populationsgenetik *f*

2260 POPULATION HOMEOSTASIS
f homéostasie *f* de population
e homeostasis *f* de población
i omeostasi *f* di popolazione
d Populationshomöostasis *f*

2261 POPULATION PRESSURE
f pressure *f* de population
e presión *f* de población
i pressione *f* di popolazione
d Populationsdruck *m*

2262 POPULATION WAVES
f ondes *f pl* de population
e ondas *f pl* de población
i onde *f pl* di popolazione
d Populationswellen *f pl*

2263 POSITIONAL ALLELES
f allèles *m pl* de position
e alelos *m pl* de posición
i alleli *m pl* di posizione

d Positionspseudoallele *n pl*

2264 POSITION CONSANGUINITY
f consanguinité *f* de position
e consanguinidad *f* de posición
i consanguinità *f* di posizione
d Positionsverwandtschaft *f* (

2265 POSITION EFFECT
f effet *m* de position
e efecto *m* de posición
i effetto *m* di posizione
d Positionseffekt *m*;
Lagewirkung *f*

2266 POST ADAPTATION
f post-adaptation *f*
e postadaptación *f*
i postadattamento *m*
d Postadaptation *f*

2267 POST DIVISION
f post-division *f*
e postdivisión *f*
i post-divisione *f*
d Postdivision *f*

2268 POSTHETEROKINESIS
f post-hétérocinèse *f*
e postheterocinesis *f*
i post-eterocinesi *f*
d Postheterokinese *f*

2269 POST-REDUCED
f post-réduit
e post-reducto
i post-ridotto
d postreduziert

2270 POST REDUCTION
f post-réduction *f*
e post-reducción *f*
i post-riduzione *f*
d Postreduktion *f*

2271 POST SPLIT ABERRATION
f aberration *f* chromatidique;
aberration *f* post-divisionne
e aberración *f* post-divisiona
i aberrazione *f* post-division
d post split aberration *f*

2272 POSTSYNDETIC INTERPHASE
    f interphase *f* postsyndétique;
       interphase *f* postsyndèse
    e interfase *f* possindética
    i interfase *f* post-sindetica
    d postsyndetische Interphase *f*

2273 POSTZYGOTIC
    f postzygotique
    e postcigótico
    i post-zigotico
    d postzygotisch

2274 POTENCE;
     POTENCY
    f potence *f*
    e potencia *f*
    i potenza *f*
    d Potenz *f*

     POTENCY, PARTIAL
     see 2084

2275 POTENCY ALLELE
    f allèle *m* de potence
    e potencia *f* alélica
    i potenza *f* allela
    d Potenzallel *n*

2276 PREADAPTATION
    f préadaptation *f*
    e preadaptación *f*
    i preadattamento *m*
    d Präadaptation *f*

2277 PRECESSION
    f précession *f*
    e precesión *f*
    i precessione *f*
    d Präzession *f*

278 PRECOCIOUS REVERSION
    f réversion *f* précoce
    e reversión *f* precoz
    i reversione *f* precoce
    d vorzeitige Rückkehr *f* in
       die Ruhephase

279 PRECOCITY
    f précocité *f*

    e precocidad *f*
    i precocità *f*
    d Präkozität *f*

     PRECOCITY, REDUCED
     see 2422

2280 PRECOCITY THEORY
    f théorie *f* de la précocité
    e teoría *f* de la precocidad
    i teoria *f* della precocità
    d Präkozitätstheorie *f*

2281 PRECONJUGATION
    f pré-conjugaison *f*
    e preconjugación *f*
    i preconiugazione *f*
    d Präkonjugation *f*

2282 PREDETERMINATION;
     DELAYED INHERITANCE
    f prédétermination *f*;
       hérédité *f* retardée
    e predeterminación *f*;
       herencia *f* retrasada
    i predeterminazione *f*;
       eredità *f* ritardata
    d Prädetermination *f*;
       Vorbestimmung *f*;
       verzögerte Vererbung *f*

2283 PREDIVISION
    f prédivision *f*
    e predivisión *f*
    i predivisione *f*
    d Prädivision *f*

2284 PREFERENTIAL REDUCTION;
     POLARIZED REDUCTION
    f réduction *f* préférentielle
    e reducción *f* preferencial
    i riduzione *f* preferenziale
    d gerichtete Reduktion *f*

2285 PREFERENTIAL SEGREGATION
    f ségrégation *f* préférentielle
    e segregación *f* preferente
    i segregazione *f* preferenziale
    d präferentielle Segregation *f*

2286   PREFORMATIONISM;
       PREFORMISM;
       PREFORMATION THEORY
       f   préformisme *m*;
           théorie *f* de la préformation
       e   teoría *f* de la preformación
       i   preformismo *m*;
           teoria *f* della preformazione
       d   Präformationstheorie *f*

2287   PREHETEROKINESIS
       f   pré-hétérokinèse *f*;
           pré-hétérocinèse *f*
       e   preheterocinesis *f*
       i   pre-eterocinesi *f*
       d   Präheterokinese *f*

2288   PREMUTATION
       f   prémutation *f*
       e   premutación *f*
       i   premutazione *f*
       d   Prämutation *f*

2289   PREPOTENCE;
       PREPOTENCY
       f   prépotence *f*;
           pouvoir raceur *m*
       e   prepotencia *f*;
           poder raceador *m*
       i   prepotenza *f*
       d   Individualpotenz *f*;
           Präpotenz *f*

2290   PREPROPHAGE
       f   préprophage *m*
       e   preprofago *m*
       i   preprofago *m*
       d   Präprophage *m*

2291   PREPROPHASE INHIBITOR
       f   inhibiteur *m* de la préprophase
       e   inhibidor *m* de la preprofase
       i   inibitore *m* della preprofase
       d   Präprophase-Inhibitor *m*

2292   PRE-REDUCED
       f   pre-réduit
       e   pre-reducto
       i   pre-ridotto
       d   präreduziert

2293   PRE-REDUCTION
       f   pré-réduction *f*
       e   pre-reducción *f*
       i   pre-riduzione *f*
       d   Präreduktion *f*

2294   PRESENCE-ABSENCE HY-
       POTHESIS;
       PRESENCE AND ABSENCE
       THEORY
       f   théorie *f* de la présence
           de l'absence
       e   teoría *f* de la presencia y
           de la ausencia
       i   teoria *f* della presenza e
           assenza
       d   Präsenz-Absenz-Hypo-
           these *f*

2295   PRESSURE
       f   pression *f*
       e   presión *f*
       i   pressione *f*
       d   Druck *m*

2296   PRESUMPTIVE REGION
       f   région *f* présumée
       e   región *f* presunta
       i   regione *f* presunta
       d   präsumptiver Organbezirk
           präsumptive Region *f*

2297   PRE-ZYGOTIC
       f   pré-zygotique
       e   pre-cigótico
       i   pre-zigotico
       d   präzygotisch

2298   PRIMARY
       f   primaire
       e   primario
       i   primario
       d   primär;
           Primär- (in Zusammen-
           setzungen)

2299   PRIMARY SEX RATIO
       f   proportion *f* primaire des
           sexes
       e   proporción *f* primaria de
           los sexos

i proporzione *f* primaria dei sessi
d primäres Geschlechterverhältnis *n*

2300 PRIMARY TARGET
f site *m* d'action primaire
e objectivo *m* primario
i obiettivo *f* d'azione primaria; posto *m* d'azione primaria
d Primärtrefferbereich *m*

2301 PRIMITIVE FORM
f forme *f* de souche; forme *f* originale
e forma *f* primitiva; forma *f* original
i forma *f* primitiva; forma *f* originale
d Stammform *f*; Urform *f*

2302 PRIMORDIUM
f primordium *m*
e primordio *m*
i primordio *m*
d Primordium *n*

2303 PROBABILITY
f probabilité *f*
e probabilidad *f*
i probabilità *f*
d Wahrscheinlichkeit *f*

2304 PROBABLE
f probable
e probable
i probabile
d wahrscheinlich

2305 PROCENTRIC
f procentrique
e procéntrico
i procentrico
d prozentrisch

2306 PROCHROMOSOME
f prochromosome *m*
e procromosoma *m*
i procromosoma *m*
d Prochromosom *n*

2307 PRODUCTIVE INFECTION
f infection *f* productive
e infección *f* productiva
i infezione *f* produttiva
d Produktivinfektion *f*

2308 PROGAMIC
f progame
e progámico
i progamico
d progam

2309 PROGAMY
f progamie *f*
e progamia *f*
i progamia *f*
d Progamie *f*

2310 PROGENE
f progène *m*
e progén *m*
i progene *m*
d Progen *n*

2311 PROGENESIS
f progénèse *f*
e progénesis *f*
i progenesi *f*
d Progenese *f*

2312 PROGENOME
f progénome *m*
e progenomio *m*
i progenomio *m*
d Progenom *n*

2313 PROGENY
f progéniture *f*; descendance *f*
e progenie *f*; descendencia *f*
i progenie *f*; discendenza *f*
d Nachkommenschaft *f*; Deszendenz *f*; Nachzucht *f*

2314 PROGENY TEST
f test *m* de descendance
e prueba *f* de la descendencia
i prova *f* della discendenza
d Nachkommenprüfung *f*; Prüfung *f* der Nachkommenschaft

2315   PROGRESSIVE
      f progressif
      e progresivo
      i progressivo
      d progressiv

2316   PROMETAPHASE
      f prométaphase f
      e prometafase f
      i prometafase f
      d Prometaphase f

2317   PROMETAPHASE STRETCH
      f élongation f prométapha-
         sique
      e elongación f prometafásica
      i allungamento m
         prometafasico
      d Prometaphasestreckung f

2318   PROMITOSE
      f promitose f
      e promitosis f
      i promitosi f
      d Promitose f

2319   PRONUCLEOLUS
      f pronucléole m
      e pronucléolo m
      i pronucleolo m
      d Pronukleolus m;
         Pronucleolus m

2320   PRONUCLEUS
      f pronucléus m
      e pronúcleo m
      i pronucleo m
      d Pronukleus m;
         Pronucleus m

2321   PROPHAGE
      f prophage m
      e profago m
      i profago m
      d Prophage m

2322   PROPHAGE RECOMBINATION
      f recombinaison f des
         prophages
      e recombinación f de los
         profagos

      i ricombinazione f dei
         profagi
      d Prophagenrekombination f

2323   PROPHASE
      f prophase f
      e profase f
      i profase f
      d Prophase f

      PROPHASE, OLD SPIRAL
      see 1970

2324   PROPHASE INDEX
      f indice m des prophases
      e índice m de las profasas
      i indice m della profase
      d Prophase-Index m

2325   PROPHASE POISONS
      f poisons mpl de la prophase
      e profase f tóxica
      i veleni mpl della profase
      d Prophasegifte npl

2326   PROPHASIC
      f prophasique
      e profásico
      i profasico
      d prophasisch;
         Prophase- (in Zusammen-
         setzungen)

2327   PROTANDRY
      f hermaphrodisme m pro-
         tandrique;
         hermaphrodisme m pro-
         térandrique
      e protandria f
      i protandria f
      d Protandrie f;
         Proterandrie f

2328   PROTERMINAL
      f proterminal
      e proterminal
      i proterminale
      d proterminal

2329   PROTOCHROMONEMA
      f protochromonema m

e protocromonema *m*
i protocromonema *m*
d Protochromonema *n*

2330 PROTOPLASM
f protoplasme *m*
e protoplasma *m*
i protoplasma *m*
d Protoplasma *n* ;
    Plasma *n*

2331 PROTOPLASMATIC;
      PROTOPLASMIC
f protoplasmique
e protoplásmico
i protoplasmico
d protoplasmatisch

2332 PROTOPLASMIC
      INCOMPATIBILITY
f incompatibilité *f* protoplasmique
e incompatibilidad *f*
   protoplásmica
i incompatibilità *f*
   protoplasmica
d protoplasmatische Inkompatibilität *f*

2333 PROTOTROPH
f prototrophe *m*
e protótrofo *m*
i prototrofo *m*
d Prototroph *m*

2334 PROTOTROPHIC
f prototrophique
e protótrofo
i prototrofico
d prototroph

2335 PROXIMAL
f proximal
e proximal
i prossimale
d proximal

2336 PSEUDOALLELE
f pseudo-allèle *m*
e pseudoalelo *m* ;
   seudoalelo *m*

i pseudoallelo *m*
d Pseudoallel *n*

2337 PSEUDOALLELISM
f pseudo-allélisme *m*
e pseudoalelismo *m* ;
   seudoalelismo *m*
i pseudoallelismo *m*
d Pseudoallelie *f*

2338 PSEUDOAMITOSIS
f pseudo-amitose *f*
e pseudoamitosis *f* ;
   seudoamitosis *f*
i pseudoamitosi *f*
d Pseudoamitose *f*

2339 PSEUDOANAPHASE
f pseudo-anaphase *f*
e pseudoanafase *f* ;
   seudoanafase *f*
i pseudoanafase *f*
d Pseudoanaphase *f*

2340 PSEUDOAPOGAMY
f pseudo-apogamie *f*
e pseudoapogamia *f* ;
   seudoapogamia *f*
i pseudoapogamia *f*
d Pseudoapogamie *f*

2341 PSEUDOASSOCIATION
f pseudo-association *f*
e pseudoasociacion *f* ;
   seudoasociacion *f*
i pseudoassociazione *f*
d Pseudoassoziation *f*

2342 PSEUDOBIVALENT
f pseudobivalent *m*
e pseudobivalente *m* ;
   seudobivalente *m*
i pseudobivalente *m*
d Pseudobivalent *n*

2343 PSEUDOCHIASMA
f pseudochiasma *m*
e pseudoquiasma *m* ;
   seudoquiasma *m*
i pseudochiasma *m*
d Pseudochiasma *n*

2344 PSEUDOCOMPATIBILITY
   f pseudo-compatibilité *f*
   e pseudocompatibilidad *f*;
      seudocompatibilidad *f*
   i pseudocompatibilità
   d Pseudokompatibilität *f*

2345 PSEUDO-CROSSING-OVER
   f pseudo-crossing-over *m*;
      pseudo-enjambement *m*
   e pseudosobrecruzamiento *m*;
      seudosobrecruzamiento *m*
   i pseudocrossing-over *m*;
      pseudoscambio *m*
   d Pseudo-Crossing-over *n*

2346 PSEUDODOMINANCE
   f pseudo-dominance *f*
   e pseudodominancia *f*;
      seudodominancia *f*
   i pseudodominanza *f*
   d Pseudodominanz *f*

2347 PSEUDOENDOMITOSIS
   f pseudo-endomitose *f*
   e pseudoendomitosis *f*;
      seudoendomitosis *f*
   i pseudoendomitosi *f*
   d Pseudoendomitose *f*

2348 PSEUDOEQUATORIAL PLATE
   f plaque *f* pseudo-équatoriale
   e placa *f* pseudoecuatorial;
      placa *f* seudoecuatorial
   i placca *f* pseudoequatoriale
   d Pseudoäquatorialplatte *f*

2349 PSEUDOEXOGENOUS
   f pseudo-exogène
   e pseudoexógeno;
      seudoexógeno
   i pseudoesogeno
   d pseudoexogen

2350 PSEUDOFERTILITY
   f pseudo-fertilité *f*
   e pseudofertilidad *f*;
      seudofertilidad *f*
   i pseudofertilità *f*
   d Pseudofertilität *f*

2351 PSEUDOFRAGMENT
   f pseudo-fragment *m*
   e pseudofragmento *m*;
      seudofragmento *m*
   i pseudofragmento *m*
   d Pseudofragment *n*

2352 PSEUDOGAMOUS
   f pseudogame
   e pseudógamo;
      seudógamo
   i pseudogamo
   d pseudogam

2353 PSEUDOGAMY
   f pseudogamie *f*
   e pseudogamia *f*;
      seudogamia *f*
   i pseudogamia *f*
   d Pseudogamie *f*

2354 PSEUDOHAPLOID (adj.)
      PSEUDOHAPLOID
   f pseudo-haploïde
      pseudo-haploïde *m*
   e pseudohaploide;
      seudohaploide
      pseudohaploide *m*;
      seudohaploide *m*
   i pseudoaploide
      pseudoaploide *m*
   d pseudohaploid
      Pseudohaploide *f*

2355 PSEUDOHAPLOIDY
   f pseudo-haploïdie *f*
   e pseudohaploidia *f*;
      seudohaploidia *f*
   i pseudoaploidismo *m*
   d Pseudohaploidie *f*

2356 PSEUDOHETEROSIS
   f pseudo-hétérosis *f*
   e pseudoheterosis *f*;
      seudoheterosis *f*
   i pseudoeterosi *f*
   d Pseudoheterosis *f*

2357 PSEUDOINCOMPATIBILITY
   f pseudo-incompatibilité *f*

e pseudoincompatibilidad *f*;
   seudoincompatibilidad *f*
i pseudoincompatibilità *f*
d Pseudoinkompatibilität *f*

58 PSEUDOMEIOSIS
    f pseudo-méiose *f*
    e pseudomeiosis *f*;
       seudomeiosis *f*
    i pseudomeiosi *f*
    d Pseudomeiose *f*

59 PSEUDOMETAPHASE
    f pseudo-métaphase *f*
    e pseudometafase *f*;
       seudometafase *f*
    i pseudometafase *f*
    d Pseudometaphase *f*

60 PSEUDOMIXIS
    f pseudomixie *f*
    e pseudomixis *f*;
       seudomixis *f*
    i pseudomixi *f*;
       pseudomissia *f*
    d Pseudomixis *f*

61 PSEUDOMONOSOMIC
    f pseudo-monosomique
    e pseudomonosómico;
       seudomonosómico
    i pseudomonosomico
    d pseudomonosom

62 PSEUDOMONOTHALLIC
    f pseudo-monothallique
    e pseudomonotálico;
       seudomonotálico
    i pseudomonotallico
    d pseudomonothallisch

63 PSEUDOMUTATION
    f pseudo-mutation *f*
    e pseudomutación *f*;
       seudomutación *f*
    i pseudomutazione *f*
    d Pseudomutation *f*

64 PSEUDOPOLYPLOID (adj.)
    PSEUDOPOLYPLOID
    f pseudo-polyploïde
       pseudo-polyploïde *m*

e pseudopoliploide;
   seudopoliploide
   pseudopoliploide *m*;
   seudopoliploide *m*
i pseudopoliploide
   pseudopoliploide *m*
d pseudopolyploid
   Pseudopolyploide *f*

2365 PSEUDOPOLYPLOIDY
      f pseudo-polyploïdie *f*
      e pseudopoliploidía *f*;
         seudopoliploidía *f*
      i pseudopoliploidismo *m*
      d Pseudopolyploidie *f*

2366 PSEUDOREDUCTION
      f pseudo-réduction *f*
      e pseudoreducción *f*;
         seudoreducción *f*
      i pseudoriduzione *f*
      d Pseudoreduktion *f*

2367 PSEUDOSATELLITE
      f pseudo-satellite *m*
      e pseudosatélite *m*;
         seudosatélite *m*
      i pseudosatellite *m*
      d Pseudosatellit *m*

2368 PSEUDOSELECTIVITY
      f pseudo-sélectivité *f*
      e pseudoselectividad *f*;
         seudoselectividad *f*
      i pseudoselettività *f*
      d Pseudoselektivität *f*

2369 PSEUDOSPINDLE
      f pseudo-fuseau *m*
      e pseudohuso *m*;
         seudohuso *m*
      i pseudofuso *m*
      d Pseudospindel *f*

2370 PSEUDOTELOPHASE
      f pseudo-télophase *f*
      e pseudotelofase *f*;
         seudotelofase *f*
      i pseudotelofase *f*
      d Pseudotelophase *f*

2371 PSEUDO WILD TYPE
f type *m* pseudo-sauvage
e tipo *m* pseudoselvaje
i falsotipo *m* selvatico
d Pseudowildtyp *m*

2372 PUFF
f boursouflure *f*;
puff *m*
e hinchazón *f* (Θ);
puff *m*
i enfiagione *f* (Θ);
puff *m*
d Puff *n*

2373 PUREBRED
f de race *f* pure
e de pura raza *f*
i di razza *f* pura
d reinrassig;
reinerbig

2374 PURE BREED
f race *f* pure
e raza *f* pura
i razza *f* pura
d reine Rasse *f*

2375 PURE LINE
f lignée *f* pure
e línea *f* pura
i linea *f* genetica pura
d reine Linie *f*

2376 PYCNOSIS
f pycnose *f*
e picnosis *f*
i picnosi *f*
d Pyknose *f*

2377 PYCNOTIC
f pycnotique
e picnótico
i picnotico
d pyknotisch

Q

2378 QUADRIPLEX;
QUADRUPLEX
f quadruplex
e cuadruplexo
i quadruplex
d quadriplex

2379 QUADRISEXUALITY
f quadrisexualité *f*
e cuadrisexualidad *f*
i quadrisessualità *f*
d Viergeschlechtigkeit *f*

2380 QUADRIVALENT (adj.)
QUADRIVALENT
f quadrivalent *m*
e cuadrivalente *m*
i quadrivalente *m*
d Quadrivalent *n*

2381 QUALITATIVE
f qualitatif
e cualitativo
i qualitativo
d qualitativ

2382 QUALITATIVE CHARACTER
f caractère *m* qualitatif
e carácter *m* cualitativo
i carattere *m* qualitativo
d qualitatives Merkmal *n*

2383 QUALITATIVE INHERITANCE
f hérédité *f* qualitative
e herencia *f* cualitativa
i eredità *f* qualitativa
d qualitative Vererbung *f*

2384 QUANTITATIVE
f quantitatif
e cuantitativo
i quantitativo
d quantitativ

2385 QUANTITATIVE CHARACTER
f caractère *m* quantitatif
e carácter *m* cuantitativo
i carattere *m* quantitativo
d quantitatives Merkmal *n*

2386 QUANTITATIVE INHERITANCE
f hérédité *f* quantitative
e herencia *f* cuantitativa
i eredità *f* quantitativa
d quantitative Vererbung *f*

2387 QUANTUM EVOLUTION
f évolution *f* du quantum (Θ)
e evolución *f* del quantum
i evoluzione *f* del quantum (Θ)
d Quantenevolution *f*

2388 QUARTET
f quartette *m*
e tetrada *f*
i tetrade *f*
d Quartett *n*;
Tetrade *f*;
Vierzellstadium *n*

2389 QUASIBIVALENT
f quasibivalent *m*
e cuasibivalente *m*
i quasibivalente *m*
d Quasibivalent *n*

2390 QUASILINKAGE
f quasilinkage *m*
e cuasiligamiento *m*
i quasilinkage *m*
d falsche Kopp(e)lung *f*

R

2391 RACE
  f race f
  e raza f
  i razza f
  d Rasse f

2392 RACE DEVELOPMENT;
  RACE HISTORY;
  PHYLOGENESIS;
  PHYLOGENY
  f phylogénèse f;
    phylogénie f
  e filogenia f;
    filogénesis f
  i filogenesi;
    filogenia;
    derivazione f
  d Phylogenie f;
    Phylogenese f;
    Stammesentwicklung f;
    Stammesgeschichte f

  RADIATION, ADAPTIVE
  see 35

2393 RADIOGENETICS
  f radiogénétique f
  e radiogenética f
  i radiogenetica f
  d Strahlengenetik f

2394 RADIOMIMETIC
  f radiomimétique
  e radiomimético
  i radiomimetico
  d radiomimetisch

2395 RAMPANT POLYPLOIDY
  f polyploïdie f étalée
  e poliploidía f exuberante
  i poliploidia f spiegata
  d rampant polyploidy f

2396 RANDOM
  f accidentel;
    aléatoire

  e fortuito;
    aleatorio;
    al azar
  i accidentale;
    aleatorio
  d zufällig;
    zufallsgemäss

2397 RANDOM DRIFT
  f dérive f fortuite
  e desviación f al azar (Θ)
  i derivazione f fortuita
  d zufallsgemässe Drift f

2398 RANDOM FIXATION
  f fixation f accidentelle
  e fixación f al azar
  i fissazione f accidentale
  d zufällige Fixierung (von
    Allelen) f

2399 RANDOMIZATION
  f randomisation f
  e randomisación f (Θ)
  i randomizzazione f;
    scompiglio m
  d Randomisation f

2400 RANDOM MATING;
  PANMIXIA;
  PANMIXIE;
  PANMIXIS
  f panmixie f
  e panmixia f
  i panmissi f
  d Panmixie f

2401 RANDOMNESS
  f accident m
  e azar m
  i accidente m
  d Zufälligkeit f;
    Zufallsgemässheit f

2402 RANGE OF VARIATION
  f étendue f de variation

e   intervalo *m* de variación
i   grado *m* di variazione
d   Variationsbreite *f*

2403   RASSENKREIS;
CIRCLE OF RACES;
RHEOGAMEON
f   rhéogaméon *m*
e   rheogameon *m*
i   reogameon *m*;
contorno *m* di razze
d   Rassenkreis *m*;
Rheogameon *n*

2404   RATE CONCEPT
f   hypothèse *f* de contrôle (⊙)
e   hipótesis *f* de control (⊙)
i   ipotesi *f* di controllo
d   Reaktionsratenhypothese *f*

2405   RATE GENE
f   gène-contrôleur *m*
(de la vitesse du processus
de développement)
e   gen *m* de control
(de la velocidad del
proceso de desarrollo)
i   gene *m* di controllo
(della velocità del
processo di sviluppo)
d   Reaktionsratengen *n*

2406   RATIO CLINE
f   "ratio cline" *m*
e   "ratio cline" *m* (⊙)
i   "ratio-cline" *m* (⊙)
d   Häufigkeitsgradient *m* (#)

2407   REACTIVATION
f   réactivation *f*
e   reactivación *f*
i   riattivazione *f*
d   Reaktivierung *f*

REACTIVATION, MULTIPLE
see 1882

2408   REARRANGEMENT
f   remaniement *m*
e   reacomodo *m*
i   ricostituzione *f*
d   Rearrangement *n*

2409   RECEPTOR
f   récepteur *m*
e   receptor *m*
i   ricevitore *m*
d   Rezeptor *m*

2410   RECESSIVE
f   récessif
e   recesivo
i   recessivo
d   rezessiv

2411   RECESSIVENESS
f   récessivité *f*
e   recesividad *f*
i   recessività *f*
d   Rezessivität *f*

2412   RECIPIENT (adj.)
RECIPIENT
f   accepteur
accepteur *m*
e   recipiente *m*
i   accettore *m*
d   Rezeptor *m*
(auch in Zusammenset-
zungen)

2413   RECIPIENT BACTERIUM
f   bactérie réceptrice *f*
e   bacteria *f* recipiente
i   batterio *m* ricettore
d   Rezeptor-Bakterium *n*

2414   RECIPROCAL
f   réciproque
e   recíproco
i   reciproco
d   reziprok

2415   RECOMBINATION
f   recombinaison *f*
e   recombinación
i   ricombinazione *f*
d   Rekombination *f*;
Neukombination *f*

2416   RECOMBINATION CLASS
f   classe *f* de recombinaison

e clase f de recombinación
i classe f di ricombinazione
d Rekombinationsklasse f

2417 RECOMBINATION FRACTION;
RECOMBINATION PER-
CENTAGE;
RECOMBINATION VALUE
f pourcentage m de recombinaison
e porcentaje m de
recombinación
i valore m di
ricombinazione
d Rekombinationsprozent-
satz m;
Rekombinationswert m

2418 RECOMBINATION INDEX
f indice m de recombinaison
e índice m de recombinación
i indice m di ricombinazione
d Rekombinationsindex m

2419 RECON
f récon m
e recón m
i recon m
d Recon n

2420 RECURRENT
f récurrent
e recurrente
i ricorrente
d zurückgreifend;
rückläufig;
wiederkehrend

2421 RECURRENT RECIPROCAL
SELECTION
f sélection f récurrente
réciproque
e selección f recurrente
recíproca
i selezione f ricorrente
reciproca
d recurrent reciprocal selection f

2422 REDUCED PRECOCITY

f précocité f réduite
e precocidad f reducida
i precocità f ridotta
d reduzierte Präkozität f

2423 REDUCTION
f réduction f
e reducción f
i riduzione f
d Reduktion f

REDUCTION, SOMATIC
see 2613

2424 REDUCTIONAL
f réductionnel
e reduccional
i riduzionale
d reduktionell;
Reduktions- (in Zusammen-
setzungen)

2425 REDUCTIONAL DIVISION;
REDUCTION DIVISION;
HETEROTYPIC DIVISION
f division f réductionnelle;
division f hétérotypique
e división f reductora;
división f reduccional;
división f heterotípica
i divisione f eterotipica;
divisione f riduzionale
d Reduktionsteilung f;
heterotypische Teilung f (

2426 REDUCTIONAL GROUPING
f groupement m réductionnel
e agrupación f reduccional
i aggruppamento m riduziona
d Reduktionsgruppierung f (

2427 REDUCTIONAL MITOSIS
f mitose f réductionnelle;
mitose f hétérotypique
e mitosis f reductora;
mitosis f reduccional;
mitosis f heterotípica
i mitosi f eterotipica;
mitosi f riduzionale
d Reduktionsmitose f

2428 REDUCTIVE INFECTION
f infection f réductive
e infección f reductiva
i infezione f riduttiva
d Reduktiv-Infektion f

2429 REDUPLICATION
f réduplication f
e reduplicación f
i reduplicazione f
d Reduplikation f

REGENERATION, MACRO-
NUCLEAR
see 1696

2430 REGRESSION
f régression f
e regresión f
i regressione f
d Regression f

REGRESSION, MULTIPLE
see 1883

REGRESSION, PARTIAL
see 2085

2431 REGRESSION COEFFICIENT
f coefficient m de régression
e coeficiente m de regresión
i coefficiente m di
regressione
d Regressionskoeffizient m

2432 REGRESSION EQUATION
f équation f de régression
e ecuación f de regresión
i equazione f di regressione
d Regressionsgleichung f

2433 REGRESSION OF DAUGHTER
(OFFSPRING) ON DAM
f régression f mère à fille
e regresión f madre a hija
i regressione f madre a
figlia
d Mutter-Tochter-Regres-
sion f

2434 REGRESSIVE
f régressif
e regresivo
i regressivo
d regressiv

2435 REGULATION DEVELOPMENT
f - -
e - -
i - -
d Regulationsentwicklung f

2436 RELATED
f apparenté ;
parent
e pariente
i imparentato
d verwandt

2437 RELATIONAL
f relatif ;
connexe
e relacional
i relazionale
d Relations- (in Zusammensetzungen)

2438 RELATIONAL COILING
f enroulement m réciproque
e arrollamiento m recíproco
i avvolgimento m reciproco
d Relationsspirale f

2439 RELATIONAL
INCOMPATIBILITY
f incompatibilité f réciproque
e incompatibilidad f
recíproca
i incompatibilità f reciproca
d reziproke Inkompatibilität f (≠)

2440 RELATIONSHIP
f parenté f ;
consanguinité f
e parentesco m ;
consanguinidad f
i consanguinità f ;
parentela f
d Verwandtschaft f

RELATIONSHIP, DIRECT
see 829

2441   RELIC SPIRAL
    f   spirale *f* rémanente
    e   espiral *f* remanente;
       vieja espiral *f* de profase
    i   spirale *f* rimanente
    d   Reliktspirale *f*

2442   RELICT
    f   survivance *f*
    e   supervivencia *f* (☉)
    i   sopravvivenza *f* (☉)
    d   Relikt *n*

2443   REMOTE EFFECT
    f   effet *m* retardé
    e   efecto *m* remoto
    i   effetto *m* remoto
    d   remote effect *m*

2444   REMOTE RELATIONSHIP
    f   consanguinité *f* lointaine
    e   consanguinidad *f* remota
    i   consanguinità *f* remota
    d   weite Verwandtschaft *f*;
       entfernte Verwandtschaft *f*

RENEWAL OF BLOOD,
PARTIAL
see 2086

2445   REPAIR
    f   réparation *f*
    e   reparación *f*
    i   riparazione *f*
    d   Reparatur *f*

2446   REPARABLE MUTANT
    f   mutant *m* réparable
    e   mutante *m* reparable
    i   mutante *m* restituibile
    d   reparable Mutante *f*

2447   REPATTERNING
    f   réarrangement *m*
    e   remodelación *f*
    i   - -
    d   (Chromosomen-)Ummuste-
       rung *f*

2448   REPEAT;
       DUPLICATION
    f   répétition *f*
    e   repetición *f*
    i   ripetizione *f*
    d   (intrachromosomale)
       Duplikation *f*

2449   REPLACEMENT HYPOTHESI
    f   hypothèse *f* du remplace-
       ment
    e   hipótesis *f* de sustitución
    i   ipotesi *f* di sostituzione
    d   "replacement"-Hypothese *f*

REPLICA HYPOTHESIS,
PARTIAL
see 2087

2450   a)REPLICATION;
       SELF-DUPLICATION
       (cytogenetics)
       b)REPLICATION (statistics)
    f   a)autoduplication *f*
         (cytogénétique)
       b)répétition *f* (statistique)
    e   a)repetición *f*
         (citogenética)
       b)repetición (estadística) *f*
    i   a)replicazione *f*;
         autoduplicazione *f*
         (citogenetica)
       b)replicazione *f*
         (statistica)
    d   a)Replikation *f*;
         Autoduplikation *f*
         (Cytogenetik)
       b)Wiederholung *f*
         (Statistik)

2451   REPRESSOR
    f   répresseur *m*
    e   represor *m*
    i   repressore *m*
    d   Repressor *m*

2452   REPRODUCTION
    f   reproduction *f*
    e   reproducción *f*
    i   riproduzione *f*
    d   Reproduktion *f*;
       Fortpflanzung *f*

REPRODUCTION, ASEXUAL
see 242

2453 REPRODUCTION RATE
f taux *m* de reproduction
e porcentaje *m* reproductor;
proporcionalidad *f*
reproductora
i tasso *m* di riproduzione
d Reproduktionsrate *f*;
Fortpflanzungsrate *f*

2454 REPRODUCTIVE
f reproductif
e reproductivo
i riproduttivo
d reproduktiv;
Fortpflanzungs- (in Zu-
sammensetzungen)

2455 REPRODUCTIVE MINIMUM
UNIT
f unité *f* minima reproduc-
trice
e unidad *f* mínima
reproductiva
i unità *f* minima
riproduttiva
d kleinste Fortpflanzungsein-
heit *f*

2456 REPULSION
f répulsion *f*
e repulsión *f*
i ripulsione *f*
d Abstossung *f*;
Repulsion *f*

2457 REPULSION PHASE
f phase *f* de répulsion
e fase *f* de repulsión
i fase *f* di ripulsione
d Repulsionsphase *f*

2458 RESIDUAL
f résiduel
e residual
i residuale
d restlich;
Rest- (in Zusammenset-
zungen)

2459 RESIDUAL CHROMOSOME
f chromosome *m* résiduel
e cromosoma *m* residual
i cromosoma *m* residuale
d Residualchromosom *n*

2460 RESIDUAL GENOTYPE
f génotype *m* résiduel
e genótipo *m* residual
i genotipo *m* residuale
d Restgenotyp *m*

2461 RESIDUAL HEREDITY
f hérédité *f* résiduelle
e herencia *f* residual
i eredità *f* residuale
d genotypisches Milieu *n*

2462 RESIDUAL HETEROZYGOTE
f hétérozygote *m* résiduel
e heterocigoto *m* residual
i eterozigote *m* residuale
d Residual-Heterozygote *f* (≠)

2463 RESIDUAL HOMOLOGY
f homologie *f* résiduelle
e homología *f* residual
i omologia *f* residuale
d residuale Homologie *f*;
Resthomologie *f* (≠)

RESISTANCE, BIOTIC
see 371

2464 RESTING NUCLEUS
f noyau *m* au repos;
noyau *m* quiescent
e núcleo *m* en reposo
i nucleo *m* in riposo
d Ruhekern *m*

2465 RESTING PHASE
f phase *f* de repos
e fase *f* de reposo
i fase *f* di riposo
d Ruhephase *f*

2466 RESTING STAGE
.f interphase *f*
e estado *m* de reposo
i stadio *m* di riposo;
resting stage
d Ruhephase *f*

2467   RESTITUTION
        f restitution f
        e restitución f
        i restituzione f
        d Restitution f

2468   RESTITUTIONAL
        f de restitution
        e restitucional
        i restituzionale
        d Restitutions- (in Zusam-
           mensetzungen)

2469   RESTORATION
        f restauration f
        e restauración f
        i restaurazione f
        d Restauration f

2470   RESTORATION BACK-CROSS
        f back-cross m de restau-
           ration
        e retrocruzamiento m de
           restauración
        i back-cross m di restauro;
           rincrocio m di restauro
        d Restaurationsrückkreu-
           zung f (#)

2471   RESTRICTION GENE
        f gène m restrictif
        e gen m restrictivo
        i gene m restrittivo
        d Restriktionsgen n

2472   RESTRICTIVE
        f restrictif
        e restrictivo
        i restrittivo
        d restriktiv

2473   RETARDATION
        f retard m
        e retardación f
        i ritardo m
        d Retardation f;
           Verzögerung f

2474   RETICULUM
        f réticulum m;
           réseau m
        e retículo m

        i reticolo m
        d Retikulum n

2475   RETRANSLOCATION
        f retranslocation f
        e retranslocación f
        i retraslocazione f
        d Retranslokation f

2476   RETURN MUTATION;
        BACK MUTATION;
        REVERSE MUTATION
        f mutation f reverse;
           mutation f en retour;
           mutation f renversée
        e mutación f invertida;
           mutación f de retroceso;
           retromutación f
        i retro-mutazione f;
           mutazione f rovescia;
           mutazione f di ritorno
        d Rückmutation f

2477   REUNION
        f réunion f
        e reunión f
        i riunione f
        d Reunion f

2478   REVERSAL HETEROPYCNOSIS
        f hétéropycnose f reverse
        e heteropicnosis f de inversió
        i eteropicnosi f invertita (Θ)
        d Heteropyknoseumkehr f

2479   REVERSE MUTATION;
        BACK MUTATION;
        RETURN MUTATION
        f mutation f reverse;
           mutation f en retour;
           mutation f renversée
        e mutación f invertida;
           mutación f de retroceso;
           retromutación f
        i retro-mutazione f;
           mutazione f rovescia;
           mutazione f di ritorno
        d Rückmutation f

2480   REVERSION
        f réversion f

e reversión *f*
i reversione *f*
d Reversion *f*

2481 RHEOGAMEON;
RASSENKREIS;
CIRCLE OF RACES
f rhéogaméon *m*
e rheogameon *m* (Ө)
i reogameon *m*
d Rassenkreis *m*;
Rheogameon *n*

2482 RIBONUCLEASE
f ribonucléase *f*
e ribonucleasa *f*
i ribonucleasi *f*
d Ribonuklease *f*

2483 RIBONUCLEIC ACID (RNA)
f acide *m* ribonucléique (ARN)
e ácido *m* ribonucléico (ARN)
i acido *m* ribonucleinico (ARN)
d Ribosenukleinsäure (RNS) *f*;
Ribonukleinsäure *f*

2484 RIBOSE
f ribose *f*
e ribosa *f*
i ribosio *m*
d Ribose *f*

2485 RIBOSOMAL
f ribosomal *m*
e ribosomal
i ribosomale
d ribosomal
Ribosomen- (in Zusammensetzungen)

2486 RIBOSOME;
MICROSOMAL PARTICLES
f ribosome *m*;
particules *f pl* microsomiques
e ribosoma *m*
i ribosoma *m*;
particelle *f pl*
microsomomali
d Ribosom *n*

2487 RING
f anneau *m*
e anillo *m*
i anello *m*
d Ring *m*

2488 RING CHROMATID
f chromatide *f* en anneau
e cromatidio *m* anular
i cromatido *m* anulare
d Ringchromatide *f*

2489 RING CHROMOSOME
f chromosome *m* en anneau
e cromosoma *m* anular
i cromosoma *m* anulare
d Ringchromosom *n*

RING CHROMOSOME, DOUBLE
SIZED
see 877

2490 RNA MESSENGER
f ARN messager *m*
e mensajero *m* ARN
i messaggero *m* RNA
d Messenger-RNS *f*;
Boten-RNS *f*

2491 ROBERTSONIAN
f robertsonien
e robertsoniano
i robertsoniano
d Robertsonsche

2492 ROTATION
f rotation *f*
e rotación *f*
i rotazione *f*
d Rotation *f*

2493 ROTATIONAL CROSS-
BREEDING
f croisement *m* en rotation
e cruzamiento *m* de rotación
i incrocio *m* di rotazione
d Rotationskreuzung *f*

S

2494 SALTANT
f mutant *m*
e saltante *m*
i saltante *m*;
mutante *m*
d Saltante *f*

2495 SAMPLE
f échantillon *m*
e muestra *f*
i campione *m*
d Stichprobe *f*;
Probe *f*

2496 SAT-CHROMOSOME
f SAT-chromosome *m*
e cromosoma *m* SAT
i SAT-cromosoma *m*
d SAT-Chromosom *n*

2497 SATELLITE
f satellite *m*;
trabant *m*
e satélite *m*
i satellite *m*;
trabante *m*
d Satellit *m*;
Trabant *m*

2498 SATELLITE NUCLEOLUS
f nucléole *m* satélite
e nucléolo *m* satélite
i nucleolo *m* satellite
d Satellitennukleolus *m*

2499 SATURATION EFFECT
f effet *m* de saturation
e efecto *m* de saturación
i effetto *m* di saturazione
d Sättigungseffekt *m*

2500 SAT-ZONE
f SAT-zone *f*
e zona *f* SAT
i SAT-zona *f*
d SAT-Zone *f*

2501 SCHIZOGENESIS ;
SCHIZOGONY;
SCISSIPARITY
f scissiparité *f*
e escisiparidad *f*
i scissiparità *f*;
schizogonia *f*;
schizogenesi *f*
d Schizogonie *f*

2502 SECONDARY
f secondaire
e secundario
i secondario
d sekundär

2503 SECONDARY CENTRIC REG
f région *f* centrique secondaire
e región *f* céntrica
secondaria
i regione *f* centrica
secondaria
d Neocentromer *n*

2504 SECONDARY CONSTRICTIO
f constriction *f* secondaire
e constricción *f* secundaria
i costrizione *f* secondaria
d sekundäre Einschnürung

2505 SECONDARY KINETOCHOR
f kinétochore *m* secondaire
e cinetócoro *m* secundario
i cinetocoro *m* secondario
d sekundäres Kinetochor *n*

2506 SECONDARY POLYPLOIDY
f polyploïdie *f* secondaire
e poliploidia *f* secundaria
i poliploidismo *m*
secondario
d sekundäre Polyploidie *f*

2507 SECONDARY SEX RATIO
f proportion *f* secondaire
des sexes

e proporción *f* secundaria
  de los sexos
i proporzione *f* secondaria
  dei sessi
d sekundäres Geschlechterverhältnis *n*

2508 SECTION
f section *f*
e sección *f*
i sezione *f*
d Sektion *f*

2509 SECTOR
f secteur *m*
e sector *m*
i settore *m*
d Sektor *m*

2510 SECTORIAL
f sectoriel
e sectorial
i settoriale
d sektorial

2511 SECTORIAL CHIMAERA
f chimère *f* sectorielle
e quimera *f* sectorial
i chimera *f* settoriale
d Sektorialchimäre *f*

2512 SEGMENT
f segment *m*
e segmento *m*
i segmento *m*
d Segment *n*

SEGMENT, DIFFERENTIAL
see 777

2513 SEGMENTAL
f segmentaire
e segmental
i segmentale
d segmental;
  Segment- (in Zusammen-
  setzungen)

2514 SEGMENTAL ALLOPOLY-
  PLOID
f allopolyploïde *m* segmentaire
e alopoliploide *m* segmental

i alloploide *m* segmentale
d Segmentallopolyploide *f*

SEGMENTS, DISLOCATED
see 837

2515 SEGREGANT
f ségrégant *m*
e segregante *m*
i segregante *m*
d Spalter *m*

2516 SEGREGATION
f ségrégation *f*
e segregación *f*
i segregazione *f*
d Segregation *f*;
  Aufspaltung *f*

SEGREGATION, NUCLEAR
see 1937

SEGREGATION,
TRANSGRESSIVE
see 2831

2517 SEGREGATIONAL LAG;
  SEGREGATION DELAY
f retard *m* de ségrégation
e retardo *m* de segregación
i ritardo *m* di segregazione
d Segregationsverzögerung *f*

2518 SEGREGATIONAL STERILITY
f stérilité *f* de ségrégation
e esterilidad *f* de segregación
i sterilità *f* di segregazione
d Aufspaltungssterilität *f*

2519 SEGREGATION DELAY
  SEGREGATIONAL LAG
f retard de ségrégation *m*
e retardo de segregación *m*
i ritardo di segregazione *m*
d Segregationsverzögerung *f*

2520 SELECTION
f sélection *f*
e selección *f*
i selezione *f*
d Selektion *f*;
  Zuchtwahl *f*

SELECTION, CANALIZING
see 422

SELECTION, DIRECTIONAL
see 828

SELECTION, DISRUPTIVE
see 849

SELECTION, GERMINAL
see 1127

SELECTION, INDIVIDUAL
see 1715

2521 SELECTION COEFFICIENT
f coefficient *m* de sélection
e coeficiente *m* de selección
i coefficiente *m* di selezione
d Selektionskoeffizient *m*

2522 SELECTION DIFFERENTIAL
f sélection *f* différentielle
e diferencial de selección *f*
i selezione *f* differenziale
d Selektionsdifferential *n*

2523 SELECTION FORCE
f force *f* de sélection
e fuerza *f* de selección
i forza *f* di selezione
d Selektionskraft *f*;
Selektionsfaktor *m*

2524 SELECTION INDEX
f index *m* de sélection
e índice *m* de selección
i indice *m* di selezione
d Selektionsindex *m*

2525 SELECTION INTENSITY
f intensité *f* de sélection
e intensidad *f* de selección
i intensità *f* di selezione
d Selektionsintensität *f*

2526 SELECTION PRESSURE;
SELECTIVE PRESSURE
f pression *f* de sélection
e presión *f* de selección
i pressione *f* di selezione
d Selektionsdruck *m*

2527 SELECTIVE
f sélectif
e selectivo
i selettivo
d selektiv

2528 SELECTIVE FACTOR
f facteur *m* sélectif
e factor *m* selectivo
i fattore *m* selettivo
d Selektionsfaktor *m*

2529 SELECTIVE PEAK
f pic *m* de sélection
e máximo *m* de selección
i punta *f* di selezione
d Selektionsgipfel *m*

2530 SELECTIVE PRESSURE;
SELECTION PRESSURE
f pression *f* de sélection
e presión *f* de selección
i pressione *f* di selezione
d Selektionsdruck *m*

2531 SELF-COMPATIBILITY
f autofertilité *f*
e autocompatibilidad *f*
i autofertilità *f*
d Selbst-Kompatibilität *f*

2532 SELF-FERTILE
f autofertile
e autofértil
i autofertile
d selbstfertil

2533 SELF-FERTILITY
f autofertilité *f*
e autofertilidad *f*
i autofertilità *f*
d Selbstfertilität *f*

2534 SELF-FERTILIZATION;
SELFING
f autofécondation *f*
e autofecundación *f*
i autofecondazione *f*
d Selbstbefruchtung *f*;
Selbstung *f*

2535 SELF-FERTILIZER
    f individu *m* à autorepro-
       duction
    e autofertilizante *m*
    i autofertilizzante *m* (☉)
    d Selbstbefruchter *m*

2536 SELF-INCOMPATIBILITY
    f autostérilité *f*
    e autoesterilidad *f*
    i autosterilità *f*
    d Selbst-Inkompatibilität *f*

2537 SELFING;
    SELF FERTILIZATION
    f autofécondation *f*
    e autofecundación *f*
    i autofecondazione *f*
    d Selbstung *f*;
       Selbstbefruchtung *f*

2538 SELF-POLLINATION
    f auto-pollinisation *f*
    e autopolinización *f*
    i autoimpollinazione *f*
    d Selbstbestäubung *f*

2539 SELF-STERILE
    f autostérile
    e autoestéril
    i autosterile
    d selbststeril

2540 SELF-STERILITY
    f autostérilité *f*
    e autoesterilidad *f*
    i autosterilità *f*
    d Selbststerilität *f*

2541 SEMEN
    f sperme *m* ;
       liquide *m* séminal;
       semence *f*
    e semen;
       líquido *m* seminal;
       esperma-seminal *m*
    i sperma *m* ;
       seme *m*
    d Samen *m* ;
       Sperma *n*;
       Samenflüssigkeit *f*

2542 SEMIALLELE
    f semi-allèle *m*
    e semialelo *m*
    i semiallelo *m*
    d Semiallel *n*

2543 SEMIALLELISM
    f semi-allélisme *m*
    e semialelismo *m*
    i semiallelismo *m*
    d Semiallelie *f*

2544 SEMIAPOSPORY
    f semi-aposporie *f*
    e semiaposporia *f*
    i semiaposporia *f*
    d Semiaposporie *f*

2545 SEMIBIVALENT (adj.)
    SEMIBIVALENT
    f semi-bivalent
       semi-bivalent *m*
    e semibivalente
       semibivalente *m*
    i semibivalente
       semibivalente *m*
    d Semibivalent *n*

2546 SEMICARYOTYPE;
    SEMIKARYOTYPE
    f semi-caryotype *m*
    e semicariotipo *m*
    i semicariotipo *m*
    d Semikaryotyp *m*

2547 SEMICHIASMA
    f semi-chiasma *m*
    e semiquiasma *m*
    i semichiasma *m*
    d Semichiasma *n*

2548 SEMIDOMINANCE
    f semi-dominance *f*
    e semidominancia *f*
    i semidominanza *f*
    d Semidominanz *f*;
       unvollständige Dominanz *f*

2549 SEMIDOMINANT (adj.)
    SEMIDOMINANT
    f semi-dominant
       semi-dominant *m*

e semidominante
   semidominante *m*
i semi-dominante
   semi-dominante *m*
d semidominant;
   unvollständig dominant

2550 SEMIHETEROTYPIC
f semi-hétérotypique
e semiheterotípico
i semieterotipico
d semiheterotypisch

2551 SEMIHOMOLOGOUS
f semi-homologue
e semihómologo
i semiomologo
d semihomolog

2552 SEMIHOMOLOGY
f semi-homologie *f*
e semihomología *f*
i semiomologia *f*
d Semihomologie *f*

2553 SEMI-INCOMPATIBILITY
f semi-incompatibilité *f*
e semi-incompatibilidad *f*
i semi-incompatibilità *f*
d Semi-Inkompatibilität *f*

2554 SEMIKARYOTYPE
   SEMICARYOTYPE
f semi-caryotype *m*
e semicariótipo *m*
i semicariotipo *m*
d Semikaryotyp *m*

2555 SEMILETHAL
f semi-léthal
e semiletal
i semiletale
d semiletal

2556 SEMILOCALIZED CEN-
   TROMERE
f centromère *m* semi-
   localisé
e centrómero *m* semi-
   localizado
i centromero *m* semi-
   localizzato

d semilokalisiertes Cen-
   tromer *n*

2557 SEMI-MATURATION
f semi-maturation *f*
e semimaduración *f*
i semimaturazione *f*
d Semi-Maturation *f*

2558 SEMIRECESSIVE
f semi-récessif
e semirrecesivo
i semirecessivo
d semirezessiv;
   unvollständig rezessiv

2559 SEMISPECIES
f semi-espèce *f*
e semiespecie *f*
i semispecie *f*
d Semispecies *f*

2560 SEMISTERILE
f semi-stérile
e semiestéril
i semisterile
d semisteril

2561 SEMISTERILITY
f semi-stérilité *f*
e semiesterilidad *f*
i semisterilità *f*
d Semisterilität *f*

2562 SEMIUNIVALENT
f semi-univalent *m*
e semiunivalente *m*
i semiunivalente *m*
d Semiunivalent *n*

2563 SENSIBLE PHASE
f phase *f* sensible
e fase *f* sensitiva
i fase *f* sensitiva
d sensible Phase *f*

2564 SENSITIVE PERIOD
f période *f* sensible
e período *m* sensitivo
i periodo *m* sensitivo
d sensible Periode *f*

565 SENSITIVITY PEAK
  f pic *m* de sensibilité
  e máximo *m* de sensibilidad
  i punta *f* di sensitività
  d Sensibilitätsmaximum *n*;
     Empfindlichkeitsgipfel *m*

566 SENSITIZING EFFECT
  f effet *m* sensibilisant
  e efecto *m* sensibilizador
  i effetto *m* sensibilizzante
  d Sensibilisierungswirkung *f*

567 SEPARATION
  f séparation *f*
  e separación *f*
  i separazione *f*
  d Trennung *f*

568 SEPTISOMIC
  f septisomique
  e septisómico
  i settisomico
  d septisom

  SERIE DE POISSON
  see 2191

569 SESQUIDIPLOID (adj.)
  SESQUIDIPLOID
  f sesquidiploïde
     sesquidiploïde *m*
  e sesquidiploide
     sesquidiploide *m*
  i sesquidiploide
     sesquidiploide *m*
  d sesquidiploid
     Sesquidiploide *f*

570 SESQUIDIPLOIDY
  f sesquidiploïdie *f*
  e sesquidiploidía *f*
  i sesquidiploidismo *m*
  d Sesquidiploidie *f*

571 SET
  f garniture *f*;
     jeu *m*
  e juego *m*
  i guarnizione *f*;
     gioco *m*

  d Satz *m*;
     Garnitur *f*

2572 SEX
  f sexe *m*
  e sexo *m*
  i sesso *m*
  d Geschlecht *n*

2573 SEX CHROMATIN
  f chromatine *f* sexuelle
  e cromatina *f* sexual
  i cromatina *f* sessuale
  d Geschlechtschromatin *n*

2574 SEX CHROMOSOME
  f chromosome *m* sexuel
  e cromosoma *m* sexual
  i cromosoma *m* sessuale
  d Geschlechtschromosom *n*

  SEX CHROMOSOMES,
  MULTIPLE
  see 1884

2575 SEX-CONTROLLED
  f contrôlé par le sexe *m*
  e controlado por el sexo *m*
  i controllato dal sesso *m*
  d geschlechtskontrolliert

2576 SEX DEGRADATION
  f dégradation *f* de la sexua-
     lité
  e degradación *f* sexual
  i degradazione *f* della
     sessualità
  d Geschlechtsabbau *m* (#)

2577 SEX DETERMINATION;
  SEXUAL DETERMINATION
  f détermination *f* du sexe;
     détermination *f* sexuelle
  e determinación *f* sexual;
     determinación *f* del sexo
  i determinazione *f* del sesso;
     determinazione *f* sessuale
  d Geschlechtsbestimmung *f*

  SEX DETERMINATION,
  BALANCE THEORY OF
  see 328

2578 SEX DIMORPHISM
  f dimorphisme *m* sexuel
  e dimorfismo *m* sexual
  i dimorfismo *m* sessuale
  d Geschlechtsdimorphismus *m*

2579 SEX-INFLUENCED
  f influencé par le sexe *m*
  e influenciado por el sexo *m*
  i influenzato dal sesso *m*
  d geschlechtsbeeinflusst

2580 SEX INHERITANCE;
  SEXUAL INHERITANCE
  f hérédité *f* sexuelle
  e herencia *f* sexual
  i eredità *f* sessuale
  d Geschlechtsvererbung *f*;
     geschlechtlichte Vererbung *f*

2581 SEX-LIMITED
  f limité à un sexe
  e limitado por el sexo
  i limitato ad un sesso
  d geschlechtsbegrenzt

2582 SEX LINKAGE
  f hérédité *f* liée au sexe
  e ligamiento *m* con el sexo;
     herencia *f* ligada al sexo
  i eredità *f* legata al sesso
  d Geschlechtskopp(e)lung *f*;
     Geschlechtsgebundenheit *f*

2583 SEX-LINKED
  f lié au sexe
  e ligado al sexo
  i legato al sesso
  d geschlechtsgekoppelt

2584 SEX RATIO
  f proportion *f* des sexes
  e proporción *f* de los sexos
  i proporzione *f* dei sessi
  d Geschlechterverhältnis *n*

  SEX RATIO, PRIMARY
  see 2299

  SEX RATIO, SECONDARY
  see 2507

2585 SEX REVERSAL
  f inversion *f* du sexe
  e sexualidad *f* invertida
  i inversione *f* sessuale
  d Geschlechtsumkehr *f*

2586 SEXUAL
  f sexuel;
     sexué
  e sexual
  i sessuale
  d geschlechtlich;
     sexual;
     sexuell

2587 SEXUAL DETERMINATION
  SEX DETERMINATION
  f détermination *f* du sexe;
     détermination *f* sexuelle
  e determinación *f* sexual;
     determinación *f* del sexo
  i determinazione *f* del ses
     determinazione *f* sessua
  d Geschlechtsbestimmung

2588 SEXUAL INHERITANCE;
  SEX INHERITANCE
  f hérédité *f* sexuelle
  e herencia *f* sexual
  i eredità *f* sessuale
  d Geschlechtsvererbung *f*;
     geschlechtliche Vererbu

2589 SEX VESICLE
  f vésicule *f* sexuelle
  e vesícula *f* sexual
  i vescicola *f* sessuale
  d sex vesicle *f*

2590 SHATTERING
  f fragmentation *f*
  e fragmentación *f*
  i frammentazione *f*
  d Fragmentation *f*

2591 a) SHIFT;
     DISPLACEMENT
  b) SHIFT
  f a) translation *f*;
        décalage *m*
     b) masquage *m*

e a) desviación *f*
  b) escamoteo *m*
i a) spostamento *m*
  b) mascheramento *m*
d a) Translokation *f;*
     Shift *n*
  b) Shift *n* (Tarnung)

2592 SIBS *pl;*
     SIBLINGS *pl*
     f frères et soeurs *pl*
     e hermanos y hermanas *pl*
     i fratelli e sorelle *pl*
     d Geschwister *pl*

2593 SIDE-BY-SIDE ASSOCIATION
     f association *f* "côte à côte"
     e asociación *f* lateral
     i associazione *f* laterale
     d seitliche Aneinanderla-
        gerung *f*

2594 SIGMA
     f sigma *m*
     e sigma *m*
     i sigma *m*
     d Sigma *n*

2595 SIMPLE CORRELATION
     f corrélation *f* simple
     e correlación *f* simple
     i correlazione *f* semplice
     d einfache Korrelation *f*

2596 SIMPLE REGRESSION
     f régression *f* simple
     e regresión *f* simple
     i regressione *f* semplice
     d einfache Regression *f*

2597 SIMPLEX (adj.)
     SIMPLEX
     f simplex
        simplex *m*
     e simplexo
        simplexo *m*
     i simplex
        simplex *m*
     d simplex
        Simplex *m*

2598 SIMULTANEOUS ADAPTATION
     f adaptation *f* simultanée
     e adaptación *f* simultánea
     i adattazione *f* simultanea
     d Simultanadaptation *f*

2599 SIMULTANEOUS MUTATION
     f mutation *f* simultanée
     e mutación *f* simultánea
     i mutazione *f* simultanea
     d Simultanmutation *f*

2600 SINGLE CHROMOSOME
     COMPENSATING SUBSTITUTION
     LINE
     f lignée *f* de substitution
        compensatoire pour un
        chromosome
     e línea *f* de substitución
        compensatoria a
        cromosoma simple
     i linea *f* di sostituzione di
        compensazione a cromosoma
        unico
     d Linie *f* mit einem substi-
        tuierten Chromosom

2601 SINGLE CROSS
     f croisement *m* unique
     e cruzamiento *m* simple
     i incrocio *m* unico
     d einmalige Kreuzung *f*

2602 SINGLE DOSE EXPRESSION
     f expression *f* d'une dose
        unique
     e expresión *f* de dosis
        simple
     i espressione *f* di dose
        singola
     d single dose expression *f;*
        Expressivität *f* eines
        einzelnen Allels

2603 SINGLE LOCUS HETEROSIS
     f hétérosis *f* pour un seul
        locus
     e heterosis *f* para un locus
        simple
     i eterosis *m* per un locus
        unico

d monogene Heterosis *f*;
durch einen einzigen Locus
bedingte Heterosis *f*

2604 SISTER (CHROMATID)
f (chromatide) soeur *f*
e (cromatidio) hermano *m*
i (cromatidio) fratello *m*
d Schwester *f* (-Chromatide)

2605 SISTER CHROMATIDS
f chromatides soeurs *f pl*
e cromatidios hermanos *mpl*
i cromatidi fratelli *mpl*
d Schwesterchromatiden *f pl*

2606 SISTER REUNION (S.R.)
f réunion *f* des chromatides
soeurs (R.S.)
e reunión *f* de los cromatidios hermanos
i riunione *f* dei cromatidi
fratelli
d Schwesterchromatiden-
Reunion *f*

2607 SISTER-STRAND CROSSING-
OVER
f crossing-over *m* entre
"sister-strand";
crossing-over *m* entre
filaments-frères
e sobrecruzamiento *m*
fraternal trenzado
i sovrincrocio *m* di filamenti
fratelli
d Schwesterstrang-Crossing-
over *n*

2608 SITE
f site *m*
e sitio *m*
i sito *m*
d Site *m*

2609 SKEW BIVALENT
f bivalent *m* asymétrique
e bivalente *m* asimétrico
i bivalente *m* asimmetrico
d asymmetrisches Bivalent *n*

2610 SKEWNESS
f asymétrie *f*
e asimetría *f*
i asimmetria *f*
d Asymmetrie *f*

2611 SOMATIC
f somatique
e somático
i somatico
d somatisch

2612 SOMATIC MUTATION
f mutation *f* somatique;
mutation *f* de bourgeon
e mutación *f* somática
i mutazione *f* somatica
d somatische Mutation *f*

2613 SOMATIC REDUCTION
f réduction *f* somatique
e reducción *f* somática
i riduzione *f* somatica
d somatische Reduktion *f*

2614 SOMATOGAMY
f somatogamie *f*
e somatogamia *f*
i somatogamia *f*
d Somatogamie *f*

2615 SOMATOPLASTIC
f somatoplastique
e somatoplástico
i somatoplastico
d somatoplastisch

2616 SPECIAL SEGMENT
f segment *m* spécial
e segmento *m* especial
i segmento speciale *m*
d Spezialsegment *n*

2617 SPECIAL SEGMENTS BRIDG
f pont *m* de segments spéci
e puente *m* de segmentos
especiales
i ponte *m* di segmenti spec
d Spezialsegmentbrücke *f*

2618   SPECIATION
       f spéciation f
       e especiación f
       i speciazione f
       d Speciation f;
          Artbildung f

2619   SPECIES
       f espèce f
       e especie f
       i specie f
       d Art f;
          Species f;
          Spezies f

2620   SPECIFIC COMBINING
       ABILITY
       f capacité f spécifique à la
          combinaison
       e aptitud f combinatoria
          específica
       i capacità f speciale di
          combinazione
       d spezifische Kombinations-
          eignung f

2621   SPECIFICITY
       f spécificité f
       e especificidad f
       i specificità f
       d Spezifität f

       SPECIFICITY, MUTAGENE
       see 1890

2622   SPECIFIC LOCUS METHOD
       f méthode f du locus
          spécifique
       e método m del locus
          específico
       i metodo m del locus
          specifico
       d specific locus-Methode f

       SPECTRUM, BIOLOGICAL
       see 355

2623   SPERMATELEOSIS
       f spermatéléose f
       e espermateleosis f
       i spermateleosi f
       d Spermateleosis f

2624   SPERMATID
       f spermatide f
       e espermatida f
       i spermatidio m
       d Spermatide f

2625   SPERMATIUM, pl -ia
       f spermatie f
       e espermacio m
       i spermazio m
       d Spermatium n, pl -ien

2626   SPERMATOCYTE
       f spermatocyte m
       e espermatócito m
       i spermatocito m
       d Spermatocyte f

2627   SPERMATOGENESIS
       f spermatogénèse f
       e espermatogénesis f
       i spermatogenesi f
       d Spermatogenese f

2628   SPERMATOGONIUM
       f spermatogonie f
       e espermatogonio m
       i spermatogonio m
       d Spermatogonium n

2629   SPERMATOZOID
       f spermatozoïde m
       e espermatozoïde m
       i spermatozoïde m
       d Spermatozoid n

2630   SPHEROME
       f sphérome m
       e esferoma m
       i sferoma m
       d Sphärom n

2631   SPHEROPLAST
       f sphéroplaste m
       e esferoplasto m
       i sferoplasto m
       d Sphäroplast m

2632   SPHEROSOME
       f sphérosome m
       e esferosoma m

i sferosoma *m*
d Sphärosom *n*

2633 SPHERULE
f sphérule *f*
e esférula *f*
i sferula *f*
d Sphaerula *f*

2634 SPINDLE
f fuseau *m*
e huso *m*
i fuso *m*
d Spindel *f*

SPINDLE, BARRELLIKE
see 334

SPINDLE, CENTRIC
see 436

SPINDLE, HOLLOW
see 1329

2635 SPINDLE ATTACHMENT
f centromère *m*
e centrómero *m*
i centromero *m*
d Spindelfaseransatzstelle *f*;
   Spindelinsertionsstelle *f*;
   Centromer *n*

2636 SPINDLE FIBRE
f fibre *f* fusoriale
e fibra *f* del huso
i fibra *f* del fuso
d Spindelfaser *f*

2637 SPINDLE POISON
f poison *m* fusorial
e veneno *m* del huso
i veleno *m* del fuso
d Spindelgift *n*

2638 SPINDLE PRECOCITY
f formation *f* précoce du
   fuseau
e precocidad *f* del huso
i precoce formazione *f* del
   fuso
d vorzeitige Spindelbildung *f*

2639 SPINDLE RESIDUE
f résidu *m* fusorial
e residuo *m* del huso
i residuo *m* del fuso
d Spindelrest *m*

SPINDLES, INTERSECTING
see 1532

SPIRAL, INTERNAL
see 1530

SPIRAL, MAJOR
see 1704

SPIRAL, MOLECULAR
see 1828

SPIRAL, RELIC
see 2441

2640 SPIRALIZATION
f spiralisation *f*
e espiralización *f*
i spiralizzazione *f*
d Spiralisation *f*;
   Spiralisierung *f*

2641 SPIRALIZATION COEFFICIE[
f coefficient *m* de spiralisation
e coeficiente *m* de
   espiralización
i coefficiente *m* di
   spiralizzazione
d Spiralisationskoeffizient *m*

2642 SPIRALIZATION CONSTANT
f constante *f* de spiralisatio[
e constante *f* de
   espiralización
i costante *f* di
   spiralizzazione
d Spiralisationskonstante *f*

2643 SPIREM
f spirème *m*
e espirema *m*
i spirema *m*
d Spirem *n*

2644 SPLIT
f fissure *f*

e separación *f*
i fessura *f*
d Spalt *m*

2645 SPLIT-SPOT
f point *m* de cassure
e punto *m* de rotura
i punto *m* di rottura
d Spaltanlage *f*

2646 SPLIT SPINDLE
f fuseau *m* éclaté
e huso *m* de rotura
i fuso *m* di rottura
d Spaltspindel *f*

2647 SPLITTING
f a. division *f*
   b. spéciation *f*
e a. división *f*
   b. especiación *f*
i a. divisione *f*
   b. speciazione *f*
d a. Spaltung *f*
   b. Artbildung *f*
   Speciation *f*

2648 SPONTANEOUS GENERATION
f génération *f* spontanée
e generación *f* espontánea;
   generación *f* primordial
i generazione *f* spontanea
d Urzeugung *f*

2649 SPORE
f spore *f*
e espora *f*
i spora *f*
d Spore *f*

2650 SPOROCYTE
f sporocyte *m*
e esporócito *m*
i sporocito *m*
d Sporocyte *f*

2651 SPOROGENESIS
f sporogénèse *f*
e esporogénesis *f*
i sporogenesi *f*
d Sporogenese *f*

2652 SPOROPHYTE
f sporophyte *m*
e esporófito *m*
i sporofito *m*
d Sporophyt *m*

2653 SPORT
f sport *m*
e sport *m*
i sport *m*
d Knospenmutation *f*

2654 SPORY
f sporie *f*
e sporia *f*
i sporia *f*
d Sporie *f*

2655 SPURIOUS ALLELOMORPH
f faux allélomorphe *m*
e falso alelomorfo *m*
i falso allelomorfo *m*
d unechtes Allelomorph *n* (#);
   Pseudoallel *n*

2656 STABILITY
f stabilité *f*
e estabilidad *f*
i stabilità *f*
d Stabilität *f*

STABILITY, DEVELOP-
MENTAL
see 743

STABILITY, MUTAGENE
see 1891

STABILITY, PHENOTYPIC
see 743

2657 STABILIZING SELECTION
f sélection *f* stabilisante
e selección *f* estabilizadora
i selezione *f* stabilizzante
d stabilisierende Selektion *f*

STAGE, DICTYATE
see 768

STANDARD
see 2881

2658 STANDARD DEVIATION
f déviation *f* standard;
écart *m* quadratique moyen;
écart-type *m*
e desviación *f* típica;
desviación *f* standard
i deviazione *f* tipica
d Standardabweichung *f*;
mittlere quadratische
Abweichung *f*

2659 STAR METAPHASE
f métaphase *f* en étoile;
métaphase *f* étoilée
e metafase *f* en estrella
i metafase *f* a stella
d Sternmetaphase *f*

2660 STARVATION
f carence *f*
e insuficiencia *f* en ácido
nucléico
i insufficienza *f* in acido
nucleico
d a. Starvation *f*;
Nukleinsäurearmut *f*
b. Hungerkultur *f*

2661 STASIGENESIS
f stasigénèse *f*
e estasigénesis *f*
i stasigenesi *f*
d Stasigenese *f*

2662 STATHMOANAPHASE
f stathmoanaphase *f*
e estatmoanafase *f*
i statmoanafase *f*
d Stathmoanaphase *f*

2663 STATHMODIERESIS
f stathmodiérèse *f*
e estatmodieresis *f*
i statmodieresi *f*
d Stathmodiärese *f*

2664 STATHMOKINESIS
f stathmocinèse *f*
e estatmocinesis *f*
i statmocinesi *f*
d Stathmokinese *f*

2665 STATHMOMETAPHASE
f stathmométaphase *f*
e estatmometafase *f*
i statmometafase *f*
d Stathmometaphase *f*

2666 STATHMOTELOPHASE
f stathmotélophase *f*
e estatmotelofase *f*
i statmotelofase *f*
d Stathmotelophase *f*

2667 STATISTICAL TEST;
TEST OF SIGNIFICANCE
f test *m* statistique
e prueba *f* estadística;
test *m* estadístico
i prova *f* statistica
d statistisches Prüfver-
fahren *n*;
Signifikanzprüfung *f*

2668 STEADY DRIFT
f dérive *f* régulière
e desviación *f* constante
i derivazione *f* regolare
d - -

2669 STEM BODY
f corps *m* propulseur
e cuerpo *m* impulsor
i corpo *m* impulsivo
d Stemmkörper *m*

2670 STEM LINE
f souche *f* principale
e línea *f* impulsora
i linea *f* di forza
d stem line *f*;
Stammlinie *f* (#)

2671 STEP ALLELOMORPHISM
f allélomorphisme *m* gradu⊕
e alelomorfismo *m* gradual
i allelomorfismo *m* gradual
d Treppenallelomorphismus⊕

2672 STERILE
f stérile
e estéril
i sterile
d steril

2673 STERILITY
 f stérilité f
 e esterilidad f
 i sterilità f
 d Sterilität f

 STERILITY, GAMETIC
 see 1030

 STERILITY, SEGREGATIONAL
 see 2518

2674 STERILITY GENE
 f gène m de stérilité
 e gen m de esterilidad
 i gene m di sterilità
 d Sterilitätsgen n

2675 STICKINESS
 f viscosité f
 e viscosidad f;
 aglutinación f
 i viscosità f
 d Klebrigkeit f

2676 STICKY ASSOCIATION
 f association f par
 agglutination
 e asociación f por
 aglutinación
 i associazione f per
 agglutinazione
 d durch Verklebung entstan-
 dener Chromosomenver-
 band m

2677 STICKY EFFECT
 f effet m d'agglutination
 e efecto m de aglutinación
 i effetto m d'agglutinazione
 d Verklebungseffekt m (#);
 chromatische
 Agglutinisierung f

2678 STRAIN;
 VARIETY
 f souche f; variété f
 e variedad f; cepa f
 i varietà f; ceppo m
 d Stamm m; Schlag m;
 Varietät f

2679 STREPSINEME
 f strepsinème m
 e estrepsinema m
 i strepsinema m
 d Strepsinema n

2680 STREPSITENE
 f strepsitène m
 e estrepsiteno m
 i strepsitene m.
 d Strepsitän n

2681 STRETCH
 f élongation f
 e elongación f
 i allungamento m
 d Streckung f

2682 STRUCTURAL
 f structural; de structure
 e estructural
 i strutturale
 d strukturell;
 Struktur- (in Zusammen-
 setzungen)

2683 STRUCTURAL CHANGE
 f modification f de structure
 e modificación f estructural
 i modificazione f strutturale
 d Strukturveränderung f

2684 STRUCTURAL HETEROZYGOTE
 f hétérozygote m de structure
 e heterocigoto m de estructura
 i eterozigote m strutturale
 d Strukturheterozygote f

2685 STRUCTURAL HYBRIDISM
 f hybridisme m de structure
 e hibridismo m de estructura
 i ibridismo m strutturale
 d Strukturhybridität f

2686 STRUCTURALLY CHANGED
 POLYPLOIDY
 f polyploïdie f modifiée dans
 sa structure
 e poliploidia f modificada en
 su estructura
 i poliploidismo m struttural-
 mente modificato

d strukturell veränderte
Polyploidie *f*

2687 STRUCTURAL MODIFICATION
f modification *f* de structure
e modificación *f* estructural
i modificazione *f* strutturale
d Strukturmodifikation *f*

2688 S-TYP POSITION EFFECT
f effet *m* de position de type
"S"
e efecto *m* de posición de
tipo "S"
i effetto *m* di posizione di
tipo "S"
d S-Typ-Positionseffekt *m*

2689 SUB-CHROMATID
f sous-chromatide *f*
e subcromátida *f*
i subcromatidio *m*
d Subchromatide *f*

2690 SUBCHROMONEMA
f sous-chromonème *m*
e subcromonema *m*
i sub-cromonema *m*
d Subchromonema *n*

2691 SUBGENE
f sous-gène *m*
e subgén *m*
i subgene *m*
d Subgen *n*

2692 SUBGENERIC
f sous-générique
e subgenérico
i subgenerico
d subgenerisch (Θ);
Untergattungs- (in
Zusammensetzungen)

2693 SUBGENUS
f sous-genre *m*
e subgénero *m*
i subgenere *m*
d Subgenus *n*;
Untergattung *f*

2694 SUBLETHAL
f subléthal

e subletal
i subletale
d subletal

2695 SUBMEDIAN
f submoyen
e submediano
i submediano
d submedian

2696 SUBMESOMITIC
f submésomitique
e submesomítico
i submesomito
d submesomitisch (#)

2697 SUB-MICROSOME
f sous-microsome *m*
e submicrosoma *m*
i submicrosoma *m*
d Submikrosom *n*

2698 SUBSEXUAL
f subsexuel;
subsexué
e subsexual
i subsessuale;
subsessuato
d subsexuell

2699 SUBSPECIES
f sous-espèce *f*
e subespecie *f*
i sottospecie *f*;
subspecie *f*
d Subspecies *f*;
Subspezies *f*;
Unterart *f*

2700 SUBSTANCE GENE
f - -
e - -
i - -
d Substanzgen *n* (#)

2701 SUBSTITUTION
f substitution *f*
e substitución *f*
i sostituzione *f*
d Substitution *f*

2702 SUBTERMINAL

f subterminal
e subterminal
i subterminale
d subterminal

2703 SUBVIABLE
f subviable
e subviable
i subviabile
d subviable;
kaum lebensfähig

2704 SUBVITAL
f subvital
e subvital
i subvitale
d subvital

2705 SUPERCONTRACTION
f supercontraction *f*
e supercontracción *f*
i supercontrazione *f*
d Superkontraktion *f*

2706 SUPERDOMINANCE
f superdominance *f*
e superdominancia *f*
i superdominanza *f*
d Superdominanz *f*

2707 SUPERDOMINANT
f superdominant
e superdominante
i superdominante
d superdominant

2708 SUPERFEMALE
f superfemelle *f*
e superhembra *f*
i superfemmina *f*
d Überweibchen *n*

2709 SUPERGENE
f supergène *m*
e supergén *m*
i supergene *m*
d Supergen *n*

2710 SUPERGENIC
f supergénique
e supergénico
i supergenico
d supergenisch

2711 SUPERMALE
f supermâle *m*
e supermacho *m*
i supermaschio *m*
d Übermannchen *n*

2712 SUPERNUMERARY (adj.);
SUPERNUMERY (adj.)
SUPERNUMERARY;
SUPERNUMERY
f supernuméraire
e supernumerario
i soprannumerario
d überzählig

2713 SUPER-RECESSIVE
f super-récessif
e super-recesivo
i super-recessivo
d super-rezessiv

2714 SUPER-RECOMBINATION
f super-recombinaison
e super-recombinación *f*
i super-ricombinazione *f*
d Super-Rekombination *f*

2715 SUPER-REDUCTION
f super-réduction *f*
e super-reducción *f*
i super-riduzione *f*
d Super-Reduktion *f*

2716 SUPERSPECIES
f super-espèce *f*
e superespecie *f*
i super-specie *f*
d Superspecies *f*;
Superspezies *f*;
Sammelart *f*

2717 SUPERSPIRAL
f super-spirale *f*
e superespiral *f*
i super-spirale *f*
d Superspirale *f*

2718 SUPERTERMINALIZATION
f superterminalisation *f*
e superterminalización *f*
i superterminalizzazione *f*
d Superterminalisation *f*

2719 SUPERVITAL
   f supervital
   e supervital
   i supervitale
   d supervital

2720 SUPPLEMENTARY GENE
   f gène *m* supplémentaire
   e gen *m* suplementario
   i gene *m* supplementare
   d Supplementärgen *n*

2721 SUPPRESSIBLE MUTANT
   f mutant *m* supprimable
   e mutante *m* supresible
   i mutante *m* sopprimibile
   d suppressible Mutante *f*

2722 SUPPRESSOR (adj.)
      SUPPRESSOR
   f suppresseur
     suppresseur *m*
   e supresor
     supresor *m*
   i soppressore
     soppressore *m*
   d Suppressor *m*

2723 SUPRASPECIES
   f supra-espèce *f*
   e supraspecie *f*
   i supra-specie *f*
   d Supraspecies *f;*
     Supraspezies *f;*
     Überart *f*

2724 SURVIVAL INDEX
   f indice *f* de survie
   e índice de supervivencia *m*
   i indice *m* di sopravvivenza
   d Erhaltungsquotient *m*

2725 SURVIVAL REDUCTION
   f diminution *f* de survie
   e reducción *f* de super-
     vivencia
   i riduzione *f* di sopra-
     vvivenza
   d Verringerung der
     Überlebensrate *f*

2726 SUSCEPTIBILITY
   f susceptibilité *f*
   e susceptibilidad *f*
   i suscettibilità *f*
   d Empfänglichkeit *f*

2727 SWITCH GENE
   f switch-gène *m*
   e - -
   i - -
   d Steuerungsgen *n* (#)

2728 SYMPATRIC
   f sympatrique
   e simpátrico
   i simpatrico
   d sympatrisch

2729 SYNAPSIS
   f synapsis *f*
   e sinapsis *f*
   i sinapsi *f*
   d Synapsis *f*

2730 SYNAPTENE (adj.)
      SYNAPTENE
   f synaptène
     synaptène *m*
   e sinaptene
     sinaptene *m*
   i sinaptene
     sinaptene *m*
   d Synaptän *n* (⊙);
     Zygotän (auch in
     Zusammensetzungen)

2731 SYNAPTIC
   f synaptique
   e sináptico
   i sinaptico
   d synaptisch

2732 SYNCHRONOUS
   f synchrone
   e sincrónico
   i sincrono
   d synchron

2733 SYNCYTE
   f syncyte *m*
   e síncito *m*
   i sincito *m*
   d Syncyte *f*

2734 SYNDESIS
      f syndèse *f*
      e síndesis *f*
      i sindesi *f*
      d Syndese *f*

2735 SYNDIPLOIDY
      f syndiploïdie *f*
      e sindiploidía *f*
      i sindiploidia *f*
      d Syndiploidie *f*

2736 SYNGAMEON
      f syngaméon *m*
      e singámeon *m*
      i singameone *m*
      d Syngameon *n*

2737 SYNGAMETY
      f syngamétie *f*
      e singametía *f*
      i singametia *f*
      d Syngametie *f*

2738 SYNGAMIC
      f singame
      e singámico
      i singamico
      d syngam

2739 SYNGAMY
      f syngamie *f*
      e singamia *f*
      i singamia *f*
      d Syngamie *f*

2740 SYNGENESIS
      f syngénèse *f*
      e singénesis *f*
      i singenesi *f*
      d Syngenese *f*

2741 SYNGENOTE
      f syngénote *m*
      e singenote *m*
      i singenote *m*
      d Syngenote *f*

2742 SYNKARYON
      f syncaryon *m*
      e sincarionte *m*
      i sincarionte *m*
      d Synkaryon *n*

2743 SYNKARYOSIS
      f syncaryose *f*
      e sincariosis *f*
      i sincariosi *f*
      d Synkaryosis *f*

2744 SYNKARYOTIC
      f syncaryotique
      e sincariótico
      i sincariotico
      d synkaryotisch

2745 SYNTELOMITIC
      f syntélomitique
      e sintelomítico
      i sintelomitico
      d syntelomitisch

2746 SYNTRIPLOIDY
      f syntriploïdie *f*
      e sintriploidía *f*
      i sintriploidismo *m*
      d Syntriploidie *f*

2747 SYNTROPH
      f syntrophe
      e síntrofo
      i sintrofo
      d syntroph

2748 SYSTEMIC MUTATION
      f mutation *f* systémique
      e mutación *f* sistémica
      i riaggruppamento cromoso-
         mico *m* (O)
      d Systemmutation *f*

T

2749   TACHYGENESIS
    f   tachygénèse *f*
    e   taquigénesis *f*
    i   tachigenesi *f*
    d   Tachygenesis *f*;
       Tachygenese *f*

2750   TACHYTELIC
    f   tachytélique
    e   taquitélico
    i   tachitelico
    d   tachytelisch

2751   TANDEM ASSOCIATION
    f   association *f* en tandem
    e   asociación *f* "tandem"
    i   associazione *f* in tandem
    d   Tandemassoziation *f*

2752   TANDEM DUPLICATION;
       TANDEM REPEATS
    f   duplication *f* en tandem
    e   duplicación *f* "tandem"
    i   duplicazione *f* in tandem
    d   Tandemduplikation *f*

2753   TANDEM FUSIONS
    f   fusions *f pl* en tandem
    e   fusiones *f pl* en tandem
    i   fusioni *f pl* in tandem
    d   Tandemfusionen *f pl*

2754   TANDEM INVERSION
    f   inversion *f* en tandem
    e   inversión *f* en tandem
    i   inversione *f* in tandem
    d   Tandeminversion *f*

2755   TANDEM REPEATS;
       TANDEM DUPLICATION
    f   duplication *f* en tandem
    e   duplicación *f* "tandem"
    i   duplicazione *f* in tandem
    d   Tandemduplikation *f*

2756   TANDEM RING
    f   anneau-tandem *m*
    e   anillo "tandem"*m*
    i   anello "tandem" *m*
    d   Tandemring *m*

2757   TANDEM SATELLITES
    f   satellites *mpl* en tandem
    e   satélites *mpl* en tandem
    i   satelliti *mpl* in tandem
    d   Tandemsatelliten *mpl*

2758   TANDEM SELECTION
    f   sélection *f* en tandem
    e   selección *f* en tandem
    i   selezione *f* in tandem
    d   Tandemselektion *f*

2759   TARGET THEORY
    f   théorie *f* de la cible
    e   teoría *f* del blanco
    i   teoria *f* del bersaglio
    d   Treffertheorie *f*

2760   TAXON
    f   taxon *m*
    e   taxón *m*
    i   taxon *m*
    d   Taxon *n*

2761   T-CHROMOSOME
    f   chromosome T *m*
    e   cromosoma T *m*
    i   cromosoma T *m*
    d   T-Chromosom *n*

2762   T-EFFECT
    f   effet T *m*
    e   efecto T
    i   effetto-T *m*
    d   T-Effekt *m*

2763   TELECHROMOMERE;
       TELOCHROMOMERE
    f   téléchromomère *n*
    e   telecromómero *m*

i   telecromomero *m*
d   Telochromomer *n*

2764   TELECHROMOSOME;
       TELOCHROMOSOME
f   téléchromosome *m*
e   telecromosoma *m*
i   telecromosoma *m*
d   Telochromosom *n*

2765   TELEGONY
f   télégonie *f*
e   telegonía *f*
i   telegonia *f*
d   Telegonie *f*

2766   TELOCENTRIC
f   télocentrique
e   telocéntrico
i   telocentrico
d   telozentrisch

2767   TELOCHROMOMERE;
       TELECHROMOMERE
f   téléchromomère *m*
e   telecromómero *m*
i   telecromomero *m*
d   Telochromomer *n*

2768   TELOCHROMOSOME;
       TELECHROMOSOME
f   téléchromosome *m*
e   telecromosoma *m*
i   telecromosoma *m*
d   Telochromosom *n*

2769   TELOGENE
f   télogène *m*
e   telogén *m*
i   telogene *m*
d   Telogen *n*

2770   TELOLECITHAL
f   télolécithe
e   telolécito
i   telolecito
d   telolezithal

2771   TELOMERE
f   télomère *m*
e   telómero *m*

i   telomero *m*
d   Telomer *n*

2772   TELOMERIC
f   télomérique
e   telomérico
i   telomerico
d   telomerisch

2773   TELOMITIC
f   télomitique
e   telomítico
i   telomitico
d   telomitisch

2774   TELOPHASE
f   télophase *f*
e   telofase *f*
i   telofase *f*
d   Telophase *f*

2775   TELOPHASE GLOBULES *pl*
f   globules *mpl* télophasiques
e   glóbulos *mpl* telofásicos
i   globuli *mpl* telofasici
d   Telophasekörperchen *npl*

2776   TELOPHASIC
f   télophasique
e   telofásico
i   telofasico
d   telophasisch;
       Telophase- (in Zusammen-
       setzungen)

2777   TELOREDUPLICATION
f   téloréduplication *f*
e   teloreduplicación *f*
i   teloreduplicazione *f*
d   Teloreduplikation *f*

2778   TELOSYNAPSIS
f   télosynapse *f*
e   telosinapsis *f*
i   telosinapsi *f*
d   Telosynapsis *f*

2779   TELOSYNDESIS
f   télosyndèse *f*
e   telosíndesis *f*
i   telosindesi *f*
d   Telosyndese *f*

2780   TELOSYNDETIC
    f télosyndétique
    e telosindético
    i telosindetico
    d telosyndetisch

2781   TEMPERATE PHAGE
    f phage *m* tempéré
    e fago *m* temperado
    i fago *m* temperato
    d temperierter Phage *m*;
       temperenter Phage *m*

2782   TEMPERATURE MUTANT
    f mutant *m* thermosensible
    e termomutante *m*
    i termomutante *m*
    d Temperaturmutante *f*

2783   T-END
    f extrémité T *m*
    e extremidad T *f*
    i estremità T *f*
    d T-Ende *n*

2784   TERMINAL
    f terminal
    e terminal
    i terminale
    d terminal

2785   TERMINAL AFFINITY
    f affinité *f* terminale
    e afinidad *f* terminal
    i affinità *f* terminale
    d Terminalaffinität *f*

2786   TERMINAL CHIASMA
    f chiasma *m* terminal
    e quiasma *m* terminal
    i chiasma *m* terminale
    d Terminalchiasma *n*

2787   TERMINAL GRANULES
    f granules *mpl* terminaux
    e granulos *mpl* terminales
    i - -
    d Telochromomeren *npl*

2788   TERMINALIZATION
    f terminalisation *f*

    e terminalización *f*
    i terminalizzazione *f*
    d Terminalisation *f*

2789   TERMONE
    f termone *f*
    e termón *m*
    i termone *m*
    d Termon *n*

2790   TERTIARY SPLIT
    f rupture *f* tertiaire
    e rotura *f* terciaria
    i rottura *f* terziaria
    d Tertiärspalt *m*

2791   TEST CROSS;
    TEST MATING
    f croisement *m* de contrôle;
       test-cross *m*;
       accouplement *m* de testage
    e cruza *f* probadora;
       acoplamiento *m* probador
    i incrocio *m* di controllo;
       incrocio *m* di paragone
    d Testkreuzung *f*;
       Testpaarung *f*

    TEST CROSS, MULTIPLE-
    POINT
    see 1881

    TEST CROSS, THREE-POINT
    see 2811

    TEST CROSS, TWO-POINT
    see 2878

2792   TESTER STRAIN
    f souche *f* de contrôle
    e cepa *f* controladora
    i ceppo *m* controllatore
    d Testlinie *f* (≠)

2793   TEST OF SIGNIFICANCE;
    STATISTICAL TEST
    f test *m* statistique
    e prueba *f* estadística;
       prueba *f* de significancia
    i test *m* statistico;
       test *m* significativo

d   Signifikanzprüfung *f*;
    statistische Prüfung *f*

2794   TETRAD
f   tétrade *f*
e   tetrada *f*
i   tetrade *f*
d   Tetrade *f*

TETRADS, DITYPE
see 855

2795   TETRADS ANALYSIS
f   analyse *f* des tétrades
e   análisis *f* de las tetradas
i   analisi *f* delle tetradi
d   Tetradenanalyse *f*

2796   TETRAHAPLOID (adj.)
TETRAHAPLOID
f   tétrahaploïde
    tétrahaploïde *m*
e   tetrahaploide
    tetrahaploide *m*
i   tetraaploide
    tetraaploide *m*
d   tetrahaploid
    Tetrahaploide *f*

2797   TETRAHAPLOIDY
f   tétrahaploïdie *f*
e   tetrahaploidía *f*
i   tetraaploidismo *m*
d   Tetrahaploidie *f*

2798   TETRAPLOID (adj.)
TETRAPLOID
f   tétraploïde
    tétraploïde *m*
e   tetraploide
    tetraploide *m*
i   tetraploide
    tetraploide *m*
d   tetraploid
    Tetraploide *f*

2799   TETRAPLOIDY;
DISOMATY
f   tétraploïdie *f*;
    disomatie *f* (☉)
e   tetraploidía *f*;
    disomatía *f* (☉)

i   tetraploidismo *m*;
    disomatia *f* (☉)
d   Tetraploidie *f*;
    Disomatie *f* (☉)

2800   TETRASOMATY
f   octoploïdie *f*
e   tetrasomatia *f*
i   tetrasomatia *f*
d   Tetrasomatie *f* (☉);
    Oktoploidie *f*

2801   TETRASOMIC
f   tétrasomique
e   tetrasómico
i   tetrasomico
d   tetrasom

2802   TETRATYPE
f   tétratype *m*
e   tetrátipo *m*
i   tetratipo *m*
d   tetratyp

2803   TETRAVALENT
f   tétravalent *m*
e   tetravalente *m*
i   tetravalente *m*
d   Tetravalent *n*

2804   THELYCARYON;
THELYKARYON
f   thélycaryon *m*
e   telicarión *m*
i   telicarion *m*
d   Thelykaryon *n*

2805   THELYCARYOTIC;
THELYKARYOTIC
f   thélycaryotique
e   telicariótico
i   telicariotico
d   thelykaryotisch

2806   THELYGENIC
f   thélygénique
e   telígeno
i   teligenico
d   thelygen

2807   THELYGENY
f   thélygénie *f*

e teligenia *f*
i teligenia *f*
d Thelygenie *f*

2808 THELYKARYON;
THELYCARYON
f thélycaryon *m*
e telicarión *m*
i telicarion *m*
d Thelykaryon *n*

2809 THELYKARYOTIC;
THELYCARYOTIC
f thélycaryotique
e telicariótico
i telicariotico
d thelykaryotisch

2810 THELYTOKY
f thélytokie *f*
e telitoquía *f*
i telitochia *f*
d Thelytokie *f*

2811 THREE-POINT TEST CROSS
f test-cross *m* à trois points
e cruza *f* probadora de tres puntos
i test *m* d'incrocio a tre vie
d Dreipunktversuch *m*

2812 THRESHOLD CHARACTER
f caractère *m* de seuil
e - -
i - -
d Schwellenmerkmal *n*

2813 THROW-BACK
f individu *m* atavique;
caractère *m* atavique
e carácter atávico *m*
i individuo *m* atavico;
carattere *m* atavico;
individuo *m* atavistico;
carattere *m* atavistico
d Rückschlag *m*

2814 THROWING-BACK
f atavisme *m*
e atavismo *m*
i atavismo *m*

d Rückschlag *m*;
Atavismus *m*

TIE, CHROMATID
see 495

2815 TOP-CROSS
f top-cross *m*
e cruzas *f pl* radiales
i top-cross *m*
d top-cross *n*

2816 TOP-CROSSING
f top-crossing *m*
e cruzamiento *m* radial
i top-crossing *m*
d top-crossing *n*

2817 TORSION . . .
TORSIONAL
f . . . de torsion
e torsional
i torsionale
d Torsions- (in Zusammensetzungen)

2818 TOUCH-AND-GO PAIRING
f appariement *m* momentané
e apareamiento *m* de toque y separación
i appaiamento *m* momentaneo
d "touch and go"-Paarung *f*

2819 TRABANT
f trabant *m*
e trabante *m*
i trabante *m*
d Trabant *m*

2820 TRACTION FIBRE
f fibre *f* de traction
e fibra *f* de tracción
i fibra *f* di trazione
d Zugfaser *f*

2821 TRAIT
f trait *m*;
caractère *m* héréditaire
e carácter *m*
i carattere *m*
d Merkmal *n*

TRAIT, ADAPTIVE
see 36

2822 TRANS-ARRANGEMENT;
TRANS-CONFIGURATION
f arrangement *m* en trans-
configuration *f*
e transacomodo *m*;
transconfiguración
i trans-configurazione (⊙)
d Transkonfiguration *f*

2823 TRANSDUCTION
f transduction *f*
e transducción *f*
i trasduzione *f*
d Transduktion *f*

TRANSDUCTION, ABORTIVE
see 9

2824 TRANSDUCTION CLONE
f clône *m* de transduction
e clon *m* de transducción
i clone *m* di trasduzione
d Transduktionsklon *m*

2825 TRANSDUCTION GROUPS
f groupes *mpl* de trans-
duction
e grupos *mpl* de transducción
i gruppi *mpl* di trasduzione
d Transduktionsgruppen *f pl*

2826 TRANSFER
f transfert *m*
e transferencia *f*
i trasferimento *m*
d Übertragung *f*;
Transfer *m*

2827 TRANSFER RNA
f ARN de transfert *m*
e transducción *f* ARN
i RNA di trasferimento *m*
d Transfer-RNS *f*;
Überträger-RNS *f*

2828 TRANSFORMATION
f transformation *f*

e transformación *f*
i trasformazione *f*
d Transformation *f*

2829 TRANSFORMING PRINCIPLE
f principe *m* transformant
e principio *m* transformante
i principio *m* trasformante
d transformierendes Agens *n*

2830 TRANSGENATION
f transgénation *f*
e transgenación *f*
i trasgenazione *f*
d Genmutation *f*;
Punktmutation *f*

2831 TRANSGRESSION;
TRANSGRESSIVE SEGRE-
GATION
f transgression *f*;
ségrégation *f* transgressive
e transgresión *f*;
segregación *f* transgresiva
i trasgressione *f*;
segregazione *f* trasgressiva
d Transgression *f*

2832 TRANSGRESSIVE
f transgressif
e transgresivo
i trasgressivo
d transgressiv

2833 TRANSIENT POLYMORPHISM
f polymorphisme *m* de
transition
e polimorfismo *m* de
transición
i polimorfismo *m* di
transizione
d transienter Polymorphismus *m*

2834 TRANSLOCATION
f translocation *f*
e translocación *f*
i traslocazione *f*
d Translokation *f*

TRANSLOCATION, IN-
SERTIONAL
see 1499

2835 TRANSLOCATION
      HETEROZYGOSIS
      f hétérozygotie f de trans-
         location
      e heterocigosis f de
         translocación
      i eterozigosi f di
         traslocazione
      d Translokationsheterozy-
         gotie f

2836 TRANSLOCATION HOMO-
      ZYGOSIS
      f homozygose f de trans-
         location
      e homocigosis f de
         translocación
      i omozigosi f di trasloca-
         zione
      d Translokationshomozygo-
         tie f

2837 TRANSLOCATION MUTANT
      f mutant m de translocation
      e mutante m de translocación
      i mutante m di traslocazione
      d Translokationsmutante f

2838 TRANSLOCATION TRISOMICS
      f trisomiques m f pl de
         translocation
      e trisómicos m pl de
         translocación;
         trisómicas f pl de
         translocación
      i trisomici m pl di
         traslocazione;
         trisomiche f pl di
         traslocazione
      d Translokationstrisome f pl

2839 TRANSMISSION
      f transmission f
      e transmisión f
      i trasmissione f
      d Transmission f

2840 TRANSMUTATION
      f transmutation f
      e transmutación f
      i trasmutazione f
      d Transmutation f

2841 TRANSPECIFIC
      f transspécifique
      e transespecífico
      i transpecifico
      d transspezifisch

2842 TRANSREPLICATION
      f transréplication f
      e transreplicación f
      i transreplicazione f
      d Transreplikation f

2843 TRANS-VECTION EFFECT
      f effet m de "trans-vection"
      e efecto m de "trans-vection"
      i effetto m di trasporto
      d Trans-Positionseffekt m

2844 TRANSVERSE EQUILIBRIUM
      f équilibre m transverse
      e equilibrio m transverso
      i equilibrio m trasverso
      d transversales Gleichge-
         wicht n (≠)

2845 TREND
      f tendance f
      e tendencia f
      i tendenza f
      d Trend m;
         Tendenz f

2846 TRIAD
      f triade f
      e triada f
      i triade f
      d Triade f

2847 TRIGENERIC
      f trigénérique
      e trigenérico
      i trigenerico
      d trigenerisch (⊙);
         Dreigattungs- (in Zusam-
         mensetzungen)

848 TRIGENIC
  f   trigénique
  e   trigénico
  i   trigenico
  d   trigenisch;
   trigen

849 TRIHAPLOID (adj.)
  TRIHAPLOID
  f   trihaploïde
   trihaploïde *m*
  e   trihaploide
   trihaploide *m*
  i   triaploide
   triaploide *m*
  d   trihaploid
   Trihaploide *f*

850 TRIHAPLOIDY
  f   trihaploïdie *f*
  e   trihaploidía *f*
  i   triaploidismo *m*
  d   Trihaploidie *f*

851 TRIHYBRID (adj.)
  TRIHYBRID
  f   trihybride
   trihybride *f*
  e   trihíbrido
   trihíbrido *m*
  i   triibrido
   triibrido *m*
  d   trihibrid
   Trihibride *f*

852 TRIHYBRIDISM
  f   trihybrydisme *m*
  e   trihibridismo *m*
  i   triibridismo *m*
  d   Trihibridie *f*

853 TRIMERICAL
  f   trimérique
  e   trimérico
  i   trimero
  d   trimer

854 TRIMERY
  f   trimérie *f*
  e   trimería *f*
  i   trimeria *f*
  d   Trimerie *f*

2855 TRIMONOECIOUS
  f   trimonoïque
  e   trimonóico
  i   trimonoico
  d   trimonoezisch

2856 TRIMORPHISM
  f   trimorphisme *m*
  e   trimorfismo *m*
  i   trimorfismo *m*
  d   Trimorphismus *m*

2857 TRIOECIOUS
  f   trioïque
  e   trióico
  i   trioico
  d   trioezisch;
   dreihäusig

2858 TRIPLE CROSS
  f   croisement *m* triple
  e   cruzamiento *m* triple;
   cruzamiento *m* de tres
   líneas
  i   incrocio *m* triplice
  e   Dreifachkreuzung *f*

2859 TRIPLE FUSION
  f   fusion *f* triple
  e   fusión *f* triple
  i   fusione *f* triplice
  d   Dreifachfusion *f*

2860 TRIPLEX (adj.)
  TRIPLEX
  f   triplex
   triplex *m*
  e   triplexo
   triplexo *m*
  i   triplex
   triplex *m*
  d   triplex

2861 TRIPLICATE GENES
  f   gènes *mpl* présents en
   trois exemplaires
  e   genes *mpl* triplicados
  i   tripletto di geni *m*
  d   - -

2862 TRIPLOID (adj.)
TRIPLOID
f triploïde
triploïde *m*
e triploide
triploide *m*
i triploide
triploide *m*
d triploid
Triploide *f*

2863 TRIPLOIDY
f triploïdie *f*
e triploidía *f*
i triploidismo *m*
d Triploidie *f*

2864 TRISOMIC
f trisomique
e trisómico
i trisomico
d trisom

2865 TRISOMY
f trisomie *f*
e trisómia *f*
i trisomia *f*
d Trisomie *f*

2866 TRIVALENT
f trivalent *m*
e trivalente *m*
i trivalente *m*
d Trivalent *n*

2867 TROPHIC NUCLEUS
f noyau *m* trophique
e núcleo *m* trófico
i nucleo *m* trofico
d Makronukleus *m*

2868 TROPHOCHROMATIN
f trophochromatine *f*
e trofocromatina *f*
i trofocromatina *f*
d Trophochromatin *n*

2869 TROPHOPLASM
f trophoplasme *m*
e trofoplasma *m*
i trofoplasma *m*
d Trophoplasma *n*

2870 TROPODIERESIS
f tropodiérèse *f*
e tropodieresis *f*
i tropodieresi *f*
d Tropodiärese *f*

2871 TROPOKINESIS
f tropocinèse *f*
e tropocinesis *f*
i tropocinesi *f*
d Tropokinese *f*

2872 TWIN (adj.)
TWIN
f jumeau
jumeau *m*
e gemelo
gemelo *m*
i gemello
gemello *m*
d Zwilling *m*

2873 TWIN GENES
f genes *mpl* jumeaux
e genes *mpl* gemelos
i geni *mpl* gemelli
d Pseudoallele *npl*

2874 TWIN HYBRIDS
f hybrides *mpl* jumeaux
e híbridos *mpl* gemelos
i ibridi *mpl* gemelli
d Zwillingsbastarde *mpl*

TWINS, IDENTICAL
see 2894

TWINS, MONOZYGOTIC
see 2894

TWINS, UNIOVULAR
see 2894

2875 TWIN SPECIES
f espèces *f pl* jumelles
e especies *f pl* gemelas
i speci *f pl* gemelle
d Zwillingsarten *f pl*

2876 TWISTING
f torsion *f*

e   torsión *f*
i   torsione *f*
d   Torsion *f*

2877   TWO-PLANE THEORY
f   théorie *f* des deux plans
e   teoría *f* de dos planos
i   teoria *f* di due piani (#)
d   Zwei-Ebenen-Theorie *f* (#)

2878   TWO-POINT TEST CROSS
f   test-cross *m* à deux points
e   cruza *f* probadora de dos puntos
i   test *m* d'incrocio a due vie
d   Zweipunktversuch *m*

2879   TYPE
f   type *m*
e   tipo *m*
i   tipo *m*
d   Typus *m*

2880   TYPE NUMBER
f   nombre type *m*
e   número tipo *m*
i   numero tipo *m*
d   Typuszahl *f* (#)

2881   TYPICAL;
STANDARD

f   typique
e   típico
i   tipico
d   typisch;
Standard- (in Zusammensetzungen)

2882   TYPOGENESIS
f   typogénèse *f*
e   tipogénesis *f*
i   tipogenesi *f*
d   Typogenese *f*

2883   TYPOLYSIS
f   typolyse *f*
e   tipólisis *f*
i   tipolisi *f*
d   Typolyse *f*

2884   TYPOSTASIS
f   typostase *f*
e   tipóstasis *f*
i   tipostasi *f*
d   Typostase *f*

2885   "TYPOSTROPHE"
f   "typostrophe" *f*
e   "tipoestrofa"
i   "tipostrofa" *f*
d   Typostrophe *f*

U

2886 ULTRAMICROSOME
  f ultramicrosome *m*
  e ultramicrosoma *m*
  i ultramicrosoma *m*
  d Ultramikrosom *n*

2887 UNBALANCED
  f déséquilibré
  e desequilibrado
  i squilibrato
  d unbalanciert

2888 UNBALANCED POLYPLOIDY
  f polyploïdie *f* déséquilibrée
  e poliploidía *f* desequilibrada
  i poliploidismo *m* squilibrato
  d unbalancierte Polyploidie *f*

2889 UNDERNUCLEATED REGION
  f zone *f* d'hétéropycnose
     négative
  e zona *f* de heteropicnosis
     negativa
  i zona *f* d'eteropicnosi
     negativa
  d negativ heteropyknotische
     Region *f*

2890 UNEQUAL BIVALENT
  f bivalent *m* inégal
  e bivalente *m* no igual;
     bivalente *m* desigual
  i bivalente *m* disuguale
  d ungleiches Bivalent *n*

2891 UNFIXABLE
  f infixable
  e infijable
  i infissabile
  d unfixierbar

2892 UNILATERAL
  f unilatéral
  e unilateral
  i unilaterale
  d einseitig

2893 UNILATERAL INHERITANCE
  f hérédité *f* unilatérale
  e herencia *f* unilateral
  i eredità *f* unilaterale
  d einseitige Vererbung *f*;
     unilaterale Vererbung *f*

2894 UNIOVULAR TWINS;
     IDENTICAL TWINS;
     MONOZYGOTIC TWINS
  f jumeaux *mpl* uniovulaires;
     jumeaux *mpl* univitellins;
     jumeaux *mpl* monozygotes
  e gemelos *mpl* idénticos;
     gemelos *mpl* uniovulares;
     gemelos *mpl* monozigóticos
  i gemelli *mpl* identici;
     gemelli *mpl* monozigotici;
     gemelli *mpl* uniovulari
  d eineiige Zwillinge (EZ) *mpl*

2895 UNISEXUAL
  f unisexuel
  e unisexual
  i unisessuale
  d unisexuell; eingeschlechtig

2896 UNISEXUALITY
  f unisexualité *f*
  e unisexualidad *f*
  i unisessualità *f*
  d Unisexualität *f*;
     Eingeschlechtigkeit *f*

2897 UNITARY
  f unitaire
  e unitario
  i unitario
  d unitär;
     unitarisch;
     je Einheit;
     Einheits- (in Zusammen-
     setzungen)

2898 UNIT CHARACTER
  f caractère *m* unitaire

e carácter *m* unitario
i carattere *m* unitario
d unifaktoriell bedingtes
   Merkmal *n*

2899   UNIVALENT
   f univalent *m*
   e univalente *m*
   i univalente *m*
   d Univalent *n*

   UNIVALENT, LOOP
   see 1684

2900   UNIVALENT SHIFT
   f - -
   e - -
   i - -
   d Univalentenwechsel *m* (#)

2901   UNSELECTED MARKERS
   f marqueurs *mpl* non
     sélectionnés

e marcadores *mpl* no
   seleccionados
i marcatori *mpl* non
   selezionati
d unselektierte Markierungs-
   gene *npl*

2902   UNSUPPRESSIBLE
   f non-suppressible
   e insuprimible
   i insopprimibile
   d unsuppressible;
     ununterdrückbar

2903   U.V.-RESTAURATION
   f restauration *f* par rayons
     U.V.
   e restauración *f* por rayos
     u.v.
   i restaurazione *f* con raggi
     U.V
   d UV-Restaurierung *f*

V

2904 VACUOLE
f vacuole *f*
e vacúolo *m*
i vacuolo *m*
d Vakuole *f*

2905 VACUOME
f vacuome *m*
e vacuoma *m*
i vacuoma *m*
d Vakuom *n*

2906 VALENCE
f valence *f*
e valencia *f*
i valenza *f*
d Valenz *f*

2907 VARIABILITY
f variabilité *f*
e variabilidad *f*
i variabilità *f*
d Variabilität *f*

2908 VARIANCE
f variance *f*
e varianza *f*
i varianza *f*
d Streuung *f*;
Varianz *f*

2909 VARIANT
f variant *m*
e variante *m*
i variante *m*
d Variante *f*

2910 VARIATION
f variation *f*
e variación *f*
i variazione *f*
d Variation *f*

VARIATION, DETERMINATE
see 738

VARIATION, DISCONTINUOUS
see 832

2911 VARIEGATE
f panaché
e variegado
i variegato
d panaschiert;
variegat

2912 VARIEGATION
f panachure *f*
e variegación *f*
i variegazione *f*
d Variegation *f*;
Panaschierung *f*
(Zustand);
Panaschüre *f* (Individuum);
Scheckung *f*

2913 VARIEGATION ALLELE
f allèle *m* de panachure
e alelo *m* de variegación
i allelo *m* di variegazione
d Scheckungsallel *n*

2914 VARIETY;
STRAIN
f souche *f*;
variété *f*
e variedad *f*;
cepa *f*
i varietà *f*;
ceppo *m*
d Varietät *f*;
Rasse *f*;
Sorte *f*

2915 VEGETATIVE
f végétatif
e vegetativo
i vegetativo
d vegetativ

VESICLE, GERMINAL
see 1129

916   VIABILITY
      f viabilité *f*
      e viabilidad *f*
      i vitalità *f*
      d Lebensfähigkeit *f*;
         Keimfähigkeit *f*

917   VIABLE
      f viable
      e viable
      i vitale
      d lebensfähig;
         keimfähig

918   VICINISM
      f hybridation *f* naturelle
      e vecinismo *m*
      i vicinismo *m*
      d Vicinismus *m*;
         Fremdbefruchtung *f*

919   VICINIST
      f hybride *f* naturelle
      e vecinista *f*
      i ibrida *f* naturale
      d Vicinist *m*;
         natürlicher Bastard *m*

920   VIRION
      f virion *m*
      e virion *m*
      i virion *m* (O)
      d - -

921   VIROGENETIC SEGMENT
      f segment *m* virogénétique
      e segmento *m* virogenético
      i segmento *m* virogenetico
      d virogenetisches Segment *n*

2922 VIRULENT PHAGE;
      INTEMPERATE PHAGE
      f phage *m* virulent;
         phage *m* intempéré
      e fago *m* virulente;
         fago *m* intemperado
      i fago *m* virulento
         fago *m* intemperato
      d virulenter Phage *m*;
         nichttemperierter Phage *m*;
         nichttemperenter Phage *m*

2923 VISIBLE
      f visible
      e visible;
         aparente
      i visibile;
         apparente
      d sichtbar

2924 VITALITY
      f vitalité *f*
      e vitalidad *f*
      i vitalità *f*
      d Vitalität *f*

2925 VITAL LOCI
      f loci *mpl* vitaux
      e loci *mpl* vitales
      i loci *mpl* vitali
      d Vitalloci *mpl*

2926 V-TYPE POSITION EFFECT
      f effet *m* de position du type V
      e efecto *m* de posición de
         tipo V
      i effetto *m* di posizione di
         tipo V
      d V-Typ-Positionseffekt *m*

W

2927 WAIST
  f étranglement *m*
  e estrangulación *f* (Θ)
  i strozzamento *m* (Θ)
  d Einschnürung *f*

2928 W-CHROMOSOME
  f chromosome W *m*
  e cromosoma W *m*
  i cromosoma W *m*
  d W-Chromosom *n*

2929 WEISMANNISM
  f Weismannisme *m*
  e Weismanismo *m*
  i weismannismo *m*
  d Weismannismus *n*

2930 WHOLE-ARM TRANSFER
  f transfert *m* du bras entier
  e transferencia *f* de todo

el brazo
  i trasferimento *m* del
    braccio intero
  d Translokation eines gan-
    zen Chromosomenarms *f*

2931 WHOLE-ARM TRANSPOSITI
  f transposition *f* du bras
    entier
  e transposición *f* de todo el
    brazo
  i trasferimento *m* del bracc
    intero
  d Translokation *f* eines
    ganzen Chromosomenarm

2932 WILD TYPE
  f type *m* sauvage
  e tipo salvaje
  i tipo selvaggio
  d Wildtyp *m*

X

2933 X (BASIC NUMBER)
  f x (nombre de base)
  e x (número básico)
  i x (numero basico)
  d x n (Basiszahl)

2934 X-CHROMOSOME
  f chromosome X m
  e cromosoma X m
  i cromosoma X m
  d X-Chromosom n

X-CHROMOSOME, ATTACHED
see 261

X-CHROMOSOMES, COMPOUND
see 598

2935 XENIA
  f xénie f

  e xenia f
  i xenia f
  d Xenie f

2936 XENOGAMY
  f xénogamie f
  e xenogamia f
  i xenogamia f
  d Xenogamie f;
    Fremdbefruchtung f

2937 XENOPLASTIC
  f xénoplastique
  e xenoplástico
  i xenoplastico
  d xenoplastisch

X-Y-CHROMOSOME,
ATTACHED
see 262

Y

Z

2939   Z (ZYGOTIC CHROMOSOME
        NUMBER)
        f   z (nombre chromosomique
            du zygote)
        e   z (número cromosómico
            zigótico)
        i   z (numero cromosomico
            zigotico)
        d   z n (zygotische Chromo-
            somenzahl)

2940   Z-CHROMOSOME
        f   chromosome Z m
        e   cromosoma Z m
        i   cromosoma Z m
        d   Z-Chromosom n

2941   ZERO POINT MUTATION
        f   mutation f de point zéro (Θ)
        e   mutación f de punto
            cero
        i   mutazione f di punto
            nullo (Θ)
        d   Nullpunktmutation f

2942   ZOOGENETICS
        f   zoogénétique f
        e   zoogenética f
        i   zoogenetica f
        d   Tiergenetik f

2943   ZYGEGENIC
        f   zygogène
        e   cigogénico
        i   zygogenetico
        d   zygogenisch (#)

2944   ZYGONEMA
        f   zygonème m
        e   zigonema m;
            cigonema m
        i   zigonema m
        d   Zygonema n

2945   ZYGOSIS
        f   zygotie f

        e   zigosis f
        i   zigosi f
        d   Zygosis f

2946   ZYGOSOME
        f   zygosome m
        e   zigosoma m
        i   zigosoma m
        d   Zygosom n

2947   ZYGOTE
        f   zygote m
        e   zigote m;
            cigota m;
            cigoto m
        i   zigote m
        d   Zygote f

948   ZYGOTENE
        f   zygotène m
        e   cigotena f;
            zigotena f
        i   zigotene m
        d   Zygotän n

2949   ZYGOTENIC
        f   zygoténique
        e   zigoteno;
            cigoteno;
            zigoténico;
            cigoténico
        i   zigotenico
        d   zygotänisch;
            Zygotän- (in Zusammen-
            setzungen)

2950   ZYGOTIC
        f   zygotique
        e   zigótico;
            cigótico
        i   zigotico
        d   zygotisch;
            Zygoten- (in Zusammen-
            setzungen)

РУССКОЕ ЧИСЛЕННОЕ ОГЛАВЛЕНИЕ

RUSSIAN NUMERICAL INDEX

1 аберрация *ж*
2 частота *ж* аберраций
3 абиогенез *м*
4 абиогенный
5 а) аборт *м*, выкидыш *м*,
   б) нарушение *ср*,
   приостановка *ж* развития
6 абортивный, недоразви-
   тый, прерванный
7 абортивная инфекция *ж*
8 абортивный митоз *м*
9 абортивная трансдукция
   ж
10 ускорение *ср*
11 добавочный
12 добавочные хромосомы *ж*
   мн
13 добавочная пластинка *ж*
14 акклиматизация *ж*
15 ацентрический
16 асимметричная
   транслокация *ж*
17 ацентрическая инверсия
   ж
18 ахиазматический
19 ахромазия *ж*
20 ахроматиновый
21 ахроматиновая фигура *ж*
22 ахроматин *м*
23 А-хромосома *ж*
24 акросиндез *м*, концевая
   конъюгация *ж*
25 приобретенный признак *м*
26 акроцентрический
27 активация *ж*
28 активатор *м*
29 система *ж* "активатор-
   диссоциация"
30 спектр *м* действия
31 приспособляемость *ж*
32 приспособление *ср*,
   адаптация *ж*

33 приспособительная
   модификация *ж*
34 адаптивный пик *м*
35 --
36 приспособительный
   признак *м*
37 приспособительное
   значение *ср*
38 сложение *ср*, добавление
   ср
39 аддитивный эффект *м*,
   эффект *м* суммирования
40 аддитивные факторы *м мн*
41 --
42 аддитивность
43 --
44 слипание *ср*, сцепление *ср*
45 сродство *ср*
46 агамета *ж*
47 агамный
48 спорофит *м*
49 агамогенез *м*
50 агамогония *ж*
51 агамный вид *м*
52 агамоспермный
53 агамоспермия
54 агамный, бесполый
55 АГ комплекс *м*
56 --
57 агматоплоидный,
   агматоплоид *м*
58 агматоплоидия *ж*
59 агматопсевдоплоидный,
   агматопсевдоплоид *м*
60 агматопсевдоплоидия *ж*
61 сертация *ж*
62 акинетический
63 --
64 аллоаутогамный
65 аллоаутогамия *ж*
66 аллель *м*, аллеломорф *м*

67   изменение *ср* частоты
      аллелей
68   аллельный,
      аллеломорфный
69   аллелизм *м*, аллелия *ж*
70   аллелобрахиальный
71   аллеломорф *м*, аллель *ж*
72   аллеломорфный,
      аллельный
73   серия *ж* аллеломорфов
      (аллелей)
74   аллеломорфизм *м*
      аллелия *ж*
75   аллелотип *м*
76   луковый тест *м* (опыты
      на корешках лука)
77   аллокарпия *ж*
78   аллохронические виды *м*
      мн
79   аллоциклический
80   аллоциклия *ж*
81   аллодиплоид *м*,
      аллодиплоидный
82   аллодиплоидия *ж*
83   аллодипломоносома *ж*
84   аллогамный
85   аллогамия *ж*, чужеродное
      опыление *ср*
86   аллоген *м*, рецессивный
      аллель *м*
87   аллогенетический
88   аллогенный
89   аллогаплоидный,
      аллогаплоид *м*
90   аллогаплоидия
91   аллогетероплоид *м*,
      аллогетероплоидный
92   аллогетероплоидия *ж*
93   --
94   аллолизогенный
95   алломорфоз *м*
96   аллопатрический

97   аллоплазма *ж*
98   аллоплоид *м*,
      аллоплоидный
99   аллоплоидия *ж*
100   аллополиплоид *м*,
      аллополиплоидный
101   аллополиплоидия *ж*
102   аллосомный
103   аллосома *ж*
104   аллосинапсис *м*,
      аллосиндез *м*
105   аллосинаптический
106   аллотетраплоид *м*,
      аллотетраплоидный
107   аллотетраплоидия *ж*
108   аллотрилоид *м*,
      аллотриплоидный
109   аллотриплоидия
110   аллотипический
111   аллозигота *ж*
112   переменная
      доминантность *ж*
113   переменная
      доминантность *ж*
114   чередование *ср* поколений
115   альтернативное
      расхождение *ср*
116   --
117   альтруистическая
      адаптация *ж*
118   альвеолярная гипотеза *ж*
119   амбисексуальный,
      двуполый, однодомный
120   амбивалентный,
      полубивалентный
121   амейоз *м*
122   амейотический
123   амиктический
124   --
125   амитоз *м*
126   амитотический
127   амиксис *м*, амиксия *ж*

128	аморфный
129	амферотокия *ж*
130	двойная звезда *ж*
131	амфиастральный митоз *м*
132	амфибивалент *м*, полубивалент *м*
133	амфидиплоид *м*, амфидиплоидный
134	амфидиплоидия *ж*
135	амфигамия *ж*
136	амфигенез *м*
137	амфигония *ж*
138	амфигаплоид *м*, амфигаплоидный
139	амфигаплоидия *ж*
140	амфикарион *м*
141	амфилепсис *м*
142	амфимиктический
143	амфимиксис *м*
144	двойная мутация *ж*
145	амфипластия *ж*
146	амфиплоид *м*, амфиплоидный
147	амфиплоидия *ж*
148	амфитена *ж*, амфитенный
149	амфиталлический
150	амфитокия *ж*
151	амфогенный
152	амфогения *ж*
153	амфогетерогония *ж*
154	анаболия *ж*
155	анахромазия *ж*
156	анагенез *м*
157	анагенетический
158	аналог *м*, аналогичный
159	анаморфный
160	анаморфоз *м*, вырождение *ср*
161	анафаза *ж*
162	анафазное движение *ср*
163	анафазный
164	анафрагмический

165	анаредупликация *ж*
166	анасхистический
167	анастомоз
168	анастральный митоз *м*
169	андроаутосома *ж*, мужская аутосома *ж*
170	андродвудомный
171	--
172	андроезия *ж*
173	андрогамия *ж*
174	андрогенез *м*
175	андрогенетический
176	андрогенный
177	андрогинный
178	андрогиния *ж*
179	андрогермафродит *м*, андрогермафродитный
180	андромерогония *ж*
181	андрооднодомный
182	андрооднодомность *ж*
183	андросома
184	андроспорогенез *м*
185	андростерильный
186	андростерильность *ж*, мужская стерильность *ж*
187	анеуплоид *м*, анеуплоидный
188	анеуплоидия *ж*
189	анеусоматия *ж*
190	анизоаутоплоидный, анизоаутоплоид *м*
191	анизоаутоплоидия *ж*
192	анизокариоз *м*
193	анизоцитоз *м*
194	анизогамета *ж*
195	анизогамия *ж*
196	анизогеномный
197	анизогеномный
198	анизогения
199	анизоплоид *м*, анизоплоидный
200	анизоплоидия *ж*

201	анизосиндез *м*
202	анизосиндетический
203	анизотрисомия *ж*
204	аннидация *ж*
205	анормогенез *м*
206	анортогенез *м*
207	анортоплоидный, анортоплоид *м*
208	анортоплоидия *ж*
209	анортоспираль *ж*
210	антефаза *ж*, препрофаза *ж*
211	опережение *ср*, упреждение *ср*
212	антиморфный
213	антимутаген *м*
214	антимутагенный
215	--
216	--
217	афазный
218	апогаметность *ср*
219	апогамный
220	апогамия
221	апогомотипный
222	апомейоз *м*
223	апомиктический
224	апомиктоз *м*
225	апомиксис *м*
226	апорогамия *ж*
227	апоспория *ж*
228	архаллаксис *м*
229	архебиоз *м*
230	архетип *м*
231	архиплазма *ж*
232	нереализация *ж*
233	ароморфоз *м*
234	арреногенетический
235	арреногения *ж*
236	арренотокия *ж*
237	артефакт *м*
238	искусственный партеногенез *м*
239	артиоплоидный, артиоплоид *м*
240	артиоплоидия *ж*
241	бесполый
242	бесполое размножение *ср*
243	ассоциация *ж*
244	ассортативное скрещивание *ср*
245	--
246	звезда *ж*
247	астроцентр *м*
248	астросфера *ж*, звезда *ж*
249	асинапсис *м*, асиндез *м*
250	асинаптический
251	асиндез *м*, асинапсис *м*
252	асингамный
253	асингамия *ж*
254	атактогамия *ж*
255	атавизм *м*
256	атавистический
257	ателомитический
258	аторзионный
259	атрактоплазма *ж*
260	атрактосома *ж*
261	сцепленные X-хромосомы *ж мн*
262	сцепленные XY-хромосомы
263	присоединение *ср*
264	разбавление *ср*
265	притяжение *ср*
266	аутоадаптация *ж*
267	аутоаллоплоид *м*, аутоаллопоидный
268	аутоаллоплоидия *ж*
269	аутобивалент *м*
270	аутогамный, самооплодотворяющийся
271	аутогамия *ж*, самооплодотворение *ср*
272	автогенез *м*
273	автогенетический

338	базидии *м*	372	биотоп *м*
339	--	373	биотип *м*
340	В-хромосома *ж*	374	двусторонний
341	пластичное поведение *ср*	375	образование *ср* мозаиков
342	межродовой	376	обладающий двойным
343	межродовое скрещивание *ср*		лучепреломлением
344	двуядерный	377	двуполый
345	биобласт *м*	378	биталлический
346	биохор *м*, жизненное пространство *ср*	379	бивалент *м*
347	биоген *м*	380	бластоцит *м*
348	биогенез *м*	381	бластогенез *м*
349	биогенетический	382	бластогенный
350	биогенетическая изоляция *ж*	383	бластомер *м*
351	возникновение *ср* жизни	384	бластовариация *ж*
352	биологический метод *м* борьбы	385	количественный признак *м*
353	биологическая изоляция *ж*	386	наследование *ср* количественных признаков
354	биологическая раса *ж*	387	блефаропласт *м*
355	биологический спектр *м*	388	блок *м*
356	биомасса *ж*	389	--
357	биомеханика *ж*	390	блок *м* генов
358	биометрический	391	блоковый эффект *м*
359	биометрика *ж*, биометрия *ж*	392	линия *ж*
360	биометрия *ж*	393	кровное родство *ср*
361	биомутант *м*	394	соматическая клетка *ж*
362	бион *м*	395	--
363	биофора *ж*	396	боттом-рецессив *м*
364	биоплазма *ж*	397	стадия *ж* букета
365	происхождение *ср* жизни	398	брахимейоз *м*
366	биосома *ж*	399	брадителический
367	биосинтез *м*	400	ветвистый, разветвленный
368	биологическая систематика *ж*	401	ветвистая хромосома *ж*
369	биотический	402	место *ср* разветвления
370	биотический потенциал *м*	403	фрагментационная гипотеза *ж*, гипотеза *ж* "сначала разрыв"
371	биотическая резистентность *ж*	404	цикл *м* "разрыв-слияние-мост"
		405	бивалент *м*, возникший в

476	хондриокинез *м*	514	хромопласт *м*
477	хондриоконт	515	--
478	хондриолиз *м*	516	хромосомный
479	хондриома *ж*	517	хромосомная
480	хондриомера *ж*		аберрация *ж*
481	хондриомит *м*	518	хромосомная химера *ж*,
482	--		хромосомный мозаик *м*
483	хондриосома *ж*	519	хромосомная нить *ж*
484	хондриосфера *ж*	520	хромосома *ж*
485	хорогамия *ж*	521	хромосомное плечо *ср*
486	хромазия *ж*	522	хромосомный мост *м*
487	хроматический,	523	хромосомный комплекс *м*
	хроматиновый	524	цикл *м* хромосомы
488	хроматидный обмен *м*	525	хромосомная карта *ж*
489	хроматида *ж*	526	--
490	хроматидная	527	число *ср* хромосом
	аберрация *ж*	528	хромосомное кольцо *ср*
491	хроматидный разлом *м*	529	хромосомный набор *м*
492	хроматидный мост *м*	530	множественные разломы
493	хроматидный		*м мн* хромосом,
494	хроматидная		сверхфрагментация *ж*
	интерференция *ж*	531	хромосомная нить *ж*
495	хроматидная петля *ж*	532	хромосомный
496	хроматин *м*	533	хромосомин *м*
497	хроматиновый мост *м*	534	--
498	хроматиновое тельце *ср*	535	хромотип *м*
499	хроматолиз *м*	536	цис-конфигурация *ж*,
500	хромидиосома *ж*		цис-положение *ср*
501	хромидия *ж*	537	цис-транс тест *м*
502	хромиоль *ж*	538	цистрон *м*
503	хромоцентр *м*	539	эффект *м* цис-положения
504	хромоцит *м*	540	клад *м*
505	хромофибрилла *ж*	541	кладогенез *м*
506	--	542	дробление *ср*
507	хромоген *м*	543	задержка *ж* дробления
508	хромомера *ж*	544	ядро *ср* дробления
509	хромомерный	545	клейстогамия *ж*
510	градиент *м* величины	546	клин *м*
	хромомер	547	клон *м*
511	хромонема *ж*	548	с-мейоз *м*,
512	хромофобный		колхициновый мейоз *м*
513	хромоплазма *ж*		

549 с-митоз *м*,
колхициновый митоз *м*
550 считывание *ср* кода
551 кодоминантность *ж*
552 кодоминантный
553 кодон *м*
554 коэффициент *м* инбридинга
555 коэффициент *м* родства
556 ценоцит *м*
557 ценогамета *ж*
558 ценогенез *м*
559 ценовид *м*
560 ценозигота *ж*
561 совпадение *ср*
562 колхицин *м*
563 колхиплоидный
564 --
565 колицин *м*
566 колициногенный
567 колициногения *ж*
568 колласома *ж*
569 коллатеральная наследственность *ж*
570 --
571 --
572 комбинационный квадрат *м*
573 комбинационная способность *ж*
574 коммискуум *м*
575 --
576 сообщество *ср*
577 компенсированная хиазма *ж*
578 компариум *м*
579 совместимость *ж*
580 совместимый
581 компенсированная хиазма *ж*
582 компенсация *ж*
583 компенсатор *м*

584 компетенция *ж*, способность *ж*
585 компетентная донорная клетка *ж*
586 конкурентоспособность *ж*
587 комплемент *м*
588 комплементарный
589 комплементарная хиазма *ж*
590 комплементарные факторы *м мн*
591 комплементация *ж*
592 --
593 полное сцепление *ср*
594 полная пенетрантность *ж*
595 комплексный признак *м*
596 множественный кроссинговер *м*
597 --
598 компаунд-Х-хромосома *ж*
599 конденсация *ж*
600 условная доминантность *ж*
601 условные факторы *м мн*
602 обусловленная доминантность *ж*
603 условная деталь *ж*
604 условные факторы *м мн*
605 конфигурация *ж*
606 конгрессия *ж*
607 конгруентное скрещивание *ср*
608 конъюгант *м*
609 конъюгация *ж*
610 --
611 перетяжка *ж*
612 --
613 контактная гипотеза *ж*
614 точка *ж* контакта
615 непрерывный
616 сокращение *ср*

683 цитогенетика *ж*
684 цитогония *ж*
685 цитокинез *м*, деление *ср* клетки
686 цитологический
687 цитолизин *м*
688 цитолиз *м*
689 цитом *м*
690 цитомера *ж*
691 микросома *ж*
692 цитомиксис *м*
693 цитоморфоз *м*
694 цитоплазма *ж*
695 цитоплазматический
696 цитоплазматическая наследственность *ж*
697 цитоплазматическая мутация *ж*
698 цитоплазмон *м*
699 цитосома *ж*
700 клеточный симбиоз *м*
701 цитотип *м*
702 дальтонический
703 дальтонизм *м*
704 мать *ж*
705 сравнение *ср* мать-дочь
706 дарвинизм *м*
707 длительная модификация *ж*
708 дочерняя клетка *ж*
709 дочерняя хроматида *ж*
710 дочерняя хромосома *ж*
711 дезадаптация *ж*
712 --
713 деконъюгация *ж*
714 дефективный
715 дефективная лизогенная бактерия *ж*
716 вторичный разрыв *м*
717 нехватка *ж*
718 дефинитивное ядро *ср*
719 дегенерация *ж*

720 дегенерация *ж*, связанная с инбридингом
721 деградация *ж*
722 задержанный
723 задержанное проявление *ср* доминантности
724 задержанное наследование *ср*, предетерминация *ж*
725 делеция *ж*
726 гетерозиготная делеция *ж*
727 дим *м*
728 денуклеинация *ж*
729 дерепрессия *ж*
730 результат *м* скрещивания двух гибридов
731 производный
732 потомство *ср*
733 потомок *м*
734 десмоны *м мн*
735 дезоксирибонуклеиновая кислота (ДНК) *ж*
736 десинапсис *м*
737 разъединение *ср*
738 направленная изменчивость *ж*
739 детерминированный
740 вредный
741 дейтеротокия *ж*
742 --
743 фенотипическая стабильность *ж*
744 отклонение *ср*
745 изменение *ср*, связанное с генами-модификаторами
746 деволюция *ж*
747 диада *ж*
748 диадельфический
749 диагенный
750 диагональная хиазма *ж*
751 диакинез *м*
752 диакинезный

753	диаллельный	786	дигаплоид *м*, дигаплоидный
754	диандрический	787	дигаплоидия *ж*
755	диафоромиксис *м*	788	дигетерозигота *ж*
756	диасхистический	789	дигибрид *м*, дигибридный
757	диастема *ж*	790	дигибридность *ж*
758	двойная звезда *ж*	791	дикарион *м*
759	двуосновный	792	дикариофаза *ж*
760	дикарион *м*	793	дикариотический
761	дикарионный	794	димегалия *ж*
762	дицентрический	795	димерный
763	дихламидный	796	димерность *ж*
764	дихогамный	797	уменьшение *ср*
765	дихогамия	798	димиксис *м*
766	дихондрический	799	двуоднодомный
767	диклинный	800	диморфный
768	ретикулярная стадия *ж* (хромосомы)	801	диморфизм *м*
769	диктиокинез *м*	802	двудомный
770	диктиосома *ж*	803	двудомность *ж*
771	ретикулярная стадия *ж* (хромосом)	804	двуяйцевые близнецы *м мн*
		805	дифилетический
772	амфитетраплоид *м*, амфитетраплоидный	806	диплобионт *м*
		807	диплобионтный
773	амфитетраплоидия *ж*	808	диплобивалент *м*
774	диэнтомофилия *ж*	809	диплохламидный
775	дифференциальный	810	диплохромосома *ж*
776	дифференциальное и интерференционное расстояния *ср*	811	диплогенотипический
		812	возникновение *ср* диплоидного и гаплоидного близнецов
777	дифференциальный сегмент *м*		
		813	двуразнодомность
778	--	814	диплоид *м*, диплоидный
779	дифференцировка *ж*	815	диплоидия *ж*
780	дигаметный, гетерогаметный	816	диплоидный набор *м* хромосом
		817	диплоидизация *ж*
781	дигаметия *ж*	818	диплокариотический, тетрагаплоидный
782	дигамия *ж*		
783	дигенез *м*, чередование *ср* поколений	819	дипломоносомный
		820	диплонема *ж*
784	дигенный	821	диплонт *м*
785	дигеномный	822	диплонтный

823	диплофаза *эс*	851	дистальный
824	диплозис *м*	852	расстояние *ср*
825	диплосома *эс*	853	распределенный
826	диплоспория *эс*		К-митоз *м*
827	диплотена *эс*,	854	дитокический
	диплотенный	855	тетрады *м мн* дитипа
828	направленный отбор *м*	856	--
829	прямое родство *ср*	857	деление *ср*
830	дизассортативное	858	дизиготический,
	спаривание *ср*,		двуяйцевый,
	спаривание *ср*		разнояйцевый
	несходных индивидуумов	859	разнояйцевые близнецы *м*
831	прерывистый		*мн*
832	прерывистая измен-	860	Д-митоз *м*
	чивость *эс*, альтерна-	861	ДНК, дезоксирибонукле-
	тивная изменчивость *эс*		иновая кислота *эс*
833	дислиэрез *м*	862	доминантность *эс*
834	комплементарный	863	доминантный
835	расхождение *ср*	864	доминиген *м*
836	альтернативное	865	донатор *м*
	распределение *ср*	866	донор *м*, донорный
837	дислоцированные	867	донорная клетка *эс*
	сегменты *м мн*	868	дозовая компенсация *эс*
838	дислокация *эс*	869	дозовая индифферент-
839	дисоматия *эс*,		ность *эс*
	тетраплоидия *эс*	870	двойное скрещивание *ср*
840	дисома *эс*	871	двойной кроссинговер *м*
841	дисомный	872	амфидиплоид *м*,
842	дисомия *эс*		амфидиплоидный
843	несопоставимый,	873	два доминантных
	несоизмеримый		комплементарных гена
844	диагональная хиазма *эс*,	874	--
	некомпенсированная	875	двойной гаплоид *м*
	хиазма *эс*	876	двойная редукция *эс*
845	диспермный	877	кольцевая хромосома *эс*
846	диспермное		удвоенного размера
	оплодотворение *ср*,	878	удвоение *ср*
	дислермия *эс*	879	дрейф *м*
847	стадия *эс* деконденсации	880	драйв *м*
848	диспирема *эс*	881	дуплексный, двойной
849	разрушительный отбор *м*	882	удвоенный
850	диссоциация *эс*	883	удвоенные гены *м мн*

884	дупликация *ж*
885	дупликационный полиплоид *м*
886	диада *ж*
887	дисгенез *м*
888	дисгенный
889	дисплоид *м*, дисплоидный
890	дисплоидный вид *м*
891	дисплоидия *ж*
892	экада *ж*
893	--
894	экобиотическая адаптация *ж*
895	экоклиматическая адаптация *ж*
896	экоклин *м*
897	экогеографическая дивергенция *ж*
898	экофен *м*
899	экофенотип *м*
900	эковид *м*
901	экотип *м*
902	эктопическая конъюгация *ж*
903	эктоплазма *ж*
904	эктосома *ж*
905	эктосфера *ж*
906	электросома *ж*
907	элиминация *ж*
908	эмаскуляция *ж*
909	эмбриальный мешок *м*
910	эндогамный
911	эндогамия *ж*
912	эндогенот *м*
913	эндогенный
914	эндомитоз *м*
915	эндомитотический
916	эндомиксис *м*
917	эндонуклеарный
918	эндоплазма *ж*
919	эндоплазматический

920	эндополиплоид *м*, эндополиплоидный
921	эндополиплоидия *ж*
922	эндосома *ж*
923	эндосперм *м*
924	эндотаксонический
925	рабочее ядро *ср*
926	рабочая стадия *ж*
927	энергида *ж*
928	перманентная гетерозиготность *ж*
929	--
930	--
931	окружение *ср*, среда *ж*
932	окружающий, средовой
933	окружающие условия *ср мн*, факторы *м мн* среды
934	корреляция *ж*, связанная со средой
935	изменчивость *ж*, связанная со средой
936	эпигамный
937	эпигенез *м*
938	эпигенетика *ж*
939	эпигенотип *м*
940	эписома *ж*
941	переходная эписомная стадия *ж*
942	эпистаз *м*, эпистазия *ж*
943	эпистатический
944	эквационный
945	эквационное деление *ср*, гомеотипическое деление *ср*
946	экваториальное тело *ср*
947	экваториальная пластинка *ж*
948	--
949	эритрофильный
950	--
951	эуцентрический
952	эухроматиновый

953 эухроматин *м*
954 эухромосома *ж*
955 евгенический
956 евгеника *ж*
957 эугетерозис *м*
958 эумейоз *м*
959 эумитоз *м*
960 эуплоидный, эуплоид *м*
961 эуплоидия *ж*
962 эволюция *ж*
963 факторы *м мн* эволюции
964 эволюционный успех *м*
965 эволютивный
966 экзаггрегация *ж*
967 проявление *ср*
968 эксмутант *м*
969 экзогамия *ж*
970 экзогенота *ж*
971 экзогенный
972 взрывной С-митоз *м*
973 взрывная эволюция *ж*,
взрывное
видообразование *ср*
974 экспоненциальная фаза *ж*
гибели
975 экспрессивность *ж*
976 ген-модификатор *м*
977 внехромосомный
978 --
979 экстрарадиальный
980 --
981 фактор *м*, ген *м*
982 генная карта *ж*
983 пара *ж* генов,
пара *ж* аллелей
984 ложное сцепление *ср*
985 ложная конъюгация *ж*
986 семейный
987 семейство *ср*
988 семейный отбор *ср*
989 оплодотворение *ср*

990 женская линия *ж*,
материнская линия *ж*
991 плодовитый, фертильный
992 оплодотворение *ср*,
фертилизин *м*
993 способность *ж* к
оплодотворению
994 расположение *ср*
хромосом цепочкой
995 нить *ж*, волокно *ср*
996 фибрилла *ж*
997 нить *ж*, волокно *ср*
998 дочернее поколение *ср*
999 первое деление *ср*
1000 расщепление *ср*,
дробление *ср*
1001 фиксация *ж*
1002 скорость *ж* фиксации
1003 фиксированный
1004 флюктуация *ж*,
колебание *ср*
1005 укрепление *ср*
1006 родоначальник *м*
1007 четырехнитевой обмен *м*,
скрещивание *ср* в четырех
направлениях
1008 фракционирование *ср*
1009 разрыв *м*, перелом *м*
1010 фрагмент *м*
1011 фрагментация *ж*
1012 братский
1013 разнояйцевые
близнецы *м мн*
1014 фримартин *м*
1015 --
1016 частота *ж*
1017 сибс *м*, родная сестра *ж*,
родной брат *м*
1018 функциональный
диплоид *м*
1019 фундаментальный
1020 основное число *ср*

1021	слияние *ср*
1022	ядро *ср* слияния
1023	гамета *ж*, половая клетка *ж*
1024	гаметический
1025	гаметическое число *ср*
1026	гаметическая несовместимость *ж*
1027	гаметический летальный фактор *м*
1028	гаметическая мутация *ж*
1029	гаметическое число *ср*
1030	гаметическая стерильность *ж*
1031	гаметобласт *м*
1032	гаметоцит *м*
1033	гаметогамия *ж*
1034	гаметогенез *м*
1035	гаметогенный
1036	гаметогенный
1037	гаметогоний *м*
1038	гаметоид *м*
1039	гаметофор *м*
1040	гаметофит *м*
1041	гаметофитный
1042	гаметный
1043	гамобий *м*
1044	гамодим *м*
1045	гамогенез *м*
1046	гамогенетический
1047	гамогенетический
1048	гамогония *ж*
1049	гамолиз *м*
1050	гамон *м*
1051	гамонт *м*
1052	гамофаза *ж*
1053	гамотропизм *м*
1054	гейтонокарпия *ж*
1055	гейтоногамный
1056	гейтоногенез *м*
1057	близнец *м*
1058	почкование *ср*
1059	размножающийся почкованием
1060	геммула *ж*
1061	ген *м*
1062	генеалогия *ж*
1063	расположение *ср* генов
1064	генное равновесие *ср*
1065	генный центр *м*
1066	генный хроматин *м*
1067	генный комплекс *м*
1068	генно-цитоплазматическая изоляция *ж*
1069	доза *ж* гена
1070	генный дрейф *м*
1071	--
1072	частота *ж* гена
1073	просачивание *ср* генов
1074	взаимодействие *ср* генов
1075	генная мутация *ж*
1076	группа *ж* генов
1077	генный ''пул'' *м*
1078	гипотеза *ж* равновесия гена и генного продукта
1079	поколение *ср*
1080	интервал *м* между поколениями
1081	генеративный
1082	генеративное ядро *ср*
1083	редупликация гена *ж*
1084	гипотеза *ж* генной реплики
1085	родовой
1086	межродовой гибрид *м*
1087	родовой тип *м*
1088	гипотеза *ж* голодания генов
1089	генная нить *ж*
1090	генетически маркированные хромосомы
1091	генетический
1092	ассортативное спаривание *ср*

№		№	
1093	генотипическая среда *ж*	1131	зародышевый путь *м*
1094	генетическая коадаптация *ж*	1132	гигантская хромосома *ж*
		1133	аппарат *м* Гольджи
1095	генетическая смерть *ж*	1134	образование *ср* аппарата Гольджи
1096	генетическое равновесие *ср*	1135	деление *ср* аппарата Гольджи
1097	генетическая нагрузка *ж*	1136	растворение *м* аппарата Гольджи
1098	генетика *ж*		
1099	генетическая ценность *ж*	1137	гонада *ж*, половая железа *ж*
1100	генетическая изменчивость *ж*	1138	гонадный
1101	--	1139	гоний *м*
1102	перенос *м* генов	1140	гонидий *м*
1103	генный	1141	гоноцит *м*
1104	генное равновесие *ср*	1142	гоногамета *ж*
1105	генная доза *ж*	1143	гоногенез *м*
1106	генная стерильность *ж*	1144	гономера *ж*
1107	геноклин *м*	1145	гономерия *ж*
1108	генокопия *ж*	1146	гономоноарренический
1109	генная дисперсия *ж*	1147	гономонотелидический
1110	геноид *м*	1148	гонофаг *м*
1111	геном *м*	1149	гоносома *ж*
1112	геномный анализ *м*	1150	гонотоконт *м*
1113	геномная мутация *ж*	1151	степень *ж*
1114	геномера *ж*	1152	градирование *ср*, расположение *ср* в определенном порядке
1115	геномный		
1116	генонема *ж*	1153	поглощающее скрещивание *ср*
1117	генономия *ж*		
1118	геносома *ж*	1154	почковый гибрид *м*
1119	генотип *м*	1155	гинандроид *м*
1120	взаимодействие *ср* генотипа со средой	1156	гинандроморф *м*
		1157	гинандроморфность *ж*
1121	генотипический	1158	гинаутосома *ж*
1122	генотипический	1159	гинодвудомный, гинодиезический
1123	геновариация *ж*		
1124	род *м*	1160	гинодвудомность *ж*, гинодиезия *ж*
1125	географическая раса *ж*		
1126	зародышевый	1161	гиноезический
1127	зародышевый отбор *м*	1162	гиноезия *ж*
1128	зародышевое пятно *ср*		
1129	зародышевый пузырек *м*		
1130	зародышевая плазма *ж*		

1163	женский фактор *м*, гинофактор *м*
1164	гиногенез *м*
1165	гиногенетический
1166	гиногенный
1167	гиногения *ж*
1168	гинооднодомный, гиномоноезический
1169	гинооднодомность *ж*, гиномоноезия *ж*
1170	гиносперм *м*
1171	гиноспора *ж*
1172	гиноспорогенез *м*
1173	спираль *ж*
1174	полукровный
1175	помесь *ж*, метис *м*, ублюдок *м*
1176	полухиазма *ж*
1177	полухроматидный разлом *м*
1178	полухроматидная фрагментация *ж*
1179	полурасхождение *ср*
1180	полумутация *ж*
1181	полураса *ж*
1182	полусибс *м*, сводная сестра *ж*, сводный брат *м*
1183	семья *ж* полусибса, семья *ж* сводной сестры
1184	гинандроморф *м*
1185	полуголодание *ср*
1186	гаплобионт *м*
1187	гаплобионтный
1188	гаплокариотип *м*
1189	гаплохламидная химера *ж*
1190	гаплодиплоид *м*, гаплодиплоидный
1191	гаплодиплоидия *ж*
1192	гаплодиплонт *м*
1193	гаплогенотипический
1194	гаплосразнодомность *ж*

1195	гаплоид *м*, гаплоидный
1196	гаплонедостаточный
1197	гаплодостаточный
1198	гаплоидия *ж*
1199	гаплом *м*
1200	гапломикт *м*
1201	гапломитоз *м*
1202	гапломиксис *м*
1203	гаплонт *м*
1204	гаплонтный
1205	гаплофаза *ж*
1206	гаплополиплоид *м*, гаплополиплоидный
1207	гаплополиплоидия *ж*
1208	гаплозис *м*
1209	гаплосомия *ж*
1210	гаплозиготный
1211	гемиаллоплоид *м*, гемиаллоплоидный
1212	гемиаллоплоидия *ж*
1213	гемиаутоплоид *м*, гемиаутоплоидный
1214	гемиаутоплоидия *ж*
1215	полухроматидный
1216	гемидиэрез *м*
1217	гемигаплоид *м*, гемигаплоидный
1218	гемигаплоидия *ж*
1219	гемиголоплоид *м*, гемиголоплоидный
1220	гемиголоплоидия *ж*
1221	гемикарион *м*
1222	гемикинез *м*
1223	гемиортокинез *м*
1224	гемиксис *м*
1225	гемизиготный
1226	гептаплоид *м*, гептаплоидный
1227	гептаплоидия *ж*
1228	геркогамия *ж*
1229	наследуемость *ж*
1230	наследственный

1306	гетеросиндез *м*	1338	гомеокинез *м*
1307	гетеротетраплоид *м*, гетеротетраплоидный	1339	гомеологичный
		1340	гомеоморфный
1308	гетеротетраплоидия *ж*	1341	гомеозис *м*
1309	гетероталлический	1342	гомеостазис *м*, гомеостаз *м*
1310	гетероталлизм *м*, гетероталлия *ж*	1343	гомеостат *м*
1311	гетероталлия *ж*, гетероталлизм *м*	1344	гомеосинапсис *м*
		1345	гомеотический
1312	гетерозисный	1346	гомеотип *м*
1313	гетеротрансформация *ж*	1347	гомеотипический
1314	гетеротропическая хромосома *ж*	1348	гомеотипическое деление *ср*, эквационное деление *ср*
1315	гетеротипическое деление *ср*	1349	гомоаллель *м*
1316	гетеротипическое деление *ср*	1350	гомоаллельный
		1351	гомобрахиальный, равноплечий
1317	гетерозиготность *ж*	1352	гомокарион *м*
1318	гетерозигота *ж*	1353	гомокариоз *м*
1319	гетерозиготный	1354	гомоцентрический
1320	гексада *ж*	1355	гомоцентричность *ж*
1321	гексаплоид *м*, гексаплоидный	1356	гомохондрический
		1357	гомохронный
1322	гексаплоидия *ж*	1358	гомодинамический
1323	гексасомный, гексасомик *м*	1359	гомеокинез *м*
		1360	гомеологичный
1324	--	1361	гомеозис *м*
1325	--	1362	гомеостазис *м*, гомеостаз *м*
1326	гистогенез *м*	1363	гомеостат *м*
1327	гистогенез *м*	1364	гомогамета *ж*
1328	голандрический	1365	гомогаметный
1329	полое веретено *ср*	1366	гомогамный
1330	голобластическое дробление *ср*	1367	гомогамия *ж*
1331	гологамный	1368	гомогенетический
1332	гологамия *ж*	1369	гомогенический
1333	гологинный	1370	гомогенота *ж*
1334	голокинетический	1371	гомогеномный
1335	голоморфоз *м*	1372	гомогения *ж*
1336	голотопия *ж*	1373	гомогетеромиксис *м*
1337	голотип *м*		

1374	гомокарион *м*
1375	гомокариоз *м*
1376	гомолецитальный
1377	гомологичный
1378	гомолог *м*
1379	гомология *ж*
1380	гомолизогенный
1381	гомомерный
1382	гомомерия *ж*
1383	гомомиксис *м*
1384	гомоморфный
1385	гомоморфный
1386	гомоморфия *ж*
1387	гомеозис *м*
1388	гомофитический
1389	гомопластический
1390	гомоплазия *ж*
1391	гомоплоид *м*, гомоплоидный
1392	гомоплоидия *ж*
1393	гомополярный
1394	--
1395	гомосомный
1396	гомостазис *м*
1397	гомосинапсис *м*
1398	гомосиндез *м*
1399	гомоталлический
1400	гомоталлизм *м*
1401	гомотрансформация *ж*
1402	гомотипическое деление *ср*
1403	гомозиготность *ж*
1404	гомозигота *ж*
1405	гомозиготизация *ж*
1406	гомозиготный
1407	хоротелический
1408	гибрид *м*, гибридный
1409	гибридная популяция *ж*
1410	--
1411	гибридизация *ж*
1412	гибридность *ж*
1413	гибридная летальность *ж*

1414	гибридогенный псевдопартогенез *м*
1415	гибридная сила *ж*, гетерозис *м*
1416	гибридная зона *ж*
1417	гипоаллеломорф *м*, гипоаллеломорфный
1418	гиперхимера *ж*
1419	гиперхромазия *ж*
1420	гиперхроматоз *м*
1421	дополнительное оплодотворение *ср* во время беременности
1422	гипердиплоид *м*, гипердиплоидный
1423	гипердиплоидия *ж*
1424	гипергаплоид *м*, гипергаплоидный
1425	гипергаплоидия *ж*
1426	гипергетеробрахиальный, сильно неравноплечий
1427	гиперморфный
1428	гиперморфоз *м*
1429	гиперплоид *м*, гиперплоидный
1430	гиперплоидия *ж*
1431	гиперполиплоид *м*, гиперполиплоидный
1432	гиперполиплоидия *ж*
1433	гиперсиндез *м*
1434	гипертелия *ж*
1435	аверсия *ж* гифов
1436	гипохромазия *ж*
1437	гипогенез *м*
1438	гипогенетический
1439	гипогаплоид *м*, гипогаплоидный
1440	гипогаплоидия *ж*
1441	гипоморфный
1442	гипоморфный

1443	гипоплоид м, гипоплоидный	1475	неполное доминирование ср
1444	гипоплоидия ж	1476	неполная пенетрантность ж
1445	гипополиплоид м, гипополиплоидный	1477	неконгруентное скрещивание ср
1446	гипополиплоидия ж	1478	независимое распределение ср генов
1447	гипостаз м, гипостазис м	1479	независимый признак м
1448	гипостатический	1480	независимая переменная ж
1449	гипосиндез м		
1450	гипотип м	1481	индекс м свободного кроссинговера
1451	гистерезис м		
1452	ида ж	1482	бактерия ж индикатор
1453	идентичный	1483	индивидуум м
1454	идентичные близнецы м мн, однояйцевые близнецы	1484	индивидуальный отбор м
		1485	индуцированный
1455	идиохроматин м	1486	индукция ж
1456	идиохромидия ж	1487	инертный
1457	идиохромосома ж	1488	инерция ж
1458	идиограмма ж	1489	наследственность ж
1459	идиомутация ж	1490	наследование ср приобретенных признаков
1460	идиоплазма ж		
1461	идиосома ж	1491	наследственый признак м
1462	идиотип м		
1463	идиовариация ж	1492	ген-ингибитор м
1464	идиосома ж	1493	подавление ср
1465	иллегитимный, незаконный	1494	--
		1495	ингибитор м
1466	коэффициент м иммиграции	1496	подавляющий
		1497	инициальный, начальный
1467	давление ср иммиграции	1498	вставка ж
1468	инбридинг м, инцухт м, родственное скрещивание ср	1499	вставочная транслокация ж
		1500	вставочный разрыв м
1469	коэффициент м инбридинга	1501	вставочный участок м, место ср вставки
1470	подавление ср, связанное с инбридингом	1502	интегрированный фактор м
1471	несовместимость ж	1503	интегрированное состояние ср
1472	фактор м несовместимости		
1473	гены м мн несовместимости	1504	интеграция ж
1474	несовместимый		

1505	вирулентный фаг *м*
1506	усиливающие факторы *м мн*
1507	--
1508	взаимодействие *ср*
1509	теория *ж* взаимодействия
1510	межплечевая конъюгация *ж*
1511	промежуток *м* (между поперечными дисками)
1512	межбивалентное соединение *ср*
1513	межцентромерная область *ж*
1514	обмен *м*, транслокация *ж*
1515	трисомик *м*, связанный с транслокацией
1516	интерхромидия *ж*
1517	интерхромомера *ж*
1518	межхромосомный
1519	интерференция *ж*
1520	интерференционное расстояние *ср*
1521	межгенный
1522	интерградация *ж*
1523	интеркинез *м*
1524	переплетение *ср*
1525	промежуточный
1526	промежуточный
1527	промежуточная наследственность *ж*
1528	интермитоз *м*
1529	внутренний
1530	внутренняя спираль *ж*
1531	интерфаза *ж*
1532	перекрещивающиеся веретена *ср мн*
1533	интерсекс *м*
1534	интерсексуальный
1535	интерсексуальность *ж*
1536	межвидовой

1537	межвидовое скрещивание *ср*, межвидовая гибридизация *ж*
1538	интерстерильность *ж*
1539	интерстициальный
1540	межзональный
1541	инбредная популяция *ж*
1542	внутрихромосомный
1543	внутриклассовая корреляция *ж*
1544	внутригенный
1545	интрагаплоид *м*, интрагаплоидный
1546	интрагаплоидия *ж*
1547	интрарадиальный
1548	интрогрессия *ж*
1549	интрогрессивный
1550	инверсия *ж*
1551	инверсионная гетерозигота *ж*
1552	инверсионный морфизм *м*
1553	инверсионная хиазма *ж*
1554	нежизнеспособность *ж*
1555	нежизнеспособный
1556	скрытый ген *м*
1557	неправильная задержка *ж*
1558	невосстановимый мутант *м*
1559	изоаллель *м*
1560	изоаллелия *ж*
1561	изоаутополиплоидный, изоаутополиплоид *м*
1562	изоаутополиплоидия *ж*
1563	равноплечий
1564	изохроматида *ж*
1565	изохроматидный разрыв *м*
1566	изохроматидный
1567	изохромосома *ж*
1568	изоцитоз *м*
1569	изодицентрический
1570	изофрагмент *м*
1571	изогамета *ж*

1572	изогамный	1611	кариоген *м*
1573	изогамия *ж*	1612	кариограмма *ж*
1574	изогенетический	1613	кариоид *м*
1575	изогенный	1614	кариокинез *м*, митоз *м*
1576	изогеномный	1615	кариокинетический,
1577	изогенный		митотический
1578	изогенность *ж*	1616	кариологический
1579	изокариоз *м*	1617	кариология *ж*
1580	изолят *м*	1618	кариолимфа *ж*
1581	изоляция *ж*	1619	кариолизис *м*
1582	изоляционные гены *м мн*	1620	кариомера *ж*
1583	изоляционный индекс *м*	1621	кариомерокинез *м*
1584	изолецитальный	1622	кариомитоз *м*
1585	--	1623	ядрышко *ср*
1586	изоморфный	1624	кариоплазмогамия *ж*
1587	изоморфизм *м*	1625	кариорексис *м*
1588	изоморфный	1626	кариосома *ж*
1589	изофен *м*	1627	кариосфера *ж*
1590	изофенный	1628	кариотека *ж*
1591	изофеногамия *ж*	1629	кариотин *м*
1592	изофенный	1630	кариотип *м*
1593	изоплоид *м*, изоплоидный	1631	катахромазия *ж*
1594	изоплоидия *ж*	1632	катафаза *ж*
1595	изополиплоид *м*,	1633	ключевой ген *м*
	изополиплоидный	1634	ключевая мутация *ж*
1596	изополиплоидия *ж*	1635	''убийца'' *ж*
1597	изопикном *м*	1636	род *м*, семья *ж*
1598	изосиндез *м*	1637	кинетическое тельце *ср*,
1599	изосиндетический		кинетохор *м*
1600	изотрансформация *ж*	1638	первичная перетяжка *ж*
1601	изотрисомия *ж*	1639	кинетическое ядро *ср*
1602	изотипия *ж*	1640	кинетохора *ж*,
1603	изозиготный		центромера *ж*
1604	изозиготность *ж*	1641	кинетоген *м*
1605	повторяющаяся мутация	1642	кинетогенез *м*
	ж	1643	кинетомера *ж*
1606	какогенический	1644	кинетонема *ж*
1607	фактор *м* ''каппа'',	1645	--
	фактор ''убийца''	1646	кинетопласт *м*
1608	--	1647	кинетоплазма *ж*
1609	кариокластический	1648	киносома *ж*
1610	кариогамия *ж*	1649	клон *м*

1650 --
1651 лабильный
1652 лабильность *ж*
1653 лакуна *ж*
1654 лаг *м*
1655 лаггард *м*, отставшая
хромосома *ж*
1656 --
1657 ламаркизм *м*
1658 "ламповые щетки"
(хроматиды)
1659 "ламповые щетки"
(хромосомы)
1660 латентный
1661 латеральный
1662 латеральная хиазма *ж*
1663 --
1664 лектотип *м*
1665 лептонема *ж*
1666 лептотена *ж*
1667 летальный
1668 летальность *ж*
1669 ограниченный
1670 линия *ж*
1671 линейный порядок *м*
1672 линейная хромосома *ж*
1673 линейная регрессия *ж*
1674 генеалогическая линия *ж*
1675 генеалогия *ж* потомства
1676 инбредная линия *ж*
1677 сцепление *ср*
1678 нагрузка *ж*, бремя *ср*
1679 локализация *ж*
1680 локализованная
центромера *ж*
1681 --
1682 локус *м*
1683 локусспецифичный
эффект *м*
1684 петля *ж* унивалента
1685 слабо конъюгирующий
1686 мутация *ж* потери

1687 перенос *м* с низкой
частотой
1688 пышный рост *м*
1689 лизогенный
1690 лизогенизация *ж*
1691 лизогения *ж*
1692 лизотип *м*
1693 макроэволюция *ж*
1694 макрогамета *ж*
1695 макромутация *ж*,
большая мутация *ж*
1696 регенерация *ж*
макронуклеуса
1697 макронуклеус *м*,
макроядро *ср*
1698 макрофилогенез *м*,
макроэволюция *ж*
1699 макроспора *ж*
1700 макроспорогенез *м*
1701 большой, главный
1702 олигоген *м*, главный ген *м*
1703 большая мутация *ж*
1704 большая спираль *ж*
1705 отцовская линия *ж*,
мужская линия *ж*
1706 проявление *ср*,
манифестация *ж*
1707 множественные
эффекты *м мн*
1708 --
1709 хромосомная карта *ж*
1710 расстояние *ср* на
хромосомной карте
1711 единица *ж* хромосомной
карты
1712 маркер *м*
1713 массовый
1714 массовая мутация *ж*
1715 индивидуальный отбор *м*,
массовый отбор *м*
1716 конъюгационный
"убийца" *м*

1717	материнский
1718	материнская наследственность *ж*
1719	совокупление *ср*, спаривание *ср*
1720	разведение *ср*
1721	--
1722	группа *ж* скрещивания
1723	система *ж* скрещивания
1724	теория *ж* скрещивания
1725	тип *м* спаривания
1726	матроклинный
1727	матрикс *м*
1728	матричный мост *м*
1729	матричное склеивание *ср*
1730	матроклинная наследственность *ж*
1731	матроклинность *ж*
1732	созревание *ср*
1733	деление *ср* созревания
1734	метод *м* наибольшего правдоподобия
1735	медиоцентрический
1736	мегаэволюция *ж*
1737	мегаспора *ж*, макроспора *ж*
1738	мегаспороцит *м*, материнская клетка *ж* эмбрионального мешка
1739	макроспорогенез *м*
1740	мегагетерохроматиновый
1741	мейоцит *м*
1742	мейомерия *ж*
1743	мейоз *м*
1744	мейосома *ж*
1745	мейоспора *ж*
1746	мейотический
1747	мейотрофный
1748	менделевский
1749	менделевское отношение *ср*
1750	менделизм *м*
1751	законы *м мн* Менделя
1752	метод *м* ментора
1753	мериклинальный
1754	мериклинальная химера *ж*
1755	меристема *ж*
1756	меробластическое дробление *ср*
1757	мерогамия *ж*
1758	мерогон *м*
1759	мерогония *ж*
1760	мерокинез *м*
1761	меромиксис *м*
1762	меростатмокинез *м*
1763	мерозигота *ж*
1764	мезомитический
1765	мезомитоз *м*
1766	метаболизирующее ядро *ср*
1767	метацентрический
1768	метагенез *м*, смена *ж* поколений
1769	метагиния *ж*
1770	метакинез *м*
1771	метамитоз *м*
1772	метандрия *ж*
1773	метафаза *ж*
1774	индекс *м* метафазной конъюгации
1775	метафазный
1776	метаплазия *ж*
1777	метаредупликация *ж*
1778	метасиндез *м*
1779	метатактический
1780	метаксения *ж*
1781	метроморфный
1782	микроцентр *м*
1783	микрохромосома *ж*
1784	микроскопление *ср*
1785	микроэволюция *ж*
1786	микрогамета *ж*
1787	микроген *м*

1788 микрогетерохроматино-
вый
1789 микромутация *ж*
1790 микронуклеус *м*,
микроядро *ср*
1791 микрофилогенез *м*
1792 микропиренный
1793 микросома *ж*
1794 микровид *м*, малый вид *м*
1795 микросфера *ж*
1796 микроспора *ж*
1797 микроспороцит *м*,
материнская пыльцевая
клетка *ж*
1798 микроподвид *м*
1799 миктон *м*
1800 миграция *ж*,
переселение *ср*
1801 мимикрия *ж*
1802 миметические гены *м мн*
1803 малый
1804 малый ген *м*
1805 неправильное деление *ср*
1806 неправильный митоз *м*
1807 митокластический
1808 митогенетический
1809 митоз *м*
1810 митосома *ж*
1811 митостатический
1812 митотический
1813 митотический индекс *м*
1814 подавление *ср* митоза
1815 митотический яд *м*
1816 миксохромосома *ж*
1817 миксоплоид *м*,
миксоплоидный
1818 миксоплоидность *ж*
1819 ложная доминантность *ж*
1820 модальное число *ср*
1821 модификация *ж*
1822 модификатор *м*

1823 комплекс *м* генов-
модификаторов
1824 модификатор *м*
1825 модуляция *ж*
1826 модулятор *м*
1827 молекулярный
1828 молекулярная спираль *ж*
1829 монада *ж*
1830 монастр *м*
1831 монида *ж*
1832 монобрахиальный,
одноплечий
1833 моноцентрический
1834 монохромосома *ж*
1835 однодомный
1836 однодомность *ж*
1837 монофакториальный
1838 моноген *м*
1839 моногенез *м*
1840 моногенный
1841 моногеномный
1842 моногеномный
1843 моногения *ж*
1844 моногония *ж*
1845 моногаплоид *м*,
моногаплоидный
1846 моногаплоидность *ж*
1847 моногибридный
1848 моногибридность *ж*
1849 монолепсис *м*
1850 мономерный
1851 мономерность *ж*
1852 монофилетический
1853 моноплоид *м*,
моноплоидный
1854 моноплоидность *ж*
1855 моноплонт *м*, гаплонт *м*
1856 --
1857 моносома *ж*
1858 моносомный, моносомик *м*
1859 моноталлический
1860 монотопия *ж*

1861	монозиготный, однояйцевый
1862	однояйцевые близнецы *м мн*, идентичные близнецы *м мн*
1863	морганида *ж*
1864	морфа *ж*
1865	морфизм *м*
1866	морфогенез *м*
1867	морфогенетический
1868	морфоз *м*
1869	мозаик *м*, мозаичная форма *ж*
1870	мозаичный
1871	мозаичность *ж*
1872	индекс *м* движения хиазм
1873	сложная перестройка *ж*
1874	мультиген *м*
1875	множественный, многократный
1876	множественные аллели *м мн*
1877	множественная центромера *ж*
1878	множественная хиазма *ж*
1879	множественная корреляция *ж*
1880	перенос *м* множественной резистентности
1881	полигенное скрещивание *ср*
1882	множественная реактивация *ж*
1883	множественная регрессия *ж*
1884	множественные половые хромосомы *ж мн*
1885	мультивалент *м*
1886	мутабильный
1887	модификатор *м* мутабильности
1888	мутаген *м*, мутагенный

1889	мутагенез *м*
1890	специфичность *ж* мутагена
1891	мутагенная стабильность *ж*
1892	мутагенный
1893	мутант *м*
1894	мутация *ж*
1895	мутационное равновесие *ср*
1896	мутационная задержка *ж*
1897	мутационный коэффициент *м*
1898	мутационная задержка *ж*
1899	мутационное давление *ср*
1900	частота *ж* мутирования
1901	ген-мутатор *м*
1902	мутаторное вещество *ср*
1903	мутон *м*
1904	взаимный
1905	"голая" центромера *ж*
1906	естественный отбор *м*
1907	--
1908	немамера *ж*
1909	неоцентромера *ж*
1910	неодарвинизм *м*
1911	неоморфный
1912	неотения *ж*
1913	новая двухплоскостная теория *ж*
1914	соединение *ср* (воссоединение)
1915	профаза *ж* на стадии рыхлого клубка
1916	удачная сочетаемость *ж*, благоприятный эффект *м* взаимодействия генов
1917	ядрышкообразующая хромосома *ж*
1918	основное число *ср*
1919	нерасхождение *ср* (хроматид, хромосом ...)

1920	негомологичная конъюгация *ж*
1921	криволейная регрессия *ж*
1922	неменделирующая наследственность *ж*
1923	отсутствие *ср* конъюгации
1924	тетрада *ж* не родительского двойного типа
1925	отсутствие *ср* редукции
1926	не сестринские хроматиды *ж мн*
1927	неоморфный
1928	конъюгация *ж* ядер
1929	ядерная шапочка *ж*
1930	ядерный диморфизм *м*
1931	кариорексис *м*, фрагментация *ж* ядра
1932	деление *ср* ядра
1933	кариорексис *м*, фрагментация *ж* ядра
1934	ядерная оболочка *ж*
1935	ядерный фенотип *м*
1936	ядерный сок *м*, кариолимфа *ж*
1937	--
1938	нуклеиновокислотное "голодание" *ср*
1939	нуклеин *м*
1940	нуклеинизация *ж*
1941	центросома *ж*, находящаяся внутри ядра
1942	нуклеогенная область *ж*
1943	нуклеоид *м*
1944	ядрышковый
1945	околоядрышковый хроматин *м*
1946	ядрышковая перетяжка *ж*
1947	фрагментация *ж* ядрышка
1948	организатор *м* ядрышка
1949	ядрышковые треки *м мн*
1950	ядрышковая зона *ж*
1951	нуклеолонема *ж*

1952	ядрышко *ср*
1953	организатор *м* ядрышка
1954	область *ж* организатора ядрышка
1955	серповидная стадия *ж* ядрышка
1956	нуклеом *м*
1957	нуклеомиксис *м*
1958	нуклеоплазма *ж*
1959	ядерно-плазменное отношение *ср*
1960	нуклеопротеин (нуклеопротеид) *м*
1961	нуклеосома *ж*
1962	нуклеотид *м*
1963	ядро *ср*
1964	чисто-рецессивный полиплоид *м*
1965	нуллисомик *м*
1966	нуллисомия *ж*
1967	числовой
1968	числовой гибрид *м*
1969	потомство *ср*
1970	профаза *ж* старой спирали
1971	олигоген *м*, главный ген *м*
1972	олигопиренный
1973	олистогетерозигота *ж*
1974	гипотеза *ж* "один диск — один ген"
1975	однодисковая тандемная дупликация *ж*
1976	мейоз *м* при одном делении
1977	гипотеза *ж* "один ген — один фермент"
1978	онтогенез *м*
1979	онтогенетический
1980	онтогенез *м*
1981	овоцентр *м*
1982	овоцит *м*
1983	овогенез *м*, оогенез *м*
1984	овокинез *м*
1985	оосома *ж*

1986	оосфера *ж*	2022	парное скрещивание *ср*
1987	овотида *ж*	2023	пара *ж* генов
1988	оперон *м*	2024	палингенез *м*
1989	оппозиционный фактор *м*	2025	паналлель *м*
1990	организатор *м*	2026	паналлоплоид *м*, паналлоплоидный
1991	центр *м* организации	2027	паналлоплоидия *ж*
1992	эффект *м* организации	2028	панаутоплоид *м*, панаутоплоидный
1993	ориентация *ж*		
1994	ортоамитоз *м*	2029	панаутоплоидия *ж*
1995	ортогенез *м*	2030	панген *м*
1996	ортокинез *м*	2031	пангенез *м*
1997	ортоплоид *м*, ортоплоидный	2032	пангеносома *ж*
1998	ортоплоидность *ж*	2033	панмиктический
1999	ортоселекция *ж*	2034	панмиксия *ж*
2000	ортоспираль *ж*	2035	парааллель *м*
2001	ортотактический	2036	парацентрический
2002	ортотеломический	2037	парахроматин *м*
2003	аутбридинг *м*	2038	бивалент *м* XY в форме парашюта
2004	ауткросс *м*		
2005	сверхдоминантность *ж*	2039	парагенеон *м*
2006	область *ж* положительного гетеропикноза	2040	парагенопласт *м*
		2041	параллельное расхождение *ср*
2007	семенной зачаток *м*		
2008	яйцо *ср*	2042	параллельная эволюция *ж*
2009	оксихроматин *м*		
2010	пахинема *ж*	2043	параллельная мутация *ж*
2011	пахитена *ж*	2044	параллельное расхождение *ср*
2012	фактор *м* упаковки		
2013	педогамия *ж*	2045	паралокус *м*
2014	педогенез *м*	2046	парамецин *м*
2015	псевдоаллели *м мн*	2047	парамейоз *м*
2016	а) конъюгация *ж*, б) спаривание *ср*	2048	парамиктический
		2049	парамитоз *м*
2017	конъюгационный блок *м*	2050	парамиксия *ж*
2018	конъюгационный коэффициент *м*	2051	паранемический
		2052	паранемическая спираль *ж*
2019	конъюгационный гетерозис *м*		
		2053	парануклеин *м*
2020	конъюгационный участок *м*	2054	парануклеоплазма *ж*
		2055	параселективность *ж*
2021	тип *м* спаривания	2056	парасексуальный

2057	парасинапсис *м*	2092	патроклинное
2058	парасиндез *м*		наследование *ср*
2059	паратактический	2093	патроклинный
2060	паратрофный	2094	партеногенез *м*
2061	паравариация *ж*	2095	--
2062	родитель *м*, родительский	2096	модификаторы *м мн* типа
2063	родительский		дифференцировки
2064	родительское	2097	тип *м* повреждения
	поколение *ср*	2098	родословная *ж*,
2065	родители *м мн*		родословное древо *ср*,
2066	партенапогамия *ж*		генеалогическая таблица
2067	партенокарпия *ж*		*ж*
2068	партеногамия *ж*	2099	пенетрантность *ж*
2069	партеногенез *м*	2100	пентаплоид *м*,
2070	партеногенетический		пентаплоидный
2071	--	2101	пентаплоидия *ж*
2072	партеномиксис *м*	2102	пентасомик *м*,
2073	партеноспермий *м*		пентасомный
2074	партеноспора *ж*	2103	процент *м* крови, часть *ж*
2075	партенот *м*		крови
2076	частичный разлом *м*	2104	перицентрический
2077	частичная хиазма *ж*	2105	периклинальный
2078	частная корреляция *ж*	2106	периферическое
2079	частичный кроссинговер *м*		движение *ср*
2080	неполное доминирование	2107	периплазма *ж*
	ср	2108	перипласт *м*
2081	частичное оплодотворение	2109	периссоплоид *м*,
	ср		периссоплоидный
2082	частичная гетероталлия *ж*	2110	периссоплоидия *ж*
2083	частичный полиплоид *м*	2111	перистазис *м*
2084	неполное доминирование	2112	постоянный
	ср	2113	допустимая доза *ж*,
2085	частная регрессия *ж*		толерантная доза *ж*
2086	частичное обновление	2114	постоянство *ср*,
	крови *ср*		стойкость *ж*
2087	гипотеза *ж* частичной	2115	постоянный, стойкий
	реплики	2116	расщепление *ср* фагов
2088	частичная реституция *ж*	2117	фазовая специфичность *ж*
2089	корпускулярная	2118	фен *м*
	наследственность *ж*	2119	феноклин *м*
2090	родительский	2120	фенокопия *ж*
2091	--	2121	феноцитология *ж*

№		№	
2122	фенодевиант *м*	2158	плазмосома *ж*
2123	феногенетика *ж*	2159	плазмотомия *ж*
2124	феном *м*	2160	плазмотип *м*
2125	феномный	2161	пласт *м*
2126	феномная задержка *ж*	2162	пластичность *ж*
2127	фенотип *м*	2163	пластида *ж*
2128	фенотипический	2164	пластидная
2129	фенотипическая		наследственность *ж*
	корреляция *ж*	2165	пластидная мутация *ж*
2130	фенотипическое	2166	пластидогенный
	смешение *ср*		комплекс *м*
2131	фотореактивация *ж*	2167	пластидом *м*
2132	фототрофный	2168	пластидотип *м*
2133	фрагмопласт *м*	2169	пластоконт *м*
2134	фрагмосома *ж*	2170	пластодесма *ж*
2135	флетический	2171	пластоген *м*
2136	филогения *ж*, филогенез *м*	2172	пластоконт *м*
2137	филогенетический	2173	пластом *м*
2138	филогенетика *ж*	2174	пластомера *ж*
2139	филогения *ж*, филогенез *м*	2175	пластосома *ж*
2140	филум *м*	2176	пластинка *ж*
2141	физиохроматин *м*	2177	плектонемный
2142	физиологическая раса *ж*	2178	плектонемная спираль *ж*
2143	фитогенетика *ж*,	2179	плейотропный
	генетика *ж* растений	2180	плейотропия *ж*
2144	плакодесмоз *м*	2181	плейотропия *ж*
2145	планогамета *ж*	2182	плейромитический
2146	планосома *ж*	2183	плоидия *ж*, плоидность *ж*
2147	бляшка *ж*	2184	плюс-модификатор *м*
2148	плазмохроматин *м*	2185	пойкилоплоидный
2149	плазмаген *м*	2186	пойкилоплоидность *ж*,
2150	плазмалемма *ж*		пойкилоплоидия *ж*
2151	плазматическая	2187	пойкилосиндез *м*
	наследственность *ж*,	2188	точечная ошибка *ж*
	цитоплазматическая	2189	точечная мутация *ж*,
	наследственность *ж*		генная мутация *ж*
2152	плазмида *ж*	2190	точечное склеивание *ср*
2153	плазмохроматин *м*	2191	закон *м* Пуассона,
2154	плазмодесма *ж*		распределение *ср*
2155	плазмодиум *м*		Пуассона
2156	плазмогамия *ж*	2192	полярный
2157	плазмон *м*		

2193	полярное тельце *ср*, полоцит *м*	2227	полигибридность *ж*
		2228	многоядерный
2194	полярная шапочка *ж*	2229	полилизогенный
2195	полярность *ж*	2230	полимерный
2196	поляризация *ж*	2231	полимерная хромосома *ж*
2197	направленная редукция *ж*	2232	полимерный
		2233	полимерия *ж*
2198	полярная пластинка *ж*	2234	полимитоз *м*
2199	полюс *м*, центр *м*	2235	полиморфизм *м*
2200	полярное поле *ср*	2236	полиморфный
2201	полярная пластинка *ж*	2237	многофазность *ж*
2202	пыльцевая леталь *ж*	2238	полифения *ж*
2203	пыльцевая стерильность *ср*	2239	полифилетический
		2240	полиплоидный, полиплоид *м*
2204	полярное тельце *ср*, полоцит *м*	2241	полиплоидный комплекс *м*
2205	полиаллельное скрещивание *ср*	2242	полиплоидия *ж*, полиплоидность *ж*
2206	многоосновный	2243	полиплотип *м*
2207	многоядерный	2244	многолучевой, полирадиальный
2208	полицентрический		
2209	полихондрический	2245	многократное удвоение *ср*
2210	полихромосома *ж*	2246	многоспутничный
2211	массовое скрещивание *ср*	2247	полисома *ж*
2212	полиэмбриония *ж*	2248	полисомия *ж*
2213	полиэнергидный	2249	полиспермный
2214	полиэргистичный	2250	полиспермия *ж*
2215	полифакториальный, многофакторный	2251	политенный
		2252	политения *ж*
2216	полигамный	2253	политопия *ж*
2217	полигамность *ж*	2254	политипический
2218	полиген *м*	2255	полиургический ген *м*
2219	межродовой	2256	популяция *ж*
2220	полигенный	2257	плотность *ж* популяции
2221	полигенная комбинация *ж*	2258	популяционное равновесие *ср*
2222	полигеноматический	2259	популяционная генетика *ж*
2223	полигенность *ж*		
2224	полигаплоидный, полигаплоид *м*	2260	популяционный гомеостаз *м*
2225	полигаплоидность *ж*	2261	популяционное давление *ср*
2226	полигибридный		

2262	популяционные волны *м* мн, волны *ж* мн жизни	2293	прередукция *ж*
2263	позиционные аллели *м* мн	2294	гипотеза *ж* присутствия-отсутствия
2264	позиционное кровное родство *ср*	2295	давление *ср*
2265	эффект *м* положения	2296	презумптивная область *ж*
2266	постадаптация *ж*	2297	предзиготический
2267	постделение *ср*	2298	первичный
2268	постгетерокинез *м*	2299	первичное соотношение *ср* полов
2269	постредуцированный	2300	первичная мишень *ж*
2270	постредукция *ж*	2301	исходная форма *ж*
2271	аберрация *ж*, вызванная после расщепление хромосом	2302	примордиум *м*, закладка *ж* органа
2272	послесиндезная интерфаза *ж*	2303	вероятность *ж*
2273	постзиготический	2304	вероятный
2274	потенция *ж*, способность *ж*	2305	процентрический
		2306	прохромосома *ж*
2275	потенциальный аллель *м*	2307	продуктивная инфекция *ж*
2276	преадаптация *ж*	2308	прогамный
2277	прецессия *ж*	2309	прогамия *ж*
2278	преждевременная реверсия *ж*	2310	проген *м*
		2311	прогенез *м*
2279	"прекосити", скороспелость *ж*	2312	прогеном *м*
		2313	потомство *ср*
2280	теория *ж* "прекосити"	2314	испытание *ср* по потомству
2281	преконъюгация *ж*		
2282	предетерминация *ж*	2315	прогрессивный
2283	предделение *ср*	2316	прометафаза *ж*
2284	направленная редукция *ж*	2317	прометафазное удлинение *ср*
2285	неслучайное расщепление *ср*	2318	промитоз *м*
		2319	пронуклеолус *м*, проядрышко *ср*
2286	преформизм *м*, теория *ж* преформизма	2320	пронуклеус *м*
		2321	профаг *м*
2287	прегетерокинез *м*	2322	рекомбинация *ж* профагов
2288	премутация *ж*		
2289	препотенция *ж*	2323	профаза *ж*
2290	препрофаг *м*	2324	профазный индекс *м*
2291	ингибитор *м* препрофазы	2325	профазные ядры *м* мн
2292	прередуцированный	2326	профазный

2327	протоандрия *ж*
2328	протерминальный
2329	протохромонема *ж*
2330	протоплазма *ж*
2331	протоплазменный, протоплазматический
2332	протоплазматическая несовместимость *ж*
2333	прототроф *м*
2334	прототрофный
2335	проксимальный
2336	псевдоаллель *м*
2337	псевдоаллелизм *м*
2338	псевдоамитоз *м*
2339	псевдоанафаза *ж*
2340	псевдоапогамия *ж*
2341	псевдоассоциация *ж*
2342	псевдобивалент *м*
2343	псевдохиазма *ж*
2344	псевдосовместимость *ж*
2345	псевдокроссинговер *м*
2346	псевдодоминантность *ж*
2347	псевдоэндомитоз *м*
2348	ложная экваториальная пластинка *ж*
2349	псевдоэкзогенный
2350	псевдофертильность *ж*
2351	ложный фрагмент *м*
2352	псевдогамный
2353	псевдогамия *ж*
2354	псевдогаплоид *м*, псевдогаплоидный
2355	псевдогаплоидность *ж*
2356	псевдогетерозис *м*
2357	ложная несовместимость *ж*
2358	псевдомейоз *м*
2359	псевдометафаза *ж*
2360	псевдомиксис *м*
2361	псевдомоносомик *м*, псевдомоносомный
2362	псевдомоноталлический

2363	ложная мутация *ж*
2364	псевдополиплоид *м*, псевдополиплоидный
2365	псевдополиплоидия *ж*
2366	псевдоредукция *ж*
2367	ложный сателлит *м*
2368	псевдоселективность *ж*, ложная избирательность *ж*
2369	ложное веретено *ср*
2370	псевдотелофаза *ж*
2371	псевдодикий тип *м*
2372	пуфф *м*
2373	чистокровный
2374	чистая расса *ж*
2375	чистая линия *ж*
2376	пикноз *м*
2377	пикнотический
2378	квадруплексный
2379	квадрисексуальность *ж*
2380	квадривалент *м*, квадривалентный
2381	качественный
2382	качественный признак *м*
2383	качественная наследственность *ж*
2384	количественный
2385	количественный признак *м*
2386	количественная наследственность *ж*
2387	квантовая эволюция *ж*
2388	тетрада *ж*
2389	квазибивалент *м*
2390	ложное сцепление *ср*
2391	раса *ж*
2392	филогенез *м*, филогения *ж*
2393	радиационная генетика *ж*
2394	радиомиметический, радиомиметик *м*
2395	--
2396	случайный

2397	случайный дрейф *м*	2431	коэффициент *м* регрессии
2398	случайная фиксация *ж*	2432	уравнение *ср* регрессии
2399	рандомизация *ж*	2433	регрессия *ж* мать-дочь
2400	панмиксия *ж*	2434	регрессивный
2401	случайность *ж*	2435	регуляционное развитие *ср*
2402	размах *м* изменчивости	2436	родственный
2403	группа подвидов *ж*	2437	относительный, взаимный
2404	гипотеза *ж* скоростей реакций	2438	взаимная спирализация *ж*
2405	ген *м*, регулирующий скорость реакции	2439	взаимная несовместимость *ж*
2406	градиент *м* скорости	2440	родство *ср*
2407	реактивация *ж*	2441	реликтовая спираль *ж*, остаточная спираль *ж*
2408	перестройка *ж*	2442	реликт *м*
2409	рецептор *м*	2443	отдаленный эффект *м*
2410	рецессивный	2444	далекое родство *ср*
2411	рецессивность *ж*	2445	репарация *ж*, восстановление *ср*
2412	реципиент *м*, реципиентный	2446	восстановимый мутант *м*
2413	бактерия *ж* реципиент	2447	--
2414	реципрокный, взаимный	2448	дупликация *ж*
2415	рекотбинация *ж*	2449	гипотеза *ж* замещения
2416	класс *м* рекотбинации	2450	а) редупликация *ж*, самовоспроизведение *ср* (в цитогенетике)
2417	процент *м* рекотбинации		
2418	индекс *м* рекотбинации	2450	в) повторность *ж* (в статистике)
2419	рекон *м*	2451	репрессор *м*
2420	возвратный	2452	репродукция *ж*, размножение *ср*
2421	возвратное реципрокное скрещивание *ср*	2453	скорость *ж* размножения
2422	редуцированное "прекосити" *ср*	2454	репродуктивный
		2455	минимальная единица *ж* репродукции
2423	редукция *ж*	2456	отталкивание *ср*
2424	редукционный	2457	фаза *ж* отталкивания
2425	редукционное деление *ср*, гетеротипическое деление *ср*	2458	остаточный
		2459	остаточная хромосома *ж*
2426	редукционная группировка *ж*	2460	остаточный генотип *м*
2427	редукционный митоз *м*	2461	остаточная наследственность *ж*
2428	редуктивная инфекция *ж*		
2429	редупликация *ж*		
2430	регрессия *ж*		

2462	остаточная гетерозигота *ж*
2463	остаточная гомология *ж*
2464	покоящееся ядро *ср*
2465	фаза *ж* покоя
2466	стадия *ж* покоя
2467	реституция *ж*
2468	реституционный
2469	восстановление *ср*
2470	восстановительное обратное скрещивание *ср*
2471	ограничительный ген *м*
2472	ограничительный, сдерживающий
2473	задержка *ж*, запаздывание *ср*, отставание *ср*
2474	ретикулум *м*, сетка *ж*
2475	ретранслокация *ж*
2476	обратная мутация *ж*
2477	соединение *ср*, воссоединение *ср*
2478	реверсия *ж* гетеропикноза
2479	обратная мутация *ж*
2480	реверсия *ж*
2481	--
2482	рибонуклеаза *ж*
2483	рибонуклеиновая кислота *ж*, РНК
2484	рибоза *ж*
2485	рибосомный
2486	рибосома *ж*
2487	кольцо *ср*
2488	кольцевая хроматида *ж*
2489	кольцевая хромосома *ж*
2490	информационная РНК *ж*
2491	робертсонов
2492	ротация *ж*, вращение *ср*
2493	ротационное скрещивание *ср*
2494	сальтант *м*
2495	выборка *ж*

2496	SAT-хромосома *ж*, хромосома *ж* с вторичной перетяжкой
2497	сателлит *м*, спутник *м*
2498	спутничное ядрышко *ср*
2499	эффект *м* насыщения
2500	SAT-зона *ж*, зона *ж* вторичной перетяжки
2501	шизогония *ж*
2502	вторичный
2503	вторичная центромерная область *ж*, неоцентро-мера *ж*
2504	вторичная перетяжка *ж*
2505	вторичная кинетохора *ж* вторичная центромера *ж*
2506	вторичная полиплоидия *ж*
2507	вторичное соотношение полов *ср*
2508	секция *ж*
2509	сектор *м*
2510	секторный
2511	секторная химера *ж*
2512	сегмент *м*
2513	сегментный
2514	сегментный аллополи-плоид *м*
2515	сегрегант *м*
2516	сегрегация *ж*, расщепление *ср*
2517	задержка *ж* расщепления
2518	сегрегационная стерильность *ж*
2519	задержка *ж* сегрегации
2520	селекция *ж*,-отбор *м*
2521	селекционный коэффициент *м*
2522	селекционный диференциал *м*
2523	селекционная сила *ж*, фактор *м* отбора
2524	селекционный индекс *м*

2525	интенсивность *ж* отбора	2558	полурецессивный
2526	давление *ср* отбора	2559	полувид *м*
2527	селективный, избирательный	2560	полустерильный, семистерильный
2528	факторы *м мн* отбора	2561	семистерильность *ж*, полустерильность *ж*
2529	селекционный пик *м*, селекционный максимум *м*	2562	семиунивалент *м*, полуунивалент *м*
2530	давление *ср* отбора	2563	чувствительная фаза *ж*
2531	самосовместимость *ж*	2564	чувствительный период *м*
2532	самофертильный	2565	пик *м* чувствительности
2533	самофертильность *ж*	2566	эффект *м* сенсибилизации
2534	самооплодотворение *ср*	2567	разделение *ср*
2535	самооплодотворяющийся организм *м*	2568	септисомный, септисомик *м*
2536	самонесовместимость *ж*	2569	сесквидиплоид *м*, сесквидиплоидный
2537	самооплодотворение *ср*	2570	сесквидиплоидность *ж*
2538	самоопыление *ср*	2571	набор *м*
2539	самостерильный	2572	пол *м*
2540	самостерильность *ж*	2573	половой хроматин *м*
2541	семя *ср*, сперма *ж*	2574	половая хромосома *ж*
2542	семиаллель *м*	2575	контролируемый полом
2543	семиаллелизм *м*	2576	деградация *ж* пола
2544	семиапоспория *ж*	2577	определение *ср* пола
2545	семибивалент *м*, полубивалент *м*	2578	половой диморфизм *м*
2546	семикариотип *м*	2579	находящийся под влиянием пола
2547	полухиазма *ж*	2580	наследование *ср* пола
2548	неполное доминирование *ср*	2581	ограниченный полом
2549	полудоминантный, с неполной доминантностью	2582	сцепление *ср* с полом
		2583	сцепленный с полом
2550	семигетеротипический	2584	соотношение *ср* полов
2551	семигомологичный	2585	превращение *ср* полов
2552	семигомология *ж*, неполная гомология *ж*	2586	половой, сексуальный
2553	полунесовместимость *ж*	2587	определение *ср* пола
2554	семикариотип *м*	2588	наследование *ср* пола
2555	семилеталь *ж*, полулеталь *ж*	2589	половой пузырек *м*
		2590	фрагментация *ж*
2556	полулокализированная центромера *ж*	2591	сдвиг *м*, смещение *ср*
2557	полусозревание *ср*	2592	братья и сестры *мн*, сибсы *м мн*

2593	ассоциация *ж* бок-о-бок
2594	сигма *ж*, среднее квадратичное отклонение *ср*
2595	простая корреляция *ж*
2596	простая регрессия *ж*
2597	симплекс *м*
2598	одновременная адаптация *ж*
2599	одновременная мутация *ж*
2600	линия *ж* с компенсаторным замещением одной хромосомы
2601	однократное скрещивание *ср*
2602	выраженность *ж* при одном аллеле
2603	моногенный гетерозис *м*
2604	сестра *ж*, сестринский
2605	сестринские хроматиды *ж мн*
2606	сестринское соединение *ср* хроматид
2607	кроссинговер *м* между сестринскими нитями
2608	участок *м*
2609	асимметричный бивалент *м*
2610	асимметрия *ж*
2611	соматический
2612	соматическая мутация *ж*
2613	соматическая редукция *ж*
2614	соматогамия *ж*
2615	соматопластический
2616	специальный участок *м*
2617	мост *м* в специальном участке
2618	видообразование *ср*
2619	вид *м*
2620	специфическая комбинационная приспособленность *ж*

2621	специфичность *ж*
2622	метод *м* специфичного локуса
2623	сперматолеозис *м*
2624	сперматида *ж*
2625	спермациум *м*
2626	сперматоцит *м*
2627	сперматогенез *м*
2628	сперматогоний *м*
2629	сперматозоид *м*, спермий *м*
2630	сфером *м*
2631	сферопласт *м*
2632	сферосома *ж*
2633	сферула *ж*
2634	веретено *ср*
2635	центромера *ж*
2636	нить *ж* веретена
2637	митотический яд *м*, действующий на веретено, метафазный яд *м*
2638	преждевременное образование *ср* веретена
2639	остаток *м* веретена
2640	спирализация *ж*
2641	коэффициент *м* спирализации
2642	константа *ж* спирализации
2643	спирема *ж*
2644	щель *ж*
2645	метод *м* расщепленных делянок
2646	расщепленное веретено *ср*
2647	а) ращепление *ср*, б) видообразование *ср*
2648	самозарождение *ср*
2649	спора *ж*
2650	спороцит *м*
2651	спорогенез *м*
2652	спорофит *м*
2653	спорт *м*, соматическая мутация *ж*

2654	спория *ж*	2685	структурная
2655	ложный аллеломорф *м*		гетерозиготность *ж*
2656	стабильность *ж*,	2686	структурно измененная
	устойчивость *ж*		полиплоидность *ж*
2657	стабилизирующий отбор *м*	2687	структурная
2658	стандартное отклонение		модификация *ж*
	ср, среднее квадратичное	2688	S-тип *м* эффекта
	отклонение *ср*		положения
2659	звездчатая метафаза *ж*	2689	субхроматида *ж*
2660	голодание *ср*	2690	субхромонема *ж*
2661	стазигенез *м*	2691	субген *м*
2662	статмоанафаза *ж*	2692	подродовой
2663	статмодиэрез *м*	2693	подрод *м*
2664	статмокинез *м*	2694	сублеталь *ж*, мутация *ж* с
2665	статмометафаза *ж*		пониженной
2666	статмотелофаза *ж*		жизнеспособностью
2667	критерий *м* значимости,	2695	субмедианный
	статистический критерий	2696	субмезомитический
	м	2697	субмикросома *ж*
2668	устойчивый дрейф *м*	2698	субсексуальный
2669	стволовое тело *ср*	2699	подвид *м*
2670	стволовая линия *ж*	2700	--
2671	ступенчатый аллелизм *м*,	2701	замещение *ср*,
	ступенчатый		субституция *ж*
	аллеломорфизм *м*	2702	субтерминальный
2672	стерильный	2703	близкий к
2673	стерильность *ж*		жизнеспособному
2674	ген *м* стерильности	2704	субвитальный
2675	клейкость *ж*	2705	сверхсокращение *ср*
2676	соединение *ср* путем	2706	сверхдоминантность *ж*
	склеивание	2707	сверхдоминантный
2677	эффект *м* склеивания	2708	сверхсамка *ж*
2678	штамм *м*,	2709	суперген *м*
	равновидность *ж*	2710	супергенный
2679	стрепсинема *ж*	2711	сверхсамец *м*
2680	стрепситена *ж*	2712	избыточный
2681	удлинение *ср*	2713	сверхрецессивный
2682	структурный	2714	сверхрекомбинация *ж*
2683	структурное изменение *ср*	2715	сверхредукция *ж*
2684	структурная гетерозигота	2716	надвид *м*
	ж	2717	сверхспираль *ж*, спираль
			ж второго порядка

2794	тетрада *ж*	2830	трансгенация *ж*,
2795	тетрадный анализ *м*		генная мутация *ж*
2796	тетрагаплоид *м*,	2831	трансгрессия *ж*
	тетрагаплоидный	2832	трансгрессивный
2797	тетрагаплоидия *ж*	2833	временный полиморфизм
2798	тетраплоид *м*,		*м*
	тетраплоидный	2834	транслокация *ж*
2799	тетраплоидия *ж*	2835	транслокационная
2800	тетрасоматия *ж*		гетерозиготность *ж*
2801	тетрасомный,	2836	транслокационная
	тетрасомик *м*		гомозиготность *ж*
2802	тетратип *м*	2837	транслокационный
2803	тетравалент *м*		мутант *м*
2804	теликарион *м*	2838	транслокационные
2805	теликариотический		трисомики *м мн*
2806	телигенический	2839	перенос *м*, передача *ж*
2807	телигения *ж*	2840	трансмутация *ж*
2808	теликарион *м*	2841	трансспецифичный,
2809	теликариотический		межвидовой
2810	телитокия *ж*	2842	трансрепликация *ж*
2811	--	2843	трансвекционный
2812	пороговый признак *м*		эффект *м*
2813	атавистический признак *м*	2844	поперечное равновесие *ср*
2814	атавизим *м*	2845	тенденция *ж*
2815	топкросс *м*	2846	триада *ж*
2816	топкроссинг *м*	2847	трехродовой
2817	торзионный	2848	трехгенный
2818	конъюгация *ж* типа ''игра	2849	тригаплоид *м*,
	в пятнашки''		тригаплоидный
2819	трабант *м*, сателлит *м*,	2850	тригаплоидия *ж*
	спутник *м*	2851	тригибридный
2820	тянущая нить *ж*	2852	тригибридность *ж*
2821	признак *м*	2853	тримерный
2822	транс-конфигурация *ж*	2854	тримерность *ж*
2823	трансдукция *ж*	2855	тримоноезический
2824	трансдукционный клон *м*	2856	триморфизм *м*
2825	трансдукционная	2857	трехдомный
	группа *ж*	2858	тройное скрещивание *ср*
2826	перенос *м*	2859	тройное слияние *ср*
2827	транспортная РНК *ж*	2860	триплекс *м*
2828	трансформация *ж*	2861	--
2829	фактор *м* трансформации	2862	триплоид *м*, триплоидный

2863	триплоидия *ж*
2864	трисомик *м*, трисомный
2865	трисомия *ж*
2866	тривалент *м*
2867	трофическое ядро *ср*
2868	трофохроматин *м*
2869	трофоплазма *ж*
2870	троподиэрез *м*
2871	тропокинез *м*
2872	близнец *м*, близнецовый
2873	псевдоаллели *м мн*
2874	близнецовые гибриды *м мн*
2875	виды-близнецы *м мн*
2876	закручивание *ср*
2877	двухплоскостная теория *ж*
2878	--
2879	тип *м*
2880	типовое число *ср*
2881	типичный, стандартный
2882	типогенез *м*
2883	типолиз *м*
2884	типостаз *м*
2885	типостроф *м*
2886	ультрамикросома *ж*
2887	несбалансированный
2888	несбалансированная полиплоидия *ж*
2889	область *ж* отрицательного гетеропикноза
2890	неравный бивалент *м*
2891	нефиксируемый
2892	односторонний
2893	одностороннее наследование *ср*
2894	однояйцевые близнецы *м мн*, идентичные близнецы *м мн*
2895	однополый
2896	однополость *ж*
2897	унитарный, единый
2898	единица *ж* наследственности
2899	унивалент *м*
2900	--
2901	не отбираемые гены-маркеры *м мн*
2902	неподавляемый
2903	восстановление *ср* от повреждений, вызванных ультрафиолетом
2904	вакуоль *ж*
2905	вакуом *м*
2906	валентность *ж*
2907	изменчивость *ж*
2908	дисперсия *ж*
2909	вариант *м*
2910	вариация *ж*
2911	мозаичный
2912	мозаичность *ж*
2913	аллель *м* пятнистости
2914	равновидность *ж*, линия *ж*, сорт *м*
2915	вегетативный
2916	жизнеспособность *ж*
2917	жизнеспособный
2918	чужеродное оплодотворение *ср*
2919	природный гибрид *м*
2920	вирион *м*
2921	вирогенетический сегмент *м*
2922	вирулентный фаг *м*
2923	видимый
2924	жизненность *ж*
2925	витальные локусы *м мн*
2926	V-тип *м* эффекта положения
2927	сужение *ср*, перетяжка *ж*
2928	W-хромосома *ж*
2929	вейсманизм *м*
2930	транслокация *ж* целого плеча хромосомы

2931 транслокация *ж* целого
      плеча хромосомы
2932 дикий тип *м*
2933 х (основное число)
2934 Х-хромосома *ж*
2935 ксения *ж*
2936 ксеногамия *ж*, чужеро-
      дное оплодотворение *ср*
2937 ксенопластический
2938 Y-хромосома *ж*
2939 z (зиготическое число
      хромосом)

2940 Z-хромосома *ж*
2941 мутация *ж* нулевой точки
2942 генетика *ж* животных
2943 зигогенный
2944 зигонема *ж*
2945 зигозис *м*
2946 зигосома *ж*
2947 зигота *ж*
2948 зиготена *ж*
2949 зиготенный
2950 зиготный, зиготический

INDEXES

FRANÇAIS

allohétéroploïde 91
allohétéroploïdie 92
alloiogénèse 93
allolysogénique 94
allomorphose 95
allopatrique 96
alloplasme 97
alloploïde 98
alloploïdie 99
allopolyploïde 100
– segmentaire 2514
allopolyploïdie 101
allosome 103
allosomique 102
allosyndèse 104
allosyndétique 105
allotétraploïde 106
allotétraploïdie 107
allotriploïde 108
allotriploïdie 109
allotypique 110
allozygote 111
alternance de générations 114
altruiste 117
ambiant 932
ambivalent 120
ambo-sexuel 119
améiose 121
améiotique 122
amfigamie 135
amictique 123
A-misdivision 124
amitose 125
amitotique 126
amixie 127
amixis 127
amorphe 128
amphérotoquie 129
amphiaster 130
amphibivalent 132
amphicarion 140
amphidiploïde 133, 872
amphidiploïdie 134
amphigénèse 136
amphigonie 137
amphihaploïde 138, 875
amphihaploïdie 139
amphilepsie 141
amphimictique 142
amphimixie 143
amphimutation 144

amphiplastie 145
amphiploïde 146
amphiploïdie 147
amphitène 148
amphithallique 149
amphitoquie 150
amphogène 151
amphogénie 152
amphohétérogonie 153
anabolie 154
anachromasie 155
anagénèse 156
anagénétique 157
analogue 158
analyse des tétrades 2795
analyse génomique 1112
anamorphique 159
anamorphisme 160
anaphase 161
anaphasique 163
anaphragmique 164
anaréduplication 165
anaschistique 166
anastomose 167
anastralmitose 168
andro-autosome 169
androdioécie 171
androdioïque 170
androécie 172
androgamie 173
androgène 176
androgénèse 174
androgénétique 175
androginie 178
androgyne 177
androhermaphrodite 179
andromérogonie 180
andromonoécie 182
andromonoïque 181
androsome 183
androsporogénèse 184
androstérile 185
androstérilité 186
aneugénique 888
aneuploïde 187
aneuploïdie 188
aneusomatie 189
animal amélioré 1151
– fondateur 1006
– tête de ligne 1006
anisoautoploïde 190

effet additif 39
- d'agglutination 2677
- de Baldwin 330
- de cis-vection 539
- de combinaison en bloc 391
- de cyanuration 668
- de position 2265
- de position de type S   2688
- de position du type V 2926
- de saturation 2499
- de "trans-vection" 2843
- d'organisation 1992
- retardé 2443
- sensibilisant 2566
effets multiples 1707
effet spécifique sur un locus 1683
- T 2762
electosome 906
élément de contrôle 617
élevage 408, 1720
élimination 907
éliminer 666
élongation 2681
- prométaphasique 2317
émasculation 908
endogamie 911
endogamique 910
endogène 913
endogénote 912
endomitose 914
endomitotique 915
endomixie 916
endonucléaire 917
endoplasme 918
endoplasmique 919
endopolyploïde 920
endopolyploïdie 921
endosome 922
endosperme 923
endotaxonique 924
énergide 927
enjambement 641
- partiel 2079
ennéaploïde 929
ennéaploïdie 930
enroulement plectonémique 2178
- réciproque 2438
entaille 1916
environnement 931
épigamique 936
épigénèse 937

épigénétique 938
épigénotype 939
épisome 940
épistasie 942
épistatique 943
équation de Poisson 2191
- de régression 2432
équationnel 944
équilibré 323
équilibre 322
- génétique 2258
- génique 1104
- mutationnel 1895
- réalisé par les mutations 1895
- transverse 2844
erreur ponctuelle 2188
erythrophile 949
espèce 2619
- agame 51
espèces jumelles 2875
état épisomique transitoire 941
- intégré 1503
étendue de variation 2402
éthéogénèse 950
étranglement 611, 2927
eucentrique 951
euchromatine 953
euchromatique 952
euchromosome 954
eugénique 955, 956
euhétérosis 957
euméiose 958
eumitose 959
euploïde 960
euploïdie 961
évolutif 965
évolution 962
- du quantum 2387
- explosive 973
- parallèle 2042
exagération 966
exhibition 967
ex-mutant 968
exogamie 969
exogène 971
exogénote 970
expression d'une dose unique 2602
expressivité 975
extrachromosomique 977
extra-radial 979
extrémité T 2783

gradient de la taille du chromomère 509
granules terminaux 2787
groupe de croisement 1722
groupement réductionnel 2426
groupes de transduction 2825
gynandroïde 1155
gynandromorphe 1156
gynandromorphisme 1157
gynautosome 1158
gynodioécie 1160
gynodioïque 1159
gynoécie 1162
gynofacteur 1163
gynogénèse 1164
gynogénétique 1165
gynogénie 1167
gynogénique 1166
gynoïque 1161
gynomonoécie 1169
gynomonoïque 1168
gynosperme 1170
gynospore 1171
gynosporogénèse 1172

haplobionte 1186
haplobiontique 1187
haplocaryotype 1188
haplodiploïde 1190
haplodiploïdie 1191
haplodiplonte 1192
haplogénotypique 1193
haplohétéroécie 1194
haploïde 1195
- dominant 1197
- semi-dominant 1196
haploïdie 1198
haplome 1199
haplomitose 1201
haplomixie 1202
haplonte 1203
haplontique 1204
haplophase 1205
haplopolyploïde 1206
haplopolyploïdie 1207
haplose 1208
haplosomie 1209
haplozygote 1210
hémialloploïde 1211
hémialloploïdie 1212
hémiautoploïde 1213

hémiautoploïdie 1214
hémicarion 1221
hémichromatidique 1215
hémicinèse 1222
hémidiérèse 1216
hémihaploïde 1217
hémihaploïdie 1218
hémiholodiploïde 1219
hémiholodiploïdie 1220
hémiorthocinèse 1223
hémixie 1224
hémizygotique 1225
heptaploïde 1226
heptaploïdie 1227
hercogamie 1234
héréditaire 1230
hérédité 1489
- à facteurs multiples 386
- alternative 634
- alternée 116
- collatérale 569
- cytoplasmique 696
- des caractères acquis 1490
- extra-chromosomique 696
- intermédiaire 1527
- liée au sexe 2582
- maternelle 1718
- matroclinale 1730
- matrocline 1730
- mendélienne 2089
- non-chromosomique 696
- non-mendélienne 1922
- patroclinale 2092
- plasmatique 2151
- plastidique 2164
- qualitative 2383
- quantitative 2386
- résiduelle 2461
- retardée 2282
- sexuelle 2588
- unilatérale 2893
héritabilité 1232
héritage 1233, 1489
hermaphrodisme 1236
- protandrique 2327
- protérandrique 2327
hermaphrodite 1235
hétérauxèse 1237
hétéroallèle 1238
hétéroallélique 1239
hétérobrachial 1240

limité 1669
- à un sexe 2581
linkage 1677
- absolu 593
liquide séminal 2541
localisation 1679
localisé 1680
loci vitaux 2925
locus 1682
loi de Poisson 2191
lois de Mendel 1751
lot de chromosomes 529
luxuriance 1688
lysogénie 1691
lysogénique 1689
lysogénisation 1690
lysotype 1692

macroévolution 1693
macrogamète 1694
macromutation 1695
macronoyau 1697
macronucléus 1697
macrophylogénèse 1698
macrospore 1699
macrosporogénèse 1700
majeur 1701
mâle ergatomorphique 948
manifestation 1706
manteau chondriosomal 482
marqueur 1712
marqueurs non sélectionnés 2901
masquage 2591
massal 1713
maternelle 1717
matrice 1727
matrix 1727
matrocline 1726
matroclinie 1731
maturation 1732
médiocentrique 1735
mégaévolution 1736
mégahétérochromatique 1740
mégaspore 1737
mégasporocyte 1738
mégasporogénèse 1739
méiocyte 1741
méiomérie 1742
méiose 1743
  en une seule division 1976
méiosome 1744

méiospore 1745
méiotique 1746
méiotrophique 1747
mélange HFC 1324
- phénotypique 2130
membrane nucléaire 1934
mendélien 1748
mendélisme 1750
mère 704
mériclinal 1753
meristème 1755
mérocinèse 1760
mérogamie 1757
mérogon 1758
mérogonie 1759
méromixie 1761
merostatmocinèse 1762
mérozygote 1763
mésomitique 1764
mésomitose 1765
métacentrique 1767
métagénèse 1768
métagynie 1769
métakinèse 1770
métamitose 1771
métandrie 1772
métaphase 1773
- en étoile 2659
- en forme de boule 331
- étoilée 2659
métaphasique 1775
métaplasie 1776
métaréduplication 1777
métasyndèse 1778
métatactique 1779
métaxénie 1780
méthode d'accouplement 1723
- du locus spécifique 2622
- du "maximum likehood" 1734
- mentor 1752
métis 1175
métissage 636
métromorphe 1781
microagglomération 1784
microcentre 1782
microchromosome 1783
micro-espèce 1794
microévolution 1785
microgamète 1786
microgène 1787
microhétérochromatique 1788

ESPAÑOL

anfogénico 151
anfoheterogonía 153
anidación 204
anillo 2487
- de Balbiani 329
- "tandem" 2756
anisoautoploide 190
anisoautoploidía 191
anisocariosis 192
anisocitosis 193
anisogameto 194
anisogamia 195
anisogenia 198
anisogenomático 196
anisogenómico 197
anisoploide 199
anisoploidía 200
anisosíndesis 201
anisosindético 202
anisotrisomía 203
anormogénesis 205
anortoespiral 209
anortogénesis 206
anortoploide 207
anortoploidía 208
antefase 210
anticipación 211
antimorfo 212
antimutagénico 214
antimutageno 213
antirrecapitulación 215
antitético 216
apareamiento 1719, 2016
- , falso 985
- análogo 244
- correctivo 625
- de oposición 830
- de pares 2022
- de semejanza 1092
- de toque y separación 2818
- ectópico 902
- flojo 1685
- interbraquial 1510
- no homólogo 1920
aparente 2923
apogametia 218
apogamia 220
apogámico 219
apohomotípico 221
apomeiosis 222
apomíctico 223

apomictosis 224
apomixia 225
aporogamia 226
aposporía 227
apparato de Golgi 1133
aptitud combinatoria 573
- combinatoria específica 2620
- por la combinación 573
aquiasmático 18
arcalaxis 228
área polar 2200
aromorfosis 233
arquebiosis 229
arquetipo 230
arquiplasma 231
arreglo en festón 994
arrenogenético 234
arrenogenía 235
arrenotoquía 236
arrollamiento plectonemático 2178
- plectonémico 2178
- recíproco 2438
artefacto 237
artioploide 239
artioploidía 240
asexual 241
asimetría 2610
asinapsis 251
asináptico 250
asíndesis 251
asingamia 253
asingámico 252
asociación 243
- de "corbata"1907
- lateral 2593
- nuclear 1928
- por aglutinación 2676
- "tandem" 2751
áster 246
astrocentro 247
astroesfera 248
atactogamia 254
atávico 256
atavismo 255, 2814
atelomítico 257
atenuación 264
atorsional 258
atracción 265, 629
- centromérica 450
atractoplasma 259
atractosoma 260

cigoteno 2949
cigótico 2950
cigoto 2947
cinetócoro 1640
- desnudo 1905
- secundario 2505
cinetogén 1641
cinetogénesis 1642
cinetómero 1643
cinetonema 1644
cinetonúcleo 1645
cinetoplasto
cinoplasma 1647
cinosoma 1648
cis-acomodo 536
cis-configuración 536
cistrón 538
citáster 671
cito 672
citoactivo 673
citoblasto 674
citocinesis 685
citodiéresis 677
citodo 676
citogamia 678
citogén 679
citogénesis 680
citogenética 683
citogenético 681
citogonia 684
citolisina 687
citólisis 688
citológico 686
citoma 689
citómero 690
citomicrosoma 691
citomixis 692
citomórfosis 693
citoplasma 694
citoplásmico 695
citoplasmón 698
citoquimera 675
citosimbiosis 700
citosoma 699
citotipo 701
cladogénesis 541
clase de recombinación 2416
clasificación 1152
cleistogamia 545
clina 546
clon 1649

clon de transducción 2824
cloroplastidio 470
cloroplasto 470
c-meiosis 548
c-mitosis 549
- distribuida 853
- explotada 972
coadaptación genética 1094
cociente híbrido extramedial 978
codominancia 551
codominante 552
codon 553
coeficiente de apareamiento 2018
- de consanguinidad 1469
- de correlación 627
- de espiralización 2641
- de inmigración 1466
- de mutación 1897
- de parentesco 555
- de regresión 2431
- de selección 2521
- de trayectoria 2091
coincidencia 561
coito 1719
colasoma 568
colchicina 562
colchiploide 563
colchiploidía 564
colicina 565
colicinogenia 567
colicinogénico 566
colocoro 610
coloración alternada de los
  segmentos cromosómicos 526
combinación poligénica 2221
comisco 574
comparación madres-hijas 705
comparium 578
compatibilidad 579
compatible 580
compensación 582
- de dosis 868
compensador 583
competencia 584
complejo AG 55
- cromosómico 523
- de genes 1067
- de poliploides 2241
- modificador 1823
- plastidiógeno 2166
complementación 591

fortificación 1005
fortuito 2396
fotoreactivación 2131
fototrófico 2132
fraccionamiento 1008
fractura 1009
fragmentación 1011, 2590
- medio-cromatidio 1178
- nuclear 1933
- nucleolar 1947
fragmento 1010
fragmoplasto 2133
fragmosoma 2134
fraternal 1012
frecuencia 1015
- de las mutaciones 1900
- de los quiasmas 464
- genética 1072
fructificación 1016
fuerza de selección 2523
fundamental 1019
fusión 1021
- céntrica 439
fusiones en tandem 2753
fusión triple 2859

gamético 1024
gameto 1023
gametoblasto 1031
gametocito 1032
gametofítico 1041
gametofito 1040
gametóforo 1039
gametogamia 1033
gametogénesis 1034
gametogénico 1036
gametogonia 1037
gametoide 1038
gametos equilibrados 324
gámico 1042
gamobio 1043
gamodemo 1044
gamófase 1052
gamogénesis 1045
gamogenético 1047
gamogonia 1048
gamólisis 1049
gamón 1050
gamonte 1051
gamotropismo 1053
geitonocarpía 1054

geitonogamia 1055
geitonogénesis 1056
gemación 1058
gemelo 2872
gemelos biovulares 804
- dicigóticos 859
- idénticos 2894
- isófanos 1013
- monozigóticos 2894
- uniovulares 2894
gémino 1057
gemíparo 1059
gemula 1060
gen 1061
- amortiguador 414
- "Buffer" 414
- de control 2405
- de esterilidad 2674
gene 1061
genealogía 1062
generación 1079
- espontánea 2648
- filial 998
- paterna 2064
- primordial 2648
generativo 1081
genérico 1085
género 1124
generotipo 1087
genes de aislamiento 1582
- complementarios 590
- de incompatibilidad 1473
- duplicados 883
- gemelos 2873
- miméticos 1802
- modelo 2095
- triplicados 2861
genética 1098
- de las poblaciones 2259
genético 1091
genetotrófico 1101
génico 1103
gen inhibidor 1492
- invisible 1556
genitor 2062
genitores 2065
gen mayor 1971
- menor 1804, 2218
- mutador 1901
genocentro 1065
genoclino 1107

homeótico 1345
homeotípico 1347
homeótipo 1346
homoalélico 1350
homoalelo 1349
homobraquial 1351
homocarionte 1374
homocaryosis 1375
homocentricidad 1355
homocéntrico 1354
homocigosis 1403
- de translocación 2836
homocigotización 1405
homocigoto 1404, 1406
- por sobrecruzamiento 646
homocóndrico 1356
homócrono 1357
homodinámico 1358
homofítico 1388
homogamético 1365
homogámeto 1364
homogamia 1367
homógamo 1366
homogenético 1368
homogenía 1372
homogénico 1369
homogenote 1370
homogenótico 1371
homoheteromixis 1373
homolecito 1376
homología 1379
- residual 2463
homólogo 1377, 1378
homolysogénico 1380
homomeria 1382
homomérico 1381
homómero 1381
homomixis 1383
homomorfia 1386
homomórfico 1384
homomorfo 1385
homoplasía 1390
homoplástico 1389
homoploide 1391
homoploidía 1392
homopolar 1393
homosinapsis 1397
homosíndesis 1398
homosis 1394
homosomal 1395
homostasis 1396

homotálico 1399
homotalismo 1400
homotransformación 1401
huevo 2008
huso 2634
- acromático 21
- central 436
- concavo 1329
- de rotura 2646
- en forma de barril 334
husos intersectores 1532
hypocromaticidad 1436

id 1452
idéntico 1453
idiocromatina 1455
idiocromidio 1456
idiocromosoma 1457
idiograma 1458
idiomutación 1459
idioplasma 1460
idiosoma 1464
idiótipo 1462
idiovariación 1463
ilegítimo 1465
incapacitación híbrida 1410
incompatibilidad 1471
- del bacteriófago 319
- gamética 1026
- protoplásmica 2332
- recíproca 2439
incompatible 1474
índice de aislamiento 1583
- de apareamiento metafásico 1774
- de las profasas 2324
- de libre entrecruzamiento 1481
- del movimiento quiasmático 1872
- de recombinación 2418
- de selección 2524
- de supervivencia 2724
- mitótico 1813
indiferencia de dosis 869
individuo 1483
inducción 1486
inducido 1485
indúctor de sobrecruzamiento 647
inercia 1488
inerte 1487
infección abortiva 7
- productiva 2307
- reductiva 2428

nucleo enérgico 925
- en reposo 2464
- generativo 1082
nucleoide 1943
nucleolar 1944
nucléolo 1952
nucleolonema 1951
nucléolo satélite 2498
nucleoma 1956
núcleo metabólico 1766
nucleomixis 1957
nucleoplasma 1958
nucleoproteína 1960
núcleos cinéticos 1639
- complementarios 618
nucleosoma 1961
nucleótida 1962
núcleo trófico 2867
nudo 1650
nueva reunión 1914
nuliplexo 1964
nulisomia 1966
nulisómico 1965
numérico 1967
número básico 337
- cromosómico 527
- fundamental 1020, 1918
- gamético 1029
- modal 1820
- tipo 2880

objectivo primario 2300
oligogén 1971
oligopirene 1972
olistheterozona 1973
ondas de población 2262
ontogenético 1979
ontogenia 1980
ontogénico 1979
occinesis 1984
oocito 1982
oogéneosis 1983
oosfera 1986
oosoma 1985
oótida 1987
operón 1988
ordenación génica 1063
orden lineal 1671
organizador 1990
- nucleolar 1953
orientación 1993

ortoamitosis 1994
ortocinesis 1996
ortoespiral 2000
ortogénesis 1995
ortoploide 1997
ortoploidía 1998
ortoselección 1999
ortotáctico 2001
ortotelomítico 2002
outbreeding 2003
outcross 2004
ovocentro 1981
ovocito 1982
ovogénesis 1983
óvulo 2007
oxicromatina 2009

paidogamia 2013
paidogénesis 2014
palingenesis 2024
palingenia 2024
panalelo 2025
panaloploide 2026
panaloploidía 2027
panautoploide 2028
panautoploidía 2029
pángene 2030
pangénesis 2031
pangenosoma 2032
panmíctico 2033
panmixia 2400
panmixis 2034
paquinema 2010
paquiteno 2011
para-alelo 2035
paracéntrico 2036
paracromatina 2037
parageneon 2039
paragenoplasto 2040
paralocus 2045
paramecina 2046
parameiosis 2047
paramíctico 2048
paramitosis 2049
paramixis 2050
paranémico 2051
paranucleína 2053
paranucléolo 1955
paranucleoplasma 2054
paraselectividad 2055
parasexual 2056

parasinapsis 2057
parasíndesis 2058
paratáctico 2059
parátrofo 2060
paravariación 2061
par de factores hereditarios 983
- de genes 2023
pareja de factores hereditarios 983
parental 2063
parentesco 2440
- directo 829
pares alelomórficos 2015
pariente 1636, 2436
parientes 2065
partenapogamia 2066
partenocarpia 2067
partenogamia 2068
partenogénesis 2069
- artificial 238
partenogenético 2070
partenogenona 2071
partenomixis 2072
partenosperma 2073
partenospermium 2073
partenospora 2074
partenote 2075
paternal 2090
patrimonio hereditario 1489
patroclino 2093
patrogénesis 2094
pedigree 2098
penetración 2099
- completa 594
- incompleta 1476
pentaploide 2100
pentaploidía 2101
pentasómico 2102
pericéntrico 2104
periclinal 2105
período sensitivo 2564
periplasma 2107
periplasto 2108
perisoploide 2109
perisoploidía 2110
peristasis 2111
permanente 2112
persistencia 2114
persistente 2115
picnosis 2376
picnótico 2377
placa 2147, 2176

placa accesoria 13
- de inhibición 1494
- ecuatorial 947
- polar 2201
- pseudoecuatorial 2348
- seudoecuatorial 2348
placo-desmosa 2144
planogameta 2145
planosoma 2146
plasmacromatina 2148
plasmagén 2149
plasma germinal 1130
plasmalema 2150
plasmidio 2152
plasmocromatina 2153
plasmodesma 2154
plasmodio 2155
plasmogamia 2156
plasmón 2157
plasmosoma 2158
plasmótipo 2160
plasmotomia 2159
plasticidad 2162
plastidio 2163
plastidiótipo 2168
plastidomio 2167
plasto 2161
plastoconto 2172
plastodesma 2170
plastogén 2171
plastoma 2173
plastomero 2174
plastosoma 2175
plectonemático 2177
plectonémico 2177
pleiotropía 2181
pleiotrópico 2179
pleuromíctico 2182
ploidia 2183
población 2256
- consanguínea 1541
- híbrida 1409
poder fecundante 993
- raceador 2289
poiquiloploide 2185
poiquiloploidía 2186
poiquilosíndesis 2187
polar 2192
polaridad 2195
polarización 2196
polibásico 2206

profase tóxica 2325
profásico 2326
progamia 2309
progámico 2308
progén 2310
progénesis 2311
progenie 2313
progenitor recurrente 309
progenomio 2312
progresivo 2315
prometafase 2316
promitosis 2318
pronúcleo 2320
pronucléolo 2319
proporcionalidad reproductora 2453
proporción de fijación 1002
- de los sexos 2584
- de sobrecruzamiento 643
- mendeliana 1749
- primaria de los sexos 2299
- secundaria de los sexos 2507
protandria 2327
proterminal 2328
protocromonema 2329
protoplasma 2330
protoplásmico 2331
protótrofo 2333, 2334
proximal 2335
prueba cis-trans 537
- de alium 76
- de la descendencia 2314
- de significancia 2793
- estadística 2793
pseudo-, sean: seudo-
puente cromatídico 492
- cromosómico 522
- de cromatina 497
- de cruzamiento 632
- de matriz 1728
- de segmentos especiales 2617
puff 2372
punto de contacto 614
- de ramificación 402
- de rotura 2645
pura raza, de 2373

quiasma 463
- complementario 589
- de compensación 581
- de inversión 1553
- diagonale 750

quiasma dispar 844
- lateral 1662
- múltiple 1878
- parcial 2077
quiasmas acordes 577
- de fusión 645
quiasma terminal 2786
quiasmatipía 468
quimera 469
- cromosómica 518
- haploclamídea 1189
- mericlinal 1754
- sectorial 2511
quimiótipo 462

radiación adaptiva 35
radiogenética 2393
radiomimético 2394
ramificado 400
randomisación 2399
"ratio cline" 2406
raza 407, 1636, 2391
- biológica 354
- fisiológica 2142
- geográfica 1125
- pura 2374
reacomodo 2408
- de rupturas múltiples 1873
reactivación 2407
- cruzada 652
- múltiple (RM) 1882
receptor 2409
recesividad 2411
recesivo 2410
- inferior 396
recipiente 2412
recíproco 2414
recombinación 2415
- de los profagos 2322
recón 2419
recurrente 2420
reducción 2423
reduccional 2424
reducción del bacteriofago 320
- de supervivencia 2725
- preferencial 2284
- somática 2613
reductor de sobrecruzamiento 648
reduplicación 2429
- de genes 1083

tetrasomatia 2800
tetrasómico 2801
tetrátipo 2802
tetravalente 2803
típico 2881
tipo 2879
- de apareamiento 1725, 2021
"tipoestrofa" 2885
tipogénesis 2882
tipólisis 2883
tipo pseudoselvaje 2371
- salvaje 2932
tipóstasis 2884
torsión 2876
torsional 2817
tóxico mitótico 1815
trabante 2819
tracto nucleolar 1949
transacomodo 2822
transconfiguración 2822
transducción 2823
- abortiva 9
- ARN 2827
transespecífico 2841
transferencia 2826
- colicinogénica de alta
  frecuencia 1325
- de baja frecuencia 1687
- de genes 1102
- de resistencia a multidroga 1880
- de todo el brazo 2930
transformación 2828
transgenación 2830
transgresión 2831
transgresivo 2832
translocación 2834
- de inserción 1499
transmisión 2839
transmutación 2840
transposición de todo el brazo 2931
transreplicación 2842
triada 2846
trigenérico 2847
trigénico 2848
trihaploide 2849
trihaploidía 2850
trihibridismo 2852
trihíbrido 2851
trimería 2854
trimérico 2853
trimonóico 2855

trimorfismo 2856
trióico 2857
triplexo 2860
triploide 2862
triploidía 2863
trisómia 2865
trisómico 2864
- de intercambio 1515
trisómicos de translocación 2838
trivalente 2866
trofocromatina 2868
trofoplasma 2869
tropocinesis 2871
tropodieresis 2870

ultramicrosoma 2886
unidad de mapa 1711
- de sobrecruzamiento 650
- mínima reproductiva 2455
unilateral 2892
unisexual 2895
unisexualidad 2896
unitario 2897
univalente 2899

vacúolo 2904
vacuoma 2905
valencia 2906
valor aditivo del genotipo 41
- de adaptación 37
valores D/I 856
valor hereditario 41, 1099
variabilidad 2907
variable independiente 1480
variación 2910
- ambiental 935
- brusca 411
- determinada 739
- discontinua 832
- discreta 832
- fenotípica correspondiente 626
- genotípica 1100
variante 2909
varianza 2908
variedad 2678, 2914
variegación 1871, 2912
variegado 1870, 2911
vecinismo 2918
vecinista 2919
vegetativo 2915
veneno del huso 2637

vesícula germinal 1129
- sexual 2589
viabilidad 2916
viable 2917
vieja espiral de profase 2441
vigor híbrido 1415
virion 2920
viscosidad 2675
- de la matriz 1729
visible 2923
vitalidad 2924

Weismanismo 2929

x (número básico) 2933
xenia 2935
xenogamia 2936
xenoplástico 2937

yema 412

z (número cromosómico zigótico)
    2939
zigonema 2944
zigosis 2945
zigosoma 2946
zigote 2947
zigotena 2948
zigoténico 2949
zigoteno 2949
zigótico 2950
zona de heteropicnosis negativa 2889
- de heteropicnosis positiva 2006
- de inserción 1501
- híbrida 1416
- intercéntrica 1513
- nucleolar 1950
- SAT 2500
zoogenética 2942

daltonico 702
daltonismo 703
dannoso 740
darvinismo 706
deficienza 717
degenerazione 719
– consanguinea 720
degenerescenza consanguinea 1470
degradazione 721
– della sessualità 2576
delezione 725
demo 727
densità della popolazione 2257
denucleizzazione 728
deriva 879
derivato 731
derivazione 2140, 2392
– fortuita 2397
– regolare 2668
desinapsi 736
determinante 738
determinazione del sesso 2587
– sessuale 2587
deuterotochia 741
deutialosoma 742
deviazione 744
– dovuta a geni modificatori 745
– tipica 2658
devoluzione 746
diacinesi 751
diacinetico 752
diade 886
diadelfo 748
diaforomissi 755
diaforomixi 755
diaginico 749
diallelo 753
diandrico 754
diaploide 786
diaploidismo 787
diascittico 756
diastema 757
diaster 758
dibasico 759
dicariofase 792
dicarion 791
dicariotico 793
dicentrico 762
diclamidio 763
diclino 767
dicogamia 765

dicogamo 764
dicondrico 766
dictiocinesi 769
dictiosoma 770
didiploide 772
didiploidismo 773
dientomofilia 774
dieterocigoto 788
difettivo 714
differenziale 775
differenziazione 779
difiletico 805
digametia 781
digametico 780
digamia 782
digenesi 783
digenico 784
digenomico 785
diibridismo 790
diibrido 789
dimeria 796
dimerico 795
diminuzione 797
dimissi 798
dimonoico 799
dimorfico 800
dimorfismo 801
– nucleare 1930
– sessuale 2578
dioicismo 803
dioico 802
diplobionte 806
diplobiontico 807
diplobivalente 808
diploclamidio 809
diplocromosoma 810
diploeteroicismo 813
diplofase 823
diplogenotipico 811
diploide 814
– funzionale 1018
diploidismo 817
diploidizzazione 815
diplomonosomico 819
diplonema 820
diplonte 821
diplontico 822
diplosi 824
diplosoma 825
diplosporia 826
diplotene 827

xenogamia 2936
xenoplastico 2937

z (numero cromosomico zigotico)
    2939
zigonema 2944
zigosi 2945
zigosoma 2946
zigote 2947
zigotene 2948
zigotenico 2949
zigotico 2950
zona d'eteropicnosi negativa 2889
– d'eteropicnosi positiva 2006
– ibrida 1416
– nucleolare 1950
zoogenetica 2942
zygogenetico 2943

Abänderung durch Modifikations-
   faktoren 745
Aberration 1
Aberrationsrate 2
abgeleitet 731
Abiogenese 3
Abiogenesis 3
abiogenetisch 4
Ablast 5
Abort 5
abortiv 6
abortive Mitose 8
- Transduktion 9
Abortiv-Infektion 7
Abstammungslinie 1675
Abstand 852
- zwischen den Generationen
   1080
Abstossung 2456
Abweichung 744
- , mittlere quadratische 2658
Acceleration 10
achiasmatisch 18
Achromasie 19
Achromatin 22
achromatisch 20
achromatische Figur 21
- Spindel 21
A-Chromosom 23
Adaptation 32
Adaptibilität 31
adaptive Modifikation 33
Addition 38
additive Faktoren 40
additiver Effekt 39
additive Wirkung 39
Additivität 42
Addospecies 43
Adhäsion 44
Affinität 45
agam 54
Agamet 46
Agamobium 48
Agamogenese 49
Agamogonie 50
Agamospecies 51

agamosperm 52
Agamospermie 53
Agens, transformierendes 2829
AG-Komplex 55
agmatoploid 57
Agmatoploide 57
Agmatoploidie 58
agmato-pseudopolyploid 59
Agmato-Pseudopolyploide 59
Agmato-Pseudopolyploidie 60
akinetisch 62
Akklimatisierung 14
Akrosyndese 26
akrozentrisch 25
Aktivator 28
Aktivator-Dissoziations-System 29
Aktivierung 27
Aktivitätsspektrum 30
Akzeleration 10
akzessorisch 11
akzessorische Metaphaseplatte 13
akzessorisches Chromosom 12
allaesthetisch 63
allel 68
Allel 66
- , rezessives 86
Allelenpaar 983
Allelie 69
allelobrachial 70
allelomorph 72
Allelomorph 71
- , unechtes 2655
allelomorphe Serie 73
Allelomorphie 74
Allelotyp 75
Allium-Test 76
allo-autogam 64
Allo-autogamie 65
Allocarpie 77
allochron 78
allodiploid 81
Allodiploide 81
Allodiploidie 82
Allodiplomonosom 83
allogam 84
Allogamie 85

Basiszahl 337, 1020
Bastard 1408
Bastardierung 1411
Bastard-Letalität 1413
Bastardnatur 1412
Bastardpopulation 1409
Bastardsterilität 1410
bathmische Kraft 339
B-Chromosom 340
bedingte Dominanz 602
bedingter Letalfaktor 603
Befruchtung 992
Befruchtungsfähigkeit 993
Begattung 1719
bigenerisch 342
begrenzt 1669
binucleat 344
binukleat 344
Bioblast 345
Biochore 346
Biogen 347
Biogenese 348
biogenetisch 349
biogenetische Isolation 350
biologische Bekämpfung 352
- Isolierung 353
- Rasse 354
biologisches Spektrum 355
Biomechanik 357
Biometrie 359, 360
biometrisch 358
Biomutante 361
Bion 362
Biophore 363
Bioplasma 364
Biosom 366
Biosynthese 367
biotisch 369
biotische Resistenz 371
biotisches Potential 370
Biotop 372
Biotyp 373
Bipartit 374
bisexuell 377
bithallisch 378
Bivalent 379
- , ungleiches 2890
Blastocyte 380
Blastogenese 381
blastogenetisch 382
Blastomere 383

Blastovariation 384
Blastozyte 380
Blepharoplast 387
Block 388
Blockmutation 389
Blutanteil 2103
Blutauffrischung 2086
Blutsverwandtschaft 393
Boten-RNS 2490
"Bottleneck"-Phänomen 395
bottom recessive 396
Brachymeiosis 398
bradytelisch 399
Brochonema 410
Bruch 1009
Bruch-Einschnürung 406
Bruch-Fusionsbrücken-Zyklus 404
Bruch-Hypothese 403
Bruch in Centromernähe 438
Bruch-Reunions-Bivalent 405
Brücken-Fragment-Konfiguration
    409
brüderlich 1012
Brut 407
Brutknospe 1060
Bukettstadium 397

Caenogenesis 418
caenogenetisch 419
Carpoxenie 423
Centriol 440
Centrochromatin 441
Centrodesmose 442
Centrogen 443
Centromer 444, 2635
Centromer- 449
- , semilokalisiertes 2556
Centromerabstand 445
Centromeranziehung 450
Centromerdislokation 448
Centromer-Interferenz 446
centromerisch 449
Centromer-Missteilung 447
Centronucleus 451
Centroplasma 452
Centroplast 453
Centrosom 434, 454
Centrosphäre 455
Centrotyp 456
cephalobrachial 457
Certation 458

fundamental 1019
funktionelle Diploide 1018
Furchung 542
- , holoblastische 1330
- , meroblastische 1756
Furchungskern 544
Furchungsverzögerung 543
Fusion 1021
Fusionskern 1022

Gamet 1023
Gameten- 1024
Gameten, balancierte 324
Gametenmutation 1028
Gametenzahl 1029
gametisch 1024
gametische Chromosomenzahl 1025
- Inkompatibilität 1026
gametischer Letalfaktor 1027
gametische Sterilität 1030
Gametoblast 1031
Gametocyte 1032
Gametogamie 1033
gametogen 1036
Gametogenese 1034
Gametogonium 1037
Gametoid 1038
Gametophor 1039
Gametophyt 1040
gametophytisch 1041
gametophytische Inkompatibilität
  1026
Gamobium 1043
Gamodeme 1044
Gamogenese 1045
gamogenetisch 1047
Gamogonie 1048
Gamolyse 1049
Gamon 1050
Gamont 1051
Gamophase 1052
Gamotropismus 1053
Garnitur 2571
Gattung 1124
Gattungen, zwei, betreffend 342
Gattungs- 1085
Gattungsbastard 1086
Gattungskreuzung 343
Gattungstyp 1087
gegenseitig 1904
Geitonocarpie 1054

Geitonogamie 1055
Geitonogenese 1056
Gemeinschaft 576
Geminus 1057
Gemmatio 1058
gemmipar 1059
Gemmula 1060
Gen 1061
Gen- 1103
- , latentes 1556
- , polyurgisches 2255
Genabstand 1710
Genbalance 1064, 1104
Genblock 390
Genchromatin 1066
Gen-Cytoplasma-Isolation 1068
Gendosis 1069, 1105
Gendrift 1070
Gene, duplikate 883
- , mimetische 1802
Genealogie 1062
Generation 1079
Generationenintervall 1080
Generationswechsel 114, 783, 1264,
  1768
generativ 1081
generativer Kern 1082
generisch 1085
gene-starvation-Hypothese 1088
Genetik 1098
genetisch 1091
genetische Coadaptation 1094
genetischer Tod 1095
genetische Variation 1100
genetisch markierte Chromosomen
  1090
genetotroph 1101
Gen-"flow" 1071
Genhäufigkeit 1072
Geninfiltration 1073
genisch 1103
genische Sterilität 1106
Genkarte 982
Genkomplex 1067
Genmilieu 1093
Genmutation 1075, 1123, 2189, 2830
Gennest 1076
Genodispersion 1109
Genoid 1110
Genokline 1107
Genokopie 1108

Halbblut- 1174
Halbbruder 1182
Halbchiasma 1176
Halbchromatiden- 1215
Halbchromatidenbruch 1177
Halbchromatidenfragmentation 1178
halbchromatidisch 1215
Halbdisjunktion 1179
Halbgeschwister 1182
Halbgeschwisterfamilie 1183
Halbmutante 1180
Halbrasse 1181
Halbschwester 1182
Halbseitenzwitter 374, 1184
Half starvation 1185
Haplobiont 1186
haplobiontisch 1187
haplochlamyde Chimäre 1189
haplodiploid 1190
Haplodiploide 1190
Haplodiploidie 1191
Haplodiplont 1192
haplogenotypisch 1193
Haploheterözie 1194
haploid 1195
Haploide 1195
Haploidie 1198
haplo-insuffizient 1196
Haplokaryotyp 1188
Haplom 1199
Haplomitose 1201
Haplomixis 1202
Haplont 1203, 1855
haplontisch 1204
Haplophase 1205
haplopolyploid 1206
Haplopolyploide 1206
Haplopolyploidie 1207
Haplosis 1208
Haplosomie 1209
haplo-suffizient 1197
haplozygotisch 1210
Häufigkeit 1015
Häufigkeitsgradient 2406
Haupt- 1701
Hauptgen 1633, 1971
hemialloploid 1211
Hemialloploide 1211
Hemialloploidie 1212
hemiautoploid 1213
Hemiautoploide 1213

Hemiautoploidie 1214
Hemidiärese 1216
hemihaploid 1217
Hemihaploide 1217
Hemihaploidie 1218
hemiholodiploid 1219
Hemiholodiploide 1219
Hemiholodiploidie 1220
Hemikaryon 1221
Hemikinese 1222
Hemiorthokinese 1223
Hemixis 1224
hemizygot 1225
Hemm- 1496
hemmend 1496
Hemmung 1493
Hemmungs- 1496
heptaploid 1226
Heptaploide 1226
Heptaploidie 1227
Herkogamie 1234
Herkunft 407
hermaphrodit 1235
Hermaphroditismus 1236
Heterauxese 1237
heteroallel 1239
Heteroallel 1238
heterobrachial 1240
heterochondrisch 1246
Heterochromatie 1250
Heterochromatin 1248
Heterochromatinosom 1249
heterochromatisch 1247
Heterochromatisierung 1251
Heterochromomer 1252
Heterochromosom 1253
heterodichogam 1254
Heterodichogamie 1255
Heteroecie 1257
heteroecisch 1256
Heteroezie 1257
heteroezisch 1256
Heterofertilisation 1258
heterogam 1262
Heterogameon 1259
Heterogamet 1260
heterogametisch 780, 1261
Heterogamie 1263
Heterogenese 1264
heterogenetisch 1265
heterogenisch 1266

Mixoploidie 1818
mock dominance 1819
modifier complex 1823
Modifikation 1821
- , abortive 33
Modifikationsgen 976, 1804, 1822
Modifikationsgenkomplex 1823
Modulation 1825
Modulator 1826
Moduszahl 1820
molekular 1827
Molekular- 1827
Molekularspirale 1828
Monade 1829
Monaster 1830
Monide 1831
monobrachial 1832
Monochromosom 1834
Monoecie 1836
monoecisch 1835
Monoezie 1836
monoezisch 1835
monofaktoriell 1837
monogen 1840
Monogen 1838
monogene Heterosis 2603
Monogenese 1839
Monogenie 1843
monogenomatisch 1842
Monogonie 1844
monohaploid 1845
Monohaploide 1845
Monohaploidie 1846
monohybrid 1847
Monohybride 1847
Monohybridie 1848
Monolepsis 1849
monomer 1850
Monomerie 1851
monophyletisch 1852
monoploid 1853
Monoploide 1853
Monoploidie 1854
Monoplont 1855
monosom 1858
Monosom 1857
monothallisch 1859
Monotopie 1860
monozentrisch 1833
Monözie 1836
monözisch 119, 1835

monozygot 1861
monozygotisch 1861
Morgan-Einheit 1863
morph 1864
Morphismus 1865
Morphogenese 1866
morphogenetisch 1867
Morphose 1868
Mosaik 1869
Mosaikbildung 375
Mosaikform 1869
mottled 1870
mottling 1871
Multibruch-Rearrangement 1873
Multigen 1874
multipel 1875
multiple Allele 1876
- Geschlechtschromosomen 1884
- Korrelation 1879
multipler Chromosomenbruch 530
multiple Regression 1883
multiples Centromer 1877
- Chiasma 1878
Multivalent 1885
Musterbildungsgene 2095
mustermodifizierende Gene 2096
mutabel 1886
mutafacient 1887
mutagen 1892
Mutagen 1888
Mutagenese 1889
Mutagenspezifität 1890
Mutagenstabilität 1891
Mutante 1893
Mutation 1894
- , cytoplasmatische 697
- , iterative 1605
- , somatische 2612
mutationsbedingte Entwicklungs-
    Abweichung 744
Mutationsdruck 1899
Mutationsgleichgewicht 1895
Mutationskoeffizient 1897
Mutationsrate 1900
Mutationsverzögerung 1898
Mutatorgen 1901
Mutatorsubstanz 1902
Muton 1903
Mutter (von Säugetieren) 704
mütterlich 1717
mütterliche Linie 990

Oozyte 1982
Operon 1988
Oppositionsfaktor 1989
Organisationseffekt 1992
Organisationszentrum 1991
Organisator 1990
Orientierung 1993
Orthoamitose 1994
Orthogenese 1995
Orthokinese 1996
orthoploid 1997
Orthoploide 1997
Orthoploidie 1998
Orthoselektion 1999
Orthospirale 2000
orthotaktisch 2001
orthotelomitisch 2002
Outbreeding 2003
Outcross 2004
Ovum 2008
Oxychromatin 2009

Paarkernphase 792
Paarung 629, 1719, 2016
- , ektopische 902
- , falsche 985
- , korrektive 625
- , nichthomologe 1920
- ähnlicher Individuen 244
Paarungsblock 2017
Paarungsheterosis 2019
Paarungskoeffizient 2018
Paarungssegment 2020
Paarungstyp 1725, 2021
Paarung unähnlicher Individuen 830
Pachynema 2010
Pachytän 2011
Packungsfaktor 2012
Paedogamie 2013
Paedogenese 2014
pair mating 2022
Palingenese 2024
Panallel 2025
pan-alloploid 2026
Pan-Alloploide 2026
Pan-Alloploidie 2027
panaschiert 1870, 2911
Panaschierung 2912
Panaschüre 2912
pan-autoploid 2028
Pan-Autoploide 2028

Pan-Autoploidie 2029
Pangen 2030
Pangenesis 2031
Pangenosom 2032
panmiktisch 2033
Panmixie 2034, 2400
Panmixis 2034
Paraallel 2035
Parachromatin 2037
Parageneon 2039
Paragenoplast 2040
Paralleldisjunktion 2041
Parallelevolution 2042
Parallelmutation 2043
Paralocus 2045
Paramaecin 2046
Parameiose 2047
paramiktisch 2048
Paramitose 2049
Paramixie 2050
paranematisch 2051
paranematische Wicklung 2052
Paranuclein 2053
Paranukleolus 1955
Paranukleoplasma 2054
Paraselektivität 2055
parasexuell 2056
Parasynapsis 2057
Parasyndese 2058
parataktisch 2059
paratroph 2060
Paravariation 2061
parazentrisch 2036
Parentalgeneration 2064
Parthenapogamie 2066
Parthenogamie 2068
Parthenogenese 2069
- , künstliche 238
parthenogenetisch 2070
Parthenokarpie 2067
Parthenomixis 2072
Parthenospermium 2073
Parthenospore 2074
Parthenote 2075
Partialbefruchtung 2081
Partialbruch 2076
Partialchiasma 2077
Partial-Crossing-over 2079
partialheterothallisch 2082
Partialrestitution 2088
partielle Korrelation 2078

plus modifier 2184
poikiloploid 2185
Poikiloploide 2185
Poikiloploidie 2186
Poikilosyndese 2187
point error 2188
point stickiness 2190
Poissonsches Gesetz 2191
Poisson-Verteilung 2191
Pol 2199
polar 2192
Polarisation 2196
Polarisierung 2196
Polarität 2195
Polfeld 2200
Polkappe 2194
Polkörperchen 2204
Pollenletalfaktoren 2202
Pollenmutterzelle 1797
Pollensterilität 2203
Polocyte 2204
Polplatte 2201
polybasisch 2206
polycaryotisch 2228
polychondrisch 2209
Polychromosom 2210
Polyembryonie 2212
polyenergid 2213
Polyenergide 2213
polyergistisch 2214
polyfaktoriell 2215
polygam 2216
Polygamie 2217
polygen 2220
Polygen 1804, 2218
polygenerisch 2219
Polygenie 2223
polygenisch 2220
Polygenkombination 2221
polygenomatisch 2222
polyhaploid 2224
Polyhaploide 2224
Polyhaploidie 2225
polyhybrid 2226
Polyhybride 2226
Polyhybridismus 2227
polykaryotisch 2228
polylysogen 2229
polymer 2232
Polymerchromosom 2231
Polymerie 2233

Polymitose 2234
polymorph 2236
Polymorphismus 2235
- , balancierter 326
- , transienter 2833
Polyphänie 2238
Polyphasie 2237
polyphyletisch 2239
polyploid 2240
Polyploide 2240
- , partielle 2083
Polyploidenkomplex 2241
Polyploidie 2242
- , balancierte 327
- , sekundäre 2506
- , strukturell veränderte 2686
- , unbalancierte 2888
Polyplotypus 2243
polyradial 2244
polysatellitisch 2246
polysom 2247
Polysomie 2248
polysperm 2249
Polyspermie 2250
polytän 2251
Polytän- 2251
Polytänie 2252
Polytopie 2253
polytypisch 2254
polyurgisches Gen 2255
polyzentrisch 2208
Population 2256
Populationsdichte 2257
Populationsdruck 2261
Populationsgenetik 2259
Populationsgleichgewicht 2258
Populationshomöostasis 2260
Populationswellen 2262
Positionseffekt 2265
Positionspseudoallele 2263
Positionsverwandtschaft 2264
positiv heteropyknotischer Abschnitt 2006
Postadaptation 2266
Postdivision 2267
Postheterokinese 2269
Postreduktion 2270
postreduziert 2269
post split aberration 2271
postsyndetische Interphase 2272
postzygotisch 2273

А-хромосома, 23
АГ комплекс, 55
аберрация, 1
абиогенез, 3
абиогенный, 4
аборт, 5
абортивная инфекция, 7
абортивная трансдукция, 9
абортивный, 6
абортивный митоз, 8
аверсия гифов, 1435
автогенез, 272
автогенетический, 273
автогенный, 274
автономный фактор, 282
агамета, 46
агамный, 47, 54
агамный вид, 51
агамогенез, 49
агамогония, 50
агамоспермия, 53
агамоспермный, 52
агматоплоид, 57
агматоплоидия, 58
агматоплоидный, 57
агматопсевдоплоид, 59
агматопсевдоплоидия, 60
агматопсевдоплоидный, 59
адаптация, 32
адаптивный пик, 34
аддитивность, 42
аддитивные факторы, 40
аддитивный эффект, 39
азигогенетический, 305
азигота, 306
акинетический, 62
акклиматизация, 14
акросиндез, 24
акроцентрический, 26
активатор, 28
активация, 27
аллелизм, 69

аллелия, 69, 74
аллелобрахиальный, 70
аллеломорф, 66, 71
аллеломорфизм, 74
аллеломорфный, 68, 72
аллелотип, 75
аллель, 66, 71
аллель пятнистости, 2913
аллельный, 68, 72
аллоаутогамия, 65
аллоаутогамный, 64
аллогамия, 85
аллогамный, 84
аллогаплоид, 89
аллогаплоидия, 90
аллогаплоидный, 89
аллоген, 76
аллогенетический, 87
аллогенный, 88
аллогетероплоид, 91
аллогетероплоидия, 92
аллогетероплоидный, 91
аллодиплоид, 81
аллодиплоидия, 82
аллодиплоидный, 81
аллодипломоносома, 83
аллозигота, 111
аллокарпия, 77
аллолизогенный, 94
алломорфоз, 95
аллопатрический, 96
аллоплазма, 97
аллоплоид, 98
аллоплоидия, 99
аллоплоидный, 98
аллополиплоид, 100
аллополиплоидия, 101
аллополиплоидный, 100
аллосинапсис, 104
аллосинаптический, 105
аллосиндез, 104
аллосома, 103

аллосомный, 102
аллотетраплоид, 106
аллотетраплоидия, 107
аллотетраплоидный, 106
аллотипический, 110
аллотриплоид, 108
аллотриплоидия, 109
аллотриплоидный, 108
аллохронические виды, 78
аллоциклический, 79
аллоциклия, 80
альвеолярная гипотеза, 118
альтернативная изменчивость, 832
альтернативное распределение, 836
альтернативное расхождение, 115
альтруистическая адаптация, 117
амбивалентный, 120
амбисексуальный, 119
амейоз, 121
амейотический, 122
амиксис, 127
амиксия, 127
амиктический, 123
амитоз, 125
амитотический, 126
аморфный, 128
амферотокия, 129
амфиастральный митоз, 131
амфибивалент, 132
амфигамия, 135
амфигаплоид, 138
амфигаплоидия, 139
амфигаплоидный, 138
амфигенез, 136
амфигония, 137
амфидиплоид, 133, 872
амфидиплоидия, 134
амфидиплоидный, 133, 872

амфикарион, 140
амфимиксис, 143
амфимиктический, 142
амфилепсис, 141
амфипластия, 145
амфиплоид, 146
амфиплоидия, 147
амфиплоидный, 146
амфиталлический, 149
амфитена, 148
амфитенный, 148
амфитетраплоид, 772
амфитетраплоидия, 773
амфитетраплоидный, 772
амфитокия, 150
амфогения, 152
амфогенный, 151
амфогетерогония, 153
анаболия, 154
анагенез, 156
анагенетический, 157
анализирующее скрещивание, 2791
аналог, 158
аналогичный, 158
анаморфный, 159
анаморфоз, 160
анаредупликация, 165
анастомоз, 167
анастральный митоз, 168
анасхистический, 166
анафаза, 161
анафазное движение, 162
анафазный, 163
анафрагмический, 164
анахромазия, 155
андроаутосома, 169
андрогамия, 173
андрогенез, 174
андрогенетический, 175
андрогенный, 176
андрогермафродит, 179

андрогермафродитный, 179
андрогиния, 178
андрогинный, 177
андродвудомный, 170
андроезия, 172
андромерогония, 180
андрооднодомность, 182
андрооднодомный, 181
андросома, 183
андроспорогенез, 184
андростерильность, 186
андростерильный, 185
анеуплоид, 187
анеуплоидия, 188
анеуплоидный, 187
анеусоматия, 189
анизоаутоплоид, 190
анизоаутоплоидия, 191
анизоаутоплоидный, 190
анизогамета, 194
анизогамия, 195
анизогения, 198
анизогеномный, 196, 197
анизокариоз, 192
анизоплоид, 199
анизоплоидия, 200
анизоплоидный, 199
анизосиндез, 201
анизосиндетический, 202
анизотрисомия, 203
анизоцитоз, 193
аннидация, 204
анормогенез, 205
анортогенез, 206
анортоплоид, 207
анортоплоидия, 208
анортоплоидный, 207
анортоспираль, 209
антефаза, 210
антиморфный, 212
антимутаген, 213
антимутагенный, 214

апогаметность, 218
апогамия, 220
апогамный, 219
апогомотипный, 221
апомейоз, 222
апомиксис, 225
апомиктический, 223
апомиктоз, 224
апорогамия, 226
апоспория, 227
аппарат Гольджи, 1133
ароморфоз, 233
арреногенетический, 234
арреногения, 235
арренотокия, 236
артефакт, 237
артиоплоид, 239
артиоплоидия, 240
артиоплоидный, 239
архаллаксис, 228
архебиоз, 229
архетип, 230
архиплазма, 231
асимметричная транслокация,
    16
асимметричный бивалент, 2609
асимметрия, 2610
асинапсис, 249, 251
асинаптический, 250
асингамия, 253
асингамный, 252
асиндез, 249, 251
ассортативное скрещивание, 244
ассортативное спаривание, 1092
ассоциация, 243
ассоциация бок-о-бок, 2593
астросфера, 248
астроцентр, 247
атавизим, 2814
атавизм, 255
атавистический, 256
атавистический признак, 2813

гетеротипическое деление, 1315, 1316, 2425
гетеротрансформация, 1313
гетеротропическая хромосома, 1314
гетерофеногамия, 1290
гетерофитный, 1291
гетерохондрический, 1246
гетерохромазия, 1250
гетерохроматизация, 1251
гетерохроматин, 1248
гетерохроматиновый, 1247
гетерохроматиносома, 1249
гетерохромомера, 1252
гетерохромосома, 1258
гетероцентрический, 1244
гетероцентричность, 1245
гибрид, 1408
гибридизация, 1411
гибридная зона, 1416
гибридная летальность, 1413
гибридная популяция, 1409
гибридная сила, 1415
гибридность, 1412
гибридный, 1408
гибридогенный псевдопартогенез, 1414
гигантская хромосома, 1132
гинандроид, 1155
гинандроморф, 1156, 1184
гинандроморфность, 1157
гинаутосома, 1158
гиногенез, 1164
гиногенетический, 1165
гиногения, 1167
гиногенный, 1166
гинодвудомность, 1160
гинодвудомный, 1159
гинодиезический, 1159
гинодиезия, 1160
гиноезический, 1161
гиноезия, 1162

гиномоноезический, 1168
гиномоноезия, 1169
гинооднодомность, 1169
гинооднодомный, 1168
гиносперм, 1170
гиноспора, 1171
гиноспорогенез, 1172
гинофактор, 1163
гипергаплоид, 1424
гипергаплоидия, 1425
гипергаплоидный, 1424
гипергетеробрахиальный, 1426
гипердиплоид, 1422
гипердиплоидия, 1423
гипердиплоидный, 1422
гиперморфный, 1427
гиперморфоз, 1428
гиперплоид, 1429
гиперплоидия, 1430
гиперплоидный, 1429
гиперполиплоид, 1431
гиперполиплоидия, 1432
гиперполиплоидный, 1431
гиперсиндез, 1433
гипертелия, 1434
гиперхимера, 1418
гиперхромазия, 1419
гиперхроматоз, 1420
гипоаллеломорф, 1417
гипоаллеломорфный, 1417
гипогаплоид, 1439
гипогаплоидия, 1440
гипогаплоидный, 1439
гипогенез, 1437
гипогенетический, 1438
гипоморфный, 1441, 1442
гипоплоид, 1443
гипоплоидия, 1444
гипоплоидный, 1443
гипополиплоид, 1445
гипополиплоидия, 1446
гипополиплоидный, 1445

считывание кода, 550

таксон, 2760
тандемная ассоциация, 2751
тандемная дупликация, 1752,
    2755
тандемная инверсия, 1754
тандемная селекция, 2758
тандемное кольцо, 2756
тандемное слияние, 1753
тандемные сателлиты, 2757
тахигенез, 2749
тахителический, 2750
телегония, 2765
телигенический, 2806
телигения, 2807
теликарион, 2804, 2808
теликариотический, 2805, 2809
телитокия, 2810
телоген, 2769
телолецитальный, 2770
теломера, 2771
теломерный, 2772
теломитический, 2773
телоредупликация, 2777
телосинапсис, 2778
телосиндез, 2779
телосиндетический, 2780
телофаза, 2774
телофазные тельца, 2775
телофазный, 2776
телохромомера, 2763, 2767
телохромосома, 2764, 2768
телоцентрический, 2766
температурный мутант, 2782
тенденция, 2845
теория взаимодействия, 1509
теория мишени, 2759
теория скрещивания, 1724
теория хиазматипии, 467
терминализация, 2788
терминальная хиазма, 2786

терминальное сродство, 2785
терминалные гранулы, 2787
терминальный, 2784
термон, 2789
тест значимости, 2793
тетравалент, 2803
тетрагаплоид, 2796
тетрагаплоидия, 2797
тетрагаплоидный, 819, 2796
тетрада, 2388, 2794
тетрадный анализ, 2795
тетрады дитипа, 855
тетраплоид, 2798
тетраплоидия, 839, 2799
тетраплоидный, 2798
тетрасоматия, 2800
тетрасомик, 2801
тетрасомный, 2801
тетратип, 2802
тип, 2879
тип повреждения, 2097
тип спаривания, 1725, 2021
типичный, 2881
типовое число, 2880
типогенез, 2882
типолиз, 2883
типостаз, 2884
типостроф, 2885
Т-конец, 2783
толерантная доза, 2113
топкросс, 2815
топкроссинг, 2816
торзионный, 2817
точечная мутация, 2189
точечная ошибка, 2188
точечное склеивание, 2190
точка контакта, 614
трабант, 2819
трансвекционный эффект, 2843
трансгенация, 2830
трансгрессивный, 2832
трансгрессия, 2831